LABORATORY HEALTH
AND SAFETY
DICTIONARY

LABORATORY HEALTH AND SAFETY DICTIONARY

W. Carl Gottschall, Ph.D.
Consultant
Littleton, Colorado

Douglas B. Walters, Ph.D.
President, KCP, Inc.
Raleigh, North Carolina

**WILEY-LISS**

A JOHN WILEY & SONS, INC., PUBLICATION

New York · Chichester · Weinheim · Brisbane · Singapore · Toronto

Douglas B. Walters contributed to this book in his private capacity. No official support or endorsement by the National Institute of Environmental Health Sciences is intended or should be inferred.

This book is printed on acid-free paper. ∞

For ordering and customer service, call 1-800-CALL-WILEY.

Library of Congress Cataloging-in-Publication Data:

Gottschall, W. Carl, 1938–
  Laboratory health and safety dictionary/by W. Carl Gottschall, Douglas B. Walters.
    p. cm.
  Includes index.
  ISBN 0-471-28317-7 (pbk.)
    I. Chemical laboratories — Safety measures — Dictionaries.   I. Walters, Douglas B. II. Title.

  QD63.5 . G67 2001
  542′.1′0289–dc21

                                                        00-068503

Printed in the United States of America

10 9 8 7 6 5 4 3 2 1

ACKNOWLEDGEMENTS

The authors wish to gratefully acknowledge Warren Kingsley for his support, review and input, especially in the early developmental stages of this manuscript.

And to our families for their love and support.

PREFACE

This dictionary is specifically designed to focus on the chemistry-related aspects of the general field of health and safety. There is considerable overlap between chemical health and safety and allied fields such as industrial hygiene, environmental health and safety, radiation safety, safety engineering, toxicology, fire safety, occupational medicine, and numerous other areas that define the rapidly expanding universe of health and safety. For this and other reasons, no dictionary of this type can be all-inclusive and this one is a constant "work-in-progress." The primary purpose of this dictionary is, for the first time, to define the terms used to describe the work of those individuals who practice chemical health and safety as a profession.

One object of the Laboratory Health and Safety Dictionary is to assist persons with limited understanding and/or varying backgrounds to better understand the vocabulary that is encountered in the field. Words, terms, and expressions are included that have grown out of documents and regulations such as OSHA's Hazard Communication Standard and the Occupational Exposure to Hazardous Chemicals in the Laboratory Standard (The Laboratory Standard), Material Safety Data Sheets (MSDSs), Right-To-Know legislation, the National Toxicology Program's Report on Carcinogens, and numerous other lists of "hazardous" agents.

In many instances Chemical Hygiene Officers (CHOs), especially those in small organizations, have collateral responsibilities. In addition, CHOs are sometimes limited in their training, background, expertise, and experience in health and safety. This book is especially designed for them.

This book is also intended to serve as a general aid to other professionals, such as fire fighters, and the general public to gain a better perspective of chemistry and the hazards that result from the improper use of chemicals.

Because many of the principal users of this dictionary seek to expand their knowledge, we purposely tried to selectively include terms that are most frequently encountered in areas related to chemical health and safety. Every effort was made to make this dictionary a user-friendly, highly readable, and, at times, even enjoyable document. For example, it is not practicable to include information on all of the potentially hazardous chemicals currently known, however, selected entries of particular concern such as dioxins or kepone are included. Similarly, basic information on the elements of the periodic table is included. An appendix of abbreviations and acronyms is included because this seems to be a trend in today's technologically advanced world. In addition, an extensive appendix of names and addresses of safety-related organizations is included. Finally, an appendix is included with brief biographies of key historical individuals whose work impacted on chemical health and safety.

Although every attempt has been made to present accurate, complete entries, it should be realized that many definitions could and have been expanded into separate books. Brevity and uniformity have been used judiciously to keep this dictionary to a manageable size.

As with every new venture, it is realized that errors and omissions may occur. It is sincerely hoped that errors are minimal; however, reader feedback is encouraged. We believe this dictionary will demonstrate that the language and jargon of chemical health and safety are exciting, dynamic, and constantly evolving.

W. CARL GOTTSCHALL
DOUGLAS B. WALTERS

CONTENTS

A

α   Lower case Greek letter alpha.

A   Atomic mass number; ampere; uppercase Greek letter alpha.

Å   Angstrom unit.

a   Atto (10^{-18}).

A&C   Abatement and control.

A&S   Accident and sickness.

A&WMA   Air and Waste Management Association.

A-E   Architect-engineer.

a.u.   Atomic units.

A/sec   Ampere/second.

AA   Atomic absorption spectroscopy; activation analysis.

AAALAC   American Association for Accreditation of Laboratory Animal Care.

AAAS   American Association for the Advancement of Science.

AAEE   American Academy of Environmental Engineers.

AAI   Alliance of American Insurers.

AAIH   American Academy of Industrial Hygienists.

AALA   American Academy for Laboratory Accreditation.

AAOHN   American Association of Occupational Health Nurses.

AAOM   American Academy of Occupational Medicine.

AAPCC   American Association of Poison Control Centers.

AARCEP   Alliance for Acid Rain Control and Energy Policy.

AAS   American Astronautical Society.

AASHTO   American Association of State Highway and Transportation Officials.

AATCC   American Association of Textile Chemists and Colorists.

Abamp   Absolute ampere.

abate   To reduce or decrease in amount, volume, or intensity; often used to designate the elimination or removal of pollution.

Abel, Sir Fredrick Augustus   English chemist (1827–1902), coinventor of cordite and the device, named after him, for determining flash points.

Abel-Pensky flash point apparatus   A device used to determine the flash point of a substance.

aberration   A deviation from the ideal or norm; used in optics to indicate a difference between the exact and apparent position of an image.

ABIH   American Board of Industrial Hygiene.

abrasion   Deterioration by physical (e.g., friction) or chemical (e.g., reaction) means; also applies to skin or mucous membranes.

abs.   Absolute.

ABSA   American Biological Safety Association.

absinthe   A specific alcoholic liqueur containing a toxic flavoring agent derived from wormwood (*Artemisia absinthium*); currently illegal in the USA.

absolute   Free from other substances, impurities, contaminants, imperfections, limitations, or interferences; the highest quality, factor, or measurement achievable; often used in conjunction with other terms, e.g., alcohol, ceiling, pressure, temperature, time, value of a real number, velocity, zero.

absolute alcohol   Pure, 100% ethyl alcohol.

absolute error   The difference between the actual and approximate value of any calculation.

absolute pressure (psia)   The pressure compared to a vacuum as opposed to gauge pressure, which has ambient pressure as the reference zero.

absolute temperature   The temperature on a scale that uses absolute zero as a reference point that can be approached but never reached; *See* Kelvin temperature scale.

absolute zero   The theoretical temperature with a complete absence of heat at which all molecular motion stops $-273.16°$ Celsius, $-459.67°$ Fahrenheit, $0°$ Kelvin.

absorb   To physically, biologically, or chemically receive, take in, or take up all, or part, of a substance (e.g., chemical) or energy (e.g., sound, radiation) and to integrate the substance into the whole.

absorbed dose   The amount of energy delivered to matter per unit mass.

absorbing medium   A substance that absorbs or takes in (e.g., scrubber, chromatographic support, sound-attenuating material).

absorption band   The frequency or wavelength range of the electromagnetic spectrum absorbed by a specific atom, molecule, radical, or functional group.

absorption coefficient   A ratio of the energy absorbed to the energy incident on the surface, as in sound or spectroscopy.

absorption spectroscopy   An instrumental technique used to detect, determine, and/or measure the electromagnetic energy received and taken in or absorbed by a substance.

absorption tower   Apparatus used to capture, trap, or purify gases by absorbing the gas or its impurities in a liquid; sometimes called a scrubber.

Ac   Actinium.

AC   Alternating current.

acaricide   A type of pesticide specifically used for acarides (e.g., mites and ticks).

ACC   American Chemistry Council.

accelerator   A substance used to reduce the time required for a reaction to occur, i.e, to speed up the reaction; also an apparatus used to speed up atomic or subatomic particles.

acceptable air quality   Air that complies with regulatory requirements of pollution contaminants and length of exposure.

acceptable risk   An agreed-upon and defined value below which adverse effects to health and/or the environment are highly unlikely to occur, i.e., a reasonable assurance of safety.

acceptance criteria   Specified limits placed on the characteristics of an item, process, or service defined in codes, standards, or other required documents.

acclimitize   To adapt to new environmental conditions, i.e., temperature, altitude, climate, or situation.

accreditation   A process to formally recognize programs as meeting established objectives and criteria.

accuracy   A term denoting the closeness of a measured value to the true value or to an accepted reference or standard. The technically accepted definition of accuracy is bias, or the difference between the observed value and the true value being estimated.

acetone   A commonly used, high-production, colorless, flammable, volatile liquid solvent with a sweetish odor.

acetylcholine   A biochemical in the brain that transmits nerve impulses in the autonomic system.

acetylcholinesterase   The enzyme, cholinesterase, hydrolyzes acetylcholine to choline and acetic acid in the body to prevent acetylcholine poisoning. *See* cholinesterase.

acetylene C_2H_2; A highly flammable, highly reactive, colorless gas used in common welding; forms explosive compounds with copper, silver, and mercury.

acfm   Actual cubic feet per minute.

ACGIH   American Conference of Governmental and Industrial Hygienists.

acid   A compound usually consisting of hydrogen plus one or more elements that, in the presence of certain solvents or water, reacts to produce hydrogen ions or a solution with pH less than neutral; acids react with basic or alkaline materials to form salts and are commonly very corrosive to eyes, skin, and mucous membranes; typically, acids are water-soluble compounds, turn litmus red, and are sour; there are several different systems for defining acids in chemistry.

acid anhydride   A compound formed when water is removed from an acid molecule; sometimes used to describe the oxide of nonmetallic elements.

acid dye   An azo, triarylmethane, or anthraquinone dye with an acid substituent; often applied in acidic solution to combine with the basic part of proteins in animal fibers.

acidimetry   The determination of the amount of acid present in a solution.

acid number   A value determined by the amount of potassium hydroxide needed to neutralize fatty acids; also, the amount of free acids in a substance.

acidosis   A decreased alkalinity of the blood and tissues characterized by sickly sweet breath, headache, nausea, vomiting, and visual disturbances; usually the result of excessive acid production resulting in the disturbance of tissue and central nervous system functions.

acid rain   Precipitation in the form of rain with an acidic pH; often caused by the presence of sulfur dioxide, nitrous oxide, and other chemicals arising from the stacks of industrial power sites or volcanic activity. Such precipitation is damaging and corrosive to fish, trees, and buildings.

ACLAM   American College of Laboratory Animal Medicine.

ACM   Asbestos-containing material.

ACOM   American College of Occupational Medicine.

Acoustical Society of America (ASA)   2 Huntington Quadrangle, Melville NY 11747-4502.

acrid   Describes a harsh, sharp, irritating, or bitter odor or taste.

acrylate   A polymer of acrylic acid or its esters used in the manufacture of thermoset plastics.

ACS   American Chemical Society; American Cancer Society.

ACSH   American Council on Science and Health.

ACT   American College of Toxicology.

ACTH   Adrenocorticotropic hormone.

actinic skin   A condition of the skin arising from exposure to ultraviolet radiation.

actinide   A member of the series of radioactive elements in the periodic table beginning with actinium and whose atomic number is 89 or higher.

actinium (Ac)   A radioactive heavy metal element, atomic number 89, atomic weight 227.03.

actinomycytes   A fungal-type microorganism pathogenic to humans and animals.

action level   The OSHA-determined level of toxicant in the environment at which medical surveillance is required; generally, one-half of the permissible exposure level or ACGIH TLV®.

activated alumina   A highly porous granular form of aluminum oxide able to absorb other substances from gases or liquids; frequently used as a catalyst, to dry gases, to purify water, and in chromatography.

activated charcoal   A form of carbon with a high capacity to adsorb gases, vapors, and colloids; frequently used in filters for removal of hazardous and noxious materials in air, in waste treatment, and in decolorization.

activation   The process of making a material radioactive by bombardment with neutrons, protons, or other nuclear particles or radiation. Also called induced radioactivity. Also, the process of improving the adsorbant properties of a substance by using steam or heat (e.g., activated charcoal or alumina).

activation analysis   A very sensitive analytical technique in which certain elements, when exposed to a high neutron flux, produce radioactive species whose characteristic emissions are detectable at extremely low concentrations and can easily be quantified.

active ingredient   The primary component in a mixture, such as a drug or pesticide, responsible for obtaining the desired result or effect.

active system   A system requiring moving parts or motion such as water flow, air flow, electric current flow, and mechanical linkage movement.

activity   The number of disintegrations occurring in a quantity of radioactive material in a given amount of time.

activity schedule   A schedule covering various time periods and detailing modifications, activities, and important events. This schedule is prepared from input by all relevant groups or individuals.

activity series   An arrangement of the metals in the periodic table according to their ability to react with water and acids.

acuity   A measure of the keenness or sharpness of perception; often used in optics.

acute   Having a short, relatively severe course, with rapid changes occurring. *See* chronic.

acute exposure   A single, brief exposure to a toxic substance whose effects become evident soon after the exposure.

acute health effect   An adverse effect with severe symptoms developing rapidly and coming quickly to a crisis. *See* chronic effect.

acute lethality   The death of test animals immediately or within 14 days of a single dose exposure to a toxic substance.

acute toxicity scale   A measurement system to rate acute toxicity in terms of lethal doses and lethal concentrations of chemicals tested on laboratory animals. Lethal doses for humans are extrapolated from the results. Toxicity ratings range from relatively harmless to extremely toxic. *See* LC_{50}, LC_{Lo}, LD_{50}, LD_{Lo}, toxic dose (TD_{Lo}).

ADA   Americans with Disabilities Act; American Dental Association.

additive   A term used to designate a substance added to another substance or mixture that does not ordinarily contain the additive. The most common examples are food additives as defined by the US Food and Drug Administration (FDA).

adduct   A nonbonded association of two or more molecules, e.g., clathrate or caged compounds.

adeno-   Prefix meaning glandular.

adenoma   A benign epithelial tumor.

adenosine diphosphate (ADP)   5′-adenylphosphoric acid; a nucleotide important in the utilization of energy in living, especially muscle, cells; contains two phosphoric groups.

adenosine triphosphate (ATP)   5′-adenylphosphoric acid; a nucleotide important in the utilization of energy in living, especially muscle, cells; contains three phosphoric groups and is the energy storage moiety.

adequate ventilation   Ventilation to reduce air contaminants below a level that may cause personal injury or illness. *See* permissible exposure limit (PEL), short-term exposure limit (STEL), threshold limit value (TLV®). Also, ventilation sufficient to prevent contaminant vapors from accumulating in air in concentrations greater than 20% of the lower flammable limit. *See* flammable limits, lower explosive limit (LEL).

adhesion   A physical or chemical state in which two surfaces are held together.

ADI Acceptable daily intake.

adiabatic Describes a reversible thermodynamic system at constant energy which neither loses nor gains heat.

adjuvant A substance that enhances the effectiveness of the primary active ingredient, e.g., drug, food; also, a substance that enhances the immune response to an antigen.

administered dose The actual amount of substance or drug given over a set time period, as opposed to the theoretical amount calculated to be delivered.

administrative check A review to determine that no irregularities appear to exist, no items are obviously missing, and there is no indication of tampering.

administrative control A management technique or policy to limit employee exposure to workplace stress, e.g., exposure time, work schedule rotation. Such rules and policies govern operations, including training programs, internal review and change programs, organizational lines of authority, and operating procedure generation, review, and approval processes.

administrative management The management of groups or departments; usually the second contact level of management for employees, but may include higher management levels.

ADP Adenosine diphosphate; automatic data processing; also known as electronic data processing (EDP).

adrenaline A hormone, also known as epinephrine, normally secreted in mammals by the adrenal gland at a constant level to assist in regulating blood pressure, heart beat, and blood sugar levels. Stress and anger temporarily increase the secretion of adrenaline.

adrenocorticotropic hormone (ACTH) A white powder polypeptide soluble in water; widely used in medicine. *See* ACTH and cortisone.

adsorption The retention, holding, or binding of other substances on a surface, e.g., activated charcoal, activated alumina.

adsorption indicator A substance used in analytical chemistry as an indicator for precipitates that changes color when an excess of the desired substance is present.

ADT Automated data terminals.

adulterate An impurity in a regulated and formulated product, e.g., food or drug, that is legally not allowed.

AEA Atomic Energy Act.

AEC   Atomic Energy Commission; original name for what are now the Department of Energy and the Nuclear Regulatory Commission.

AECF   Association of Environmental Consulting Firms.

AED   Aerodynamic equivalent diameter.

aeolian   Pertaining to the wind.

aerobic   Describes an environment with oxygen or air present, e.g., chemical reactions or organisms that require oxygen to occur or to grow.

aerodynamic diameter (AED)   The diameter of a sphere with a density of one, having the same settling velocity as the particle being studied.

aerosol   A suspension of fine solid particles, i.e., fumes, dust, or smoke, or liquid particles, i.e., mist or fog, in gas. *See* smoke, fog, mist.

af   Audio frequency.

AFFF   Aqueous film-forming foam.

AFL-CIO   American Federation of Labor-Congress of Industrial Organizations.

aflatoxin   A highly toxic, stable mold or mycotoxin derived from *Aspergillus flavus* that is a common contaminant of moldy foods.

AFS   American Foundrymen's Society.

AFSHP   Association of Federal Safety and Health Professionals.

Ag   *Argentum* (Latin); silver.

AGA   American Gas Association.

agar   A colloid, also known as agar-agar, derived from algae in specific types of seaweed; used as a medium for culturing microorganisms and as a food thickener.

Agency for Toxic Substances and Disease Registry (ATSDR)   1600 Clifton Rd., NE, Atlanta GA 30333.

agent   A biological, chemical, physical, or radiological substance that may be hazardous to humans or the environment.

Agent Orange   A highly toxic herbicide, used for defoliation in the Vietnam War, that contains 2,4,5-trichlorophenoxyacetic acid (2,4,5,-T), 2,4-dichlorophenoxyacetic acid (2,4-D), and trace amounts of dibenzo-*p*-dioxins.

agglomeration   The growth of colloidal particles in suspended precipitations that causes the resultant cluster or floc to increase in size.

aggregate   A mixture of crushed stone, gravel, or sand used in sewage treatment and construction.

agitate   To mix, stir, or move as a result of violent action.

Agricola, Georgius   Latin name of Georg Bauer (1494–1555), a German mineralogist, metallurgist, physician, and author of *De Re Metallica* (1556), a record of sixteenth century mining and metalworking in which the use of protective masks for miners and smelters was advocated.

AGSE   Association of Groundwater Scientists and Engineers.

AGST   Above-ground storage tank(s).

AHA   American Hospital Association.

AHERA   Asbestos Hazard Emergency Response Act.

AHJ   Authority having jurisdiction.

AIA   American Institute of Architects; Asbestos Information Association.

AIC   American Institute of Chemists.

AICCPU   American Institute of Chartered Casualty Property Underwriters.

AIChE   American Institute of Chemical Engineers.

AIDS   Acquired immunodeficiency syndrome.

AIEEE   American Institute of Electrical and Electronics Engineers.

AIHA   American Industrial Hygiene Association.

AIHF   American Industrial Hygiene Foundation.

AIME   American Institute of Mining, Metallurgical and Petroleum Engineers.

AIP   American Institute of Physics.

air   A mixture of gases that surrounds the earth comprised of about 78% nitrogen, 21% oxygen, remainder primarily carbon dioxide, noble gases, hydrogen, methane, ozone, etc.

Air and Waste Management Association (AWMA)   1 Gateway Center, Pittsburgh PA 15222.

air balance   A system to ensure that the amount of air supplied to a facility is equal to or greater than the air exhausted from the facility.

air barrier   The unidirectional movement of air past and parallel to the plane of an opening at a velocity greater than that on either side, which creates resistance to movement of airborne particulates through the opening. Also known as an air curtain.

airborne particulate matter   A designation by the ACGIH that classifies material likely to be deposited in the respiratory tract.

air changes   Calculated by dividing the air flow rate in a room by the size of the room divided by time; usually expressed in air changes per hour.

air cleaner   A device for the treatment or removal of air contaminants.

air conditioning   The control of temperature and humidity of air in the home and workplace.

air conduction   The process by which sound is transmitted through air.

air convection   The movement of air as a result of gravity and differences in density.

air curtain   *See* air barrier.

air duct   A system used to transport conditioned air from one location to another.

airheads   Integrating air samplers.

air line respirator   A device worn by workers that provides a sufficient amount of clean, breathable air, using a tube of specified construction, from an external source to the potentially contaminated area.

airlock   An intermediate chamber between volumes (building areas) that have unequal pressure or temperature.

air monitor   A device to sample and measure the contaminants in the air.

air pollution   The contamination of air by undesired and hazardous or noxious substances that may be comprised of gases, vapors, mists, fumes, or dust; regulated by the US Environmental Protection Agency.

Air Pollution Control Association (APCA)   PO Box 2861, Pittsburgh PA 15230.

air-purifying respirator   A device worn by workers that contains filters that remove harmful air contaminants.

air quality   A legally regulated level of pollutants that may not be exceeded for a specific time in a specific location.

air sampling   A method used to sample, collect, and analyze air for the presence of contaminants or pollutants.

Air Transport Association of America (ATAA)   1301 Pennsylvania Ave., NW, Washington DC 20004-1707.

AISG   American Insurance Services Group.

AISS   American Iron and Steel Society.

AL   Acceptable level; action level.

Al   Aluminum.

ALAP   As low as practicable; *See* ALARA, which is preferred.

ALARA   As low as reasonably achievable; a basic concept of radiation protection that specifies that the radioactive discharges and radiation exposure to personnel be kept as low as practical.

alarm   An energized sound or light that serves to alert personnel that action must be taken to prevent damage to equipment or to indicate that there is a hazard to personnel or other undesirable conditions.

albumin   Water-soluble, naturally occurring proteins occurring in milk, blood serum, and eggs.

albuminuria   The presence of albumin in urine; a symptom of kidney disease.

ALCA   American Leather Chemists' Association.

alchemy   A medieval mixture of chemical science and philosophy that attempted to change common metals into gold, discover a universal cure for disease, and prolong life.

alcohol   A broad class of naturally occurring organic compounds containing a hydroxyl group; denotes 95% ethyl alcohol in chemistry laboratories; colloquially used to denote alcoholic beverages usually containing less than 50% ethyl alcohol.

alcoholysis   A reaction between alcohol and an organic compound to form a new compound and water.

ALD   Average lethal dose.

aldehyde   A reactive organic functional group with the general formula RCHO characterized by the presence of a carbonyl group, $C{=}O$; often volatile and used in flavoring and fragrances.

aldrin   A broad-spectrum chlorinated insecticide banned from use in the USA.

algae   Photosynthetic organisms occurring in fresh and salt water, which include seaweed and kelp; used in foods and food supplements as well as for the treatment of sewage; may adversely affect water quality by lowering the dissolved oxygen content.

-algia   Suffix meaning pain.

algicide   A chemical, such as copper sulfate, added to a water supply to control algae growth.

algorithm   A step-by-step procedure for solving a problem, often mathematical problems solved by computers.

ALI   Annual limits/intake.

alicyclic   A group of nonaromatic organic compounds possessing closed ring structures.

aliphatic   A straight or branch-chained organic group comprised of three **subgroups**   alkanes, alkenes, alkynes.

aliquot   A defined part of the whole; used in analytical chemistry to define the precise part or fraction of the total sample that is tested or analyzed.

alkali   A substance with the ability to neutralize an acid and form a salt. Sometimes referred to as a caustic or base. Alkalis can be very corrosive to the eyes, skin, and mucous membranes. *See* base.

alkali metal   A strongly electropositive **metal** found in group IA of the periodic table of elements, e.g., lithium, potassium, sodium, that reacts violently with water.

alkalimetry   A method of analysis for the concentration of free, unbound bases present in a solution.

alkaline earth metals   Members of group IIA of the periodic table of elements; less reactive than the alkali metals.

alkaloid   An organic compound, occurring naturally in certain plants, that contains nitrogen and frequently a ring structure and is often highly toxic. Examples include strychnine, caffeine, morphine, atropine, nicotine, quinine, cocaine.

alkalosis   A condition resulting from the loss of acid or accumulation of base in the human body; sometimes caused by exposure to high altitude.

alkane   A saturated, i.e., containing only C—C single bonds, hydrocarbon with the general formula C_nH_{2n+2}; also called paraffin.

alkene   An unsaturated hydrocarbon containing one or more carbon to carbon double bonds with the general formula C_nH_{2n}; also called olefins.

alkyl   A paraffinic hydrocarbon group derived from an alkane by removing one hydrogen atom possessing the general formula C_nH_{2n+1}.

alkylate   The process of adding one or more alkyl groups to a compound. Used in the petroleum industry to increase octane ratings.

alkyne   Unsaturated hydrocarbons containing one or more carbon to carbon triple bonds with the general formula C_nH_{2n-2}, e.g., acetylene.

allergen   A substance that produces an unusual response in persons particularly sensitive to that substance. *See* sensitizer.

allergy   The response of a hypersensitive person to an allergen in his/her environment. Allergies are estimated to occur in about 10% of the population.

Alliance for Acid Rain Control and Energy Policy (AARCEP)   444 Capitol St., NW, Washington DC 20001.

Alliance of American Insurers (AAI)   1501 Woodfield Rd., Schaumburg IL 60173-4980.

allotrope   A substance capable of existing in one or more forms; polymorphic; examples include carbon, diamond, graphite; oxygen, and ozone.

alloy   A solid or liquid mixture of two or more metals whose properties are different, and more specifically desirable, than any of the components.

alnico   A powerful, permanently magnetic alloy of iron, aluminum, nickel, and cobalt.

alopecia   Loss of hair or feathers; baldness.

alpha, α   A Greek symbol, meaning the first, used to designate the position closest to another group or position in an organic compound; optical rotation; an allotrophic form; radiation in the form of the helium nucleus.

alpha emitter   A substance that gives off radiation in the form of alpha particles or helium nuclei.

alpha particle   A positively charged radioactive particle with a charge of two, consisting of a helium nucleus made up of two protons and two neutrons, with low penetrating power and short range that cannot penetrate the skin but is extremely dangerous if inhaled or ingested; also, an alpha ray.

alpha radiation   Alpha particle radiation with low penetration, stopped by paper; a health hazard only inside the body.

ALR   Allergenic effects.

alternating current (AC)   Electric current that reverses its direction at regular intervals.

Altman, Sidney   Canadian chemist (1939–) who shared the 1989 Nobel Prize in Chemistry for studies of catalytic properties of RNA.

alum   Aluminum ammonium sulfate, aluminum potassium sulfate, aluminum sulfate; often used as an emetic, astringent, or styptic.

alumina   A porous form of aluminum oxide used for its adsorbant properties in gas and water purification and chromatography.

aluminosis   A form of pneumoconiosis caused by the presence of aluminum-containing dust in the lungs.

Aluminum (Al)   A light group IIIA metallic element, atomic number 13, atomic weight 26.98; a common metal with many commercial uses.

aluminum hydride   A white to grayish white compound with the formula AlH_3, which produces hydrogen in the presence of water and presents high fire and explosive risks.

alveolitis   Inflammation of the alveoli of the lungs.

alveolus   A terminal air sac in the lung with capillaries that enable gases to be absorbed; plural is alveoli.

Am   Americium.

AMA   American Medical Association.

amalgam   A mixture of mercury and another metal; most commonly used in dentistry.

ambient   Describes existing, present or surrounding environmental conditions as in ambient temperature.

amebiasis   Infection by ameba.

amended water   Water containing a wetting agent to enhance penetration. *See* surfactant.

American Academy for Laboratory Accreditation (AALA)   656 Quince Orchard Rd., Gaithersburg MD 20878-1409.

American Academy of Environmental Engineers (AAEE)   130 Holiday Ct., Annapolis MD 21401.

American Academy of Industrial Hygienists (AAIH)   302 South Waverly Rd., Lansing MI 48917.

American Academy of Occupational Medicine (AAOM) 2340 Arlington Heights Rd., Suite 400, Arlington Heights IL 80005.

American Association for Accreditation of Laboratory Animal Care (AAALAC) 11300 Rockville Pike, Rockville MD 21401.

American Association for the Advancement of Science (AAAS) 1333 H St., NW, Washington DC 20005.

American Association of Occupational Health Nurses (AAOHN) 50 Lennox Pointe, Atlanta GA 30324.

American Association of Poison Control Centers (AAPCC) 20 North Pine, Baltimore MD 21201.

American Association of State Highway and Transportation Officials (AASHTO) 444 Capitol St., NW, Washington DC 20001.

American Association of Textile Chemists and Colorists (AATCC) PO PO PO Box 12215, 1 Davis Dr., Research Triangle Park NC 27709.

American Astronautical Society (AAS) 6852 Rolling Mill Pl., Springfield VA 22152.

American Biological Safety Association (ABSA) 1202 Allanson Rd., Mundelein IL 60060.

American Board of Industrial Hygiene (ABIH) 6015 W. St. Joseph, Suite 102, Lansing MI 48917-3980.

American Cancer Society (ACS) 1599 Clifton Rd., NE, Atlanta GA 30329.

American Chemical Society (ACS) 1155 Sixteenth St., NW, Washington DC 20036.

American Chemistry Council (ACC) 1300 Wilson Blvd., Arlington VA 22209.

American College of Laboratory Animal Medicine (ACLAM) 96 Chester St., Chester NH 03036.

American College of Occupational Medicine (ACOM) 55 West Seegers Rd., Arlington Heights IL 60005.

American College of Toxicology (ACT) 9650 Rockville Pike, Bethesda MD 20814.

American Conference of Governmental Industrial Hygienists (ACGIH) 1330 Kemper Meadow Dr., Cincinnati OH 45240.

American Council on Science and Health (ACSH) 1995 Broadway, New York NY 10023-5860.

American Dental Association (ADA)   211 East Chicago Ave., Chicago IL 60611.

American Federation of Labor-Congress of Industrial Organizations (AFL-CIO)   815 16th St., NW, Washington DC 20006.

American Foundrymen's Society (AFS)   505 State St., Des Plaines IL 60016-8399.

American Gas Association (AGA)   1515 Wilson Blvd., Arlington VA 22209-2470.

American Hospital Association (AHA)   325 Seventh St., NW, Washington DC 20004.

American Industrial Hygiene Association (AIHA)   2700 Prosperity Ave., Suite 250, Fairfax VA 22031.

American Institute of Architects (AIA)   1735 New York Ave., NW, Washington DC 20006.

American Institute of Chartered Casualty Property Underwriters (AICPCU)   720 Providence Rd., Malvem PA 19355.

American Institute of Chemical Engineers (AIChE)   345 E. 47th St., New York NY 10017.

American Institute of Chemists (AIC)   7315 Wisconsin Ave., Bethesda MD 20814.

American Institute of Electrical and Electronics Engineers (AIEEE)   345 E. 47th St., New York NY 10017.

American Institute of Mining, Metallurgical, and Petroleum Engineers (AIME)   345 E. 47th St., New York NY 10017.

American Institute of Physics (AIP)   335 E. 47th St., New York NY 10017.

American Insurance Services Group (AISG)   85 John St., New York NY 10038.

American Iron and Steel Society (AISS)   1133 15th St., NW, Washington DC 20005.

American Leather Chemists' Association (ALCA)   University of Cincinnati, Cincinnati OH 45221.

American Medical Association (AMA)   535 North Dearborn St., Chicago IL 60610.

American Meteorological Society (AMS)   45 Beacon St., Boston MA 02108.

American National Standards Institute (ANSI)   1819 L St., NW, Washington DC 20036.

American Nuclear Society (ANS)   555 N. Kensington Ave., LaGrange Park IL 60525.

American Occupational Medical Association (AOMA)   2340 S. Arlington Heights Rd., Arlington Heights IL 60005.

American Oil Chemists Society (AOCS)   PO Box 3489, Champaign IL 61826-3489.

American Petroleum Institute (API)   1220 L St., NW, Washington DC 20005.

American Pharmaceutical Association (APhA)   2215 Constitution Ave., NW, Washington, DC 20037-2985.

American Pollution Control Association (APCA)   PO Box 2861, Pittsburgh PA 15230.

American Public Health Association (APHA)   1015 15th St., NW, Washington DC 20005.

American Red Cross (ARC)   430 17th St., NW, Washington DC 20006.

American Society for Heating, Refrigeration and Air Conditioning Engineers, Inc. (ASHRAE)   1791 Tullie Circle, N.E., Atlanta GA 30329.

American Society for Mechanical Engineers (ASME)   Three Park Ave., New York NY 10016-5990.

American Society for Metals International (ASM)   9639 Kinsman Rd., Materials Park OH 44073-0002.

American Society for Nondestructive Testing (ASNT)   1711 Arlingate Ln., PO Box 28518, Columbus OH 43228-0518.

American Society for Testing Materials (ASTM)   100 Barr Harbor Dr., West Conshohocken PA 19428-2959.

American Society for Training & Development (ASTD)   PO Box 1443, 1640 King St., Alexandria VA 22313.

American Society of Civil Engineers (ASCE)   1015 15 St., NW, Suite 600, Washington DC 20005.

American Society of Microbiology (ASM)   1752 N St., NW, Washington DC 20036-2804.

American Society of Safety Engineers (ASSE)   1800 E. Oakton, Des Plaines IL 60018.

American Society of Sanitary Engineers (ASSE)   PO Box 40362, Bay Village OH 44140.

American Standard Code for Information Interchange (ASCII)   A binary alphanumeric code used throughout the computer industry developed by the International Standards Organization (ISO) and computer manufacturers. It is the widely accepted standard for encoding text and is commonly used as the basis for e-mail messages.

American Standards Association (ASA)   11 W. 42nd St., New York NY 10036.

American Supply Association (ASA)   222 Merchandise Mart, Suite 1360, Chicago IL 60654.

Americans with Disabilities Act (ADA)   A US law established in 1991 that prohibits discrimination in public or the workplace on the basis of disability.

American Vacuum Society (AVS)   335 E. 45th St., New York NY 10017.

American Veterinary Medical Association (AVMA)   1931 North Meacham Rd., Schaumburg IL 60173.

American Water Works Association (AWWA)   6666 West Quincy Ave., Denver CO 80235.

American Welding Society (AWS)   PO Box 351040, Miami FL 33135.

americium (Am)   A radioactive, transuranic element, atomic number 95, atomic weight 243.

Ames test   A quick laboratory test for mutagenicity using Salmonella typhimurium.

amide   An organic compound containing the CONH2 group.

amine   An organic compound which is similar to NH_3 with other functional groups replacing some or all of the H. Primary amines have 2 H and 1 alkyl, secondary have 2 alkyl and 1 H, and tertiary have 3 alkyl groups.

amino acids   The chemical compounds containing an amino group as well as a carboxyl group; considered the building blocks of proteins.

amorphous   Having no structure; usually refers to a solid with no crystal structure, but the other states of matter are also amorphous.

amosite   A form of asbestos.

ampere   SI unit measuring electric current.

amphetamine   1-phenyl-2-aminopropane; a colorless, flammable liquid with a strong odor; usually associated with central nervous system drugs.

amphibole   A form of asbestos.

ampicillin   6, D-α-Aminophenyl-acetamido penicillanic acid; an antibiotic.

amplitude   Size or magnitude; the maximum value of a periodically varied quantity.

ampoule, ampule   A sealed container, usually glass, for transfer of a small amount of substance without contamination or exposure to potential reactants.

AMS   American Meteorological Society; aerial measuring system.

amu   Atomic mass unit.

amylase   A class of enzymes that changes starches into sugars.

amyl nitrite   An antidote for cyanide poisoning.

anaerobic   Literally, "without air"; generally meaning oxygen free.

analgesia   The loss of sensitivity to pain.

analgesic   A compound or mixture used to relieve pain.

analog   A compound of the same structural type.

anaphylaxis   A severe allergic reaction to a foreign substance with serious consequences including death.

androgen   A male sex hormone.

androsterone $C_{19}H_{30}O_2$; a male steroid; metabolic product of testosterone.

anemia   A symptom characterized by a reduced number of red blood cells in the blood of the organism.

anemometer   An instrument to measure gas velocity.

anesthesia   The loss of sensation or feeling.

anesthetic   A compound that induces unconsciousness or lack of feeling in a part of an organism.

Anfinsen, Christian Boehmer   American chemist (1916–) who shared the 1972 Nobel Prize in Chemistry for work on ribonuclease.

angiosarcoma   A malignant neoplasm originating in blood vessels.

Ångstrom, Anders Jonas   Swedish astronomer and physicist (1814–1874) who studied light, made spectral analyses in the solar system, and discovered hydrogen in the solar atmosphere; the Ångstrom unit used to measure the length of light waves is named after him.

angstrom unit   Å; a very small unit of measure; 10^{-10} meters or defined as the red line of cadmium $= 6438.4696$ Å.

anhydride   A compound derived from another compound (e.g., an acid) by removing water.

anhydrite   A marblelike form of calcium sulfate.

anhydrous   No water molecules are present in the form of a hydrate or as water of crystallization; literally, "without water."

aniline   Phenylamine; an important starting compound for the synthesis of dyes, drugs, etc.

aniline dye   A large class of mostly toxic dyes derived from aniline.

anion   Negatively charged ion; an ion that is attracted to the electrolytic anode. *See* ionization.

ANL   Argonne National Laboratory.

anneal   To temper a substance by gradual change of temperature after maintenance at a given temperature; the process relieves internal stress and strain or impurities.

anode   The electrode at which oxidation occurs. In a galvanic cell, the anode is negative, but in an electrolytic cell, the anode is positive.

anodizing   An electrolytic treatment of certain reactive metals to form a tightly adhering oxide coating that protects and can sometimes color the metal.

anorexia   Loss of appetite.

anosmia   Loss of the sense of smell.

anoxia   Loss or deficit of oxygen; literally, "without oxygen."

ANS   American Nuclear Society.

ANSI   American National Standards Institute.

antacid   A substance used to counteract or neutralize an acid condition; usually used in the context of the human stomach.

antagonist   A counteracting or opposing compound, object, or organism.

anthracite   A form of coal; hard coal as opposed to bituminous.

anthracosilicosis   A form of pneumoconiosis in which carbon and silica deposits occur in the lungs; associated with coal dust environments.

anthracosis   Carbon accumulated in the lungs; caused by inhalation of coal dust or smoke.

anthrax   An acute infectious disease, generally associated with sheep or cattle but can occur in humans and sometimes prove fatal.

anthropometry   The science of measuring the human body.

anti-   A prefix meaning against or counter; in geometric isomers, both groups are on the same side of a C=N bond.

antibiotics   Chemotherapeutic agents that can interfere directly with the proliferation of microorganisms at concentrations that are tolerated by the host.

antibody   A protein formed in the body in response to an antigen, which may subsequently defend the body against that antigen.

anticaking agent   A substance added to a finely divided matrix to maintain the physical status; prevents agglomeration or reaction of the matrix with, e.g., water in the air.

anticancer agent   An agent that is anticarcinogenic; prevents or delays tumors.

anticipation   The first responsibility of industrial hygienists in the practice of assuring a safe environment is to anticipate what problems might occur; *See* recognition, evaluation, and control; also, occurring before the normal onset time of a given disease or illness.

anticoagulant   A substance used to prevent coagulation; usually used for anticlotting blood.

anticontamination (anti-C) clothing   More rigorously protective clothing that may be required for specific jobs and/or operations. Full anti-C clothing consists of paper coveralls, hoods, gloves, shoe covers, and/or boots. Worn in addition to normal precautionary clothing for the purpose of radiological/ contamination control; not worn outside controlled areas.

anticorrosive   A substance used to prevent corrosion.

antidote   A specific therapeutic agent or remedy that counteracts an injurious effect.

antiemetic   A substance used to prevent vomiting.

antifertility   A substance used to suppress ovulation or sperm production; contraceptives may be administered orally or by injection.

antifoam agent   A substance used to prevent the formation of foam.

antifreeze   A substance used to prevent freezing; usually a chemical that lowers the freezing point of water or the solvent of interest.

antigen   Any object that is capable of stimulating the production of an immune response.

antihistamine   A compound used to inhibit or counteract histamines; usually associated with colds or allergies.

antihypertensive   A compound that lowers the blood pressure of an organism.

anti-inflammatory   A compound used to prevent or minimize inflammation.

antiknock agent   A compound that prevents knocking; usually associated with automobile engines.

antimalarial agent   A compound used to prevent or combat malaria.

antimildew agent   A compound used to prevent mildew.

antimony (Sb)   *Stibium* (Latin); a group VA metallic element, atomic number 51, atomic weight 121.75; widely used in alloys.

antineoplastic agent   A compound that prevents or inhibits the formation of neoplasms or cancer. *See* anticancer agent.

antioxidant   A substance that retards alteration by oxidation.

antiozonate   A compound used to inhibit ozone oxidation of a substrate; inhibits strong or severe oxidizing agents.

antipsychotic agent   A compound used to inhibit or prevent psychotic malfunctions.

antipyretic   A compound used to prevent or reduce fever or inflammation.

antiseptic   A compound that prevents the multiplication of microorganisms.

antistatic   A compound or process used to prevent static charge buildup; important for certain plastics in explosive or flammable environments.

antitoxin   An antibody produced to counter or neutralize a given biological toxin.

anuria   The absence or insufficient excretion of urine.

AOAC   Association of Official Analytical Chemists.

AOCS   American Oil Chemists Society.

AOM   Annual operating and maintenance cost.

APCA   Air Pollution Control Association.

APCD   Air Pollution Control Division.

APEN   Air pollution emission notice.

APhA   American Pharmaceutical Association.

APHA   American Public Health Association.

API   American Petroleum Institute.

aplastic anemia   A drastic reduction in the number of red blood cells.

apnea, apneic   The transient lack of breathing.

applied dose   The dose given as contrasted to the dose received.

approved storage facility   A place in which materials are stored, consisting of one or more approved containers, conforming to the requirements and covered by a license or permit issued by the appropriate authority. Also called above-ground magazine, service magazine.

aprotic solvent   A solvent that does not accept or supply hydrogen ions.

AQRV   Air quality-related values.

AQTX   Aquatic toxicity.

aqua fortis   Weak nitric acid.

aqua regia   A mixture of 1 part nitric and 3 or 4 parts hydrochloric acids that is a strong dissolving liquid, capable of dissolving even "noble metals" like gold.

aquatic toxicity   The hazardous effect observed in a specific form of fish or sea life as a result of exposure to a toxic substance.

aqueous   Describes a water-based solution or suspension.

Ar   Argon.

ARAR   Applicable or relevant and appropriate requirement.

ARC   American Red Cross.

arc furnace   A furnace that uses an electric arc to generate the heat; electrodes are usually carbon.

arc welding   Welding performed utilizing electrode arcs.

archiving   Saving, as a history record (e.g., past effective information).

area sampling   Determinations or measurements made in a specified place.

arene   *See* aromatic.

argon (Ar)   A group VIII noble gas element, atomic number 18, atomic weight 39.94.

Argonne National Laboratory (ANL)   9700 S. Cass Ave., Argonne IL 60440.

argyria   The impregnation of the body (skin) tissue with silver salts, resulting in a local or generalized gray-blue coloration.

arithmetic mean   The sum of a group of values divided by the number of values.

aromatic   Describes hydrocarbons characterized by the presence of the benzene ring structure (e.g., toluene, xylene).

aromaticity   *See* resonance.

Arrhenius, Svante August   Swedish scientist (1859–1927) awarded the 1903 Nobel Prize in Chemistry for his studies of compound dissociation in solvents; noteworthy is his equation. *See* Arrhenius equation.

Arrhenius equation $d \ln k/dT = A/RT^2$; where A is the energy of activation, k is the reaction rate constant, R is the gas law constant, and T is the absolute temperature.

arsenic (As)   A group V element, atomic number 33, atomic weight 74.92, compounds of which are associated with poisoning. They are also clastogens and mutagens.

arsine AsH_3; a colorless toxic gas often used in the electronics industry.

ART   Aqueous recycle technologies.

arthritis   A condition of inflammation of a joint, usually with pain and swelling.

arthro-   Pertaining to the joints.

arthropod   A member of the largest animal phylum; invertebrate animals with a hard jointed exoskeleton, segmented bodies, and jointed, paired appendages.

arthus phenomenon   A severe local inflammation where repeated injections have occurred.

ARU   Acid recovery unit.

aryl   A group of compounds containing the benzene or multiple aromatic ring structure.

As   Arsenic.

AS   Atmosphere, standard.

ASA   American Standards Association; American Supply Association; Acoustical Society of America.

asbestos   A group of naturally occurring minerals that separate into fibers and were used extensively for insulation. The great majority of asbestos in the US is chrysotile (95%) and, to a lesser extent, amosite. Certain man-made fibers such as refractory ceramic fibers (RCF) are considered asbestos for control purposes.

Asbestos Information Association (AIA)   1745 Jefferson Davis Highway, Arlington VA 22202.

asbestosis   A disease of the respiratory system caused by inhalation of asbestos fibers.

asbestos vacuum   A high-efficiency particulate air (HEPA)-filtered vacuum device for asbestos use only.

asbestos worker   A worker whose 8-hour personal asbestos air sample exceeds 0.1 fibers/cm^3, as determined by the Occupational Safety and Health Administration (OSHA) reference method, or who is routinely at risk of exposure to airborne asbestos.

as-built   Describes a drawing or other final design output that incorporates all approved changes and is the final accepted configuration of a project, system component, or item.

ascariasis   An infestation of nematode worms; usually in the intestines, but also can occur in the liver, lungs, or stomach.

ASCE   American Society of Civil Engineers.

ASCII   American Standard Code for Information Interchange.

ascorbic acid $C_6H_8O_6$; vitamin C; white crystals soluble in water and essential in the human diet to prevent scurvy; a common dietary supplement.

-ase   A suffix used in biochemistry to denote an enzyme, e.g., cholinesterase; however, there are also enzymes ending with "in," e.g., pepsin.

aseptic   Sterile or free from germs.

ASHAA   Asbestos in Schools Hazard Abatement Act.

ASHRAE   American Society of Heating, Refrigerating, and Air Conditioning Engineers, Inc.

ASI   Aviation Safety Institute.

askarel   The generic name for synthetic dielectric material that evolves only nonexplosive gases when decomposed by an electric arc.

ASLAP   American Society of Laboratory Practitioners.

ASM   American Society for Metals; American Society of Microbiology.

ASME   American Society of Mechanical Engineers.

ASNT   American Society for Nondestructive Testing.

aspect ratio   The ratio of one dimension to another.

aspergilli   Molds and fungi that ferment carbohydrates into organic acids, e.g., citric acid.

asphalt   A black or dark brown semisolid or solid rich in high-molecular-weight organic compounds; used for paving, roofing, etc.

asphyxia   The lack of oxygen and interference with the oxygenation of the blood.

asphyxiation   Suffocation resulting from being deprived of oxygen. Chemical asphyxiants act chemically to prevent oxygen from reaching the tissue or prevent the tissue from using it even though the blood is well oxygenated. (e.g., carbon monoxide combining with hemoglobin in the blood to reduce the capacity for transporting oxygen). Simple asphyxia results when a substance has displaced the oxygen in the air. *See* oxygen deficiency.

aspiration hazard   The danger of drawing into the lungs substances (e.g., petroleum-based solvents) that may result in chemical pneumonitis.

aspiration pneumonia   Pneumonia caused by the aspiration of a foreign substance, i.e., chemical or biological.

aspirin   Acetylsalicylic acid; a white, amorphous crystalline compound; a widely used analgesic, antipyretic medicine.

assay   The weight (%) of material in a given item. Usually used for ores.

ASSE   American Society of Safety Engineers; American Society of Sanitary Engineers.

assessment   An appraisal to evaluate the effectiveness of an activity/operation. To determine the extent of compliance with required procedures and practices. To perform an evaluation of a material control and accounting anomaly or material discrepancy.

assigned protection factor   The level of respiratory protection afforded by properly fitted and functioning respirators worn by trained personnel.

Association of Environmental Consulting Firms (AECF)   1 E. Wacker Dr., Chicago IL 60601.

Association of Federal Safety and Health Professionals (AFSHP)   7549 Wilhelm Dr., Lanham MD 20706.

Association of Groundwater Scientists and Engineers (AGSE)   PO Box 182039, Columbus OH 43218.

Association of Official Analytical Chemists (AOAC)   111 N. 19th St., Arlington VA 22209.

AST   Above-ground storage tank.

astatine (At)   A radioactive group VII element, atomic number 85, atomic weight 210.

ASTD   American Society for Training and Development.

asthma   A disease characterized by recurrent attacks of dyspnea, wheezing, and perhaps coughing caused by spasmodic contraction of the bronchioles in the lungs.

ASTM   American Society for Testing Materials.

Aston, Francis William   English chemist (1877–1945) awarded the 1922 Nobel Prize in Chemistry for isotope studies with his mass spectrograph.

asymmetry   Lack of symmetry; e.g., in organic chemistry, four different groups attached to a tetrahedral carbon atom.

asymptomatic   Neither causing nor exhibiting symptoms.

At   Astatine.

ATAA   Air Transport Association of America.

at%   Atomic percent.

at. no.   Atomic number.

at. wt   Atomic weight.

ataxia   Loss of muscular coordination.

-ate   A suffix used in chemistry to denote the highest oxidation state of the metal in an inorganic salt.

atm   Atmosphere, a unit of pressure; 760 millimeters of mercury pressure is called the standard atmospheric pressure. *See* pressure.

atmosphere   The gas phase environment or surroundings; most commonly of the earth but can also be generally used; a unit of pressure = 760 mm of mercury = 1 atmosphere.

atmospheric pollution   The unhealthy additive(s) to natural or clean atmospheres.

ATMX   Atomic materials rail transport car.

atom   The smallest part of an element that remains unchanged during chemical reaction, consisting of a nucleus of protons and neutrons surrounded by electrons; the fundamental building block of chemical elements.

atomic absorption spectroscopy (AA)   An analytical technique that vaporizes a substance and determines the absorbance at given wavelengths to quantitatively determine the amount of the absorbing atom(s).

atomic mass unit (amu)   The mass of the carbon-12 atom is defined to equal 12 amu.

atomic number (Z)   The number of positively charged protons in the nucleus of an atom, which is also equal to the number of electrons on an electrically neutral atom. Each chemical element has its own atomic number.

atomic radius   The radius of an (assumed spherical) atom, generally in Å units.

atomic weight (A)   The mass of a specific number of atoms of an element (approximately 6×10^{23}) in grams; the total number of protons and neutrons in the nucleus of an atom of the element; the mass of an atom. *See* mass number.

atopic dermatitis   A skin disease characterized by the degeneration or wasting away of skin cells, characterized by a lack of accent; a hypersensitivity usually associated with heredity.

ATP   Adenosine triphosphate.

atrophy   A wasting or reduction in size or function.

atropine $C_{17}H_{23}NO_3$; a white crystalline compound used in medicine as an antidote for cholinesterase inhibitors such as nerve gases.

ATSDR   Agency for Toxic Substances and Disease Registry.

attenuation   The process by which a beam of radiation is reduced in intensity when passing through material; a combination of absorption and scattering processes.

atto-(a)   A prefix $= 10^{-18}$; as instrumental detection limits decrease, a term encountered more often.

Au   *aurum* (Latin); gold.

audible range   The frequency range over which normal ears hear, approximately 20 hertz through 20,000 hertz.

audiogram   The written record of the audiometer test.

audiologist   Someone who specializes in testing hearing and correcting any disorder(s) found.

audiometer   An instrument used to test hearing.

audio testing program   Part of a written hearing conservation program used to document the testing, evaluation, and prevention of worker exposure to hazardous noise.

audit   A planned and documented activity performed to determine, by investigation, examination, or evaluation of objective evidence, the adequacy of and compliance with established procedures, instruction, drawings, and other applicable documents and the effectiveness of implementation. Not to be confused with surveillance or inspection activities performed for the sole purpose of process control or product acceptance. A documented activity performed in accordance with written procedures or checklists to verify by examination and evaluation of objective evidence that applicable elements of a quality assurance or safety program have been developed, documented, and effectively implemented in accordance with specified requirements.

auditor   An individual such as a technical specialist, a management representative, or an auditor-in-training, who performs any portion of a quality or safety audit.

auditory   Having to do with hearing.

Auger, Pierre-Victor   French physicist (1899–1931) who discovered the photoelectric effect.

Auger electron   The electron that falls from a higher atomic energy level to a lower one after incident energy has ejected the electron from the lower energy level. This secondary process emits a photon whose energy is determined by the difference in energy between the levels and leaves an ionized atom. The process is common in low Z atoms.

auric   Golden; having to do with gold compounds.

authority having jurisdiction (AHJ)   Federal, state, and local agencies that have the authority to regulate.

authorized entry   A confined space, as defined by OSHA in 29 CFR 1910.146, which requires a permit to enter.

autoclave   An apparatus for sterilization by steam under pressure.

auto-ignition point   The temperature at which a substance will ignite without introduction of an initiating source.

auto-ignition temperature   The lowest temperature at which a flammable gas or vapor-air mixture will ignite from a heat source without the necessity of spark or flame. *See* ignition, temperature.

autoimmunity   A condition in which an organism produces antibodies against its own tissues.

autolysis   Self-digestion.

autolyte   A substance that conducts electric current.

automatic actuation logic   The matrix of permissives, interlocks, and any other conditions that must be satisfied to generate an automatic actuation.

automatic control logic   The matrix of permissives, interlocks, and any other conditions that must be satisfied to generate a control function.

autooxidation   The oxidation of a substance by air without need of other compounds or abnormal conditions.

aversion time   The time required for the human eye to blink when exposed to a stimulus like light.

avian pneumocephalitis   A synonym for Newcastle disease.

Aviation Safety Institute (ASI)   PO Box 304, Worthington OH 43085.

AVMA   American Veterinary Medical Association.

AVO   Avoid verbal orders.

Avogadro, Amedeo   Italian chemist (1776–1856) noted for stating the principle that equal volumes of gases at the same temperature and pressure contain the same number of molecules, i.e., 6.023×10^{23} for 22.4 liters of a gas.

Avogadro's number (N) 6.023×10^{23}; the number of molecules in a mole of that compound; the number is for 22.414 liters of gas at standard temperature and pressure.

avoid verbal orders (AVO)   A written authorization is necessary.

AVS   American Vacuum Society.

awareness barrier   Attachment/device that by physical and visual means warns a person of an approaching or present hazard.

AWG   American wire gauge.

AWMA   American Waste Management Association.

AWQC   Ambient water quality criteria.

AWS   American Welding Society.

awu   Atomic mass unit.

AWWA   American Water Works Association.

AWWARF   American Water Works Association Research Foundation.

axial fan   A propeller-type fan used to move large volumes of air with little resistance.

axillary   Having to do with the armpits.

axis of rotation   An imaginary line or point about which motion occurs, e.g., limb rotation.

axis of symmetry   An imaginary line about which rotation may occur and that will produce two identical parts.

azeotrope   (or azeotropic mixture); A liquid mixture of two or more compounds with a constant boiling point having the same composition in the vapor and liquid phases.

azide   Any compound in which the nitrogen present is in a -3 oxidation state; heavy metal azides are explosive, as are many organic azides.

azo dye   The group of dyes that contain the $-N{=}N-$ chromophore.

B

β Lower case Greek letter beta; beta particle (radiation).

B   Boron; blower.

b   Barn; unit of measure for neutron capture cross section; 1 barn = 10^{-24} cm^2.

Ba   Barium.

Babbitt   An alloy used for making bearings containing tin, copper, antimony, and arsenic.

BAC   Blood alcohol concentration; budgeted cost at completion; biologically activated.

bacillus   A type of aerobic, Gram-positive rod-shaped bacteria.

background level   The naturally occurring amount of a substance or contaminant being measured; the blank or reference level.

background radiation   The radiation that comes from the environment, consisting of radiation from cosmic rays and from the naturally radioactive elements of the earth, including that from within the body. Radiation occurring naturally everywhere. An average individual exposure from background radiation is 125 millirem per year in midlatitudes at sea level.

backpressure   The difference between the pressure flowing into a system and the actual pressure in the system. If the back pressure in a system is higher than the supply pressure in a water system, contamination of potable water may result.

BACT   Best available control technology.

bacteria   A small, unicellular microorganism commonly found in plants, animals, and the environment certain species of which may cause disease in humans.

bactericide   An agent that can kill bacteria.

bacteriostat   An agent that prevents, slows, or stops bacteria growth.

baffle   A device that that deflects, slows, stops, or changes the flow of a fluid, i.e., gases or liquids.

bagasse   The fiber of sugar cane stalks that remains after the juice is extracted; composed mainly of cellulose with a content of approximately 4% ash, 2% protein. and <0.5% silica. It rapidly turns into dust when dry.

bagassosis   A fungal disease of the lung caused by exposure to airborne dried sugar cane stalks, or bagasse; similar to farmer's lung.

baghouse   A part of an air-purifying system consisting of bags made of textile or other fibers through which air containing dust and/or harmful particulates is drawn.

baking powder   A leavening agent widely used in baking; typically composed of sodium bicarbonate, tartaric acid, cornstarch, and potassium tartrate. When heated, carbon dioxide is produced, which raises the dough.

baking soda $NaHCO_3$; sodium bicarbonate.

balance   To equal. Chemical equations are said to be balanced when the number of atoms reacting yield an equal number of product atoms and charges, if any, balance. Also, an instrument used to weigh substances; often used in laboratories.

ball mill   A steel device used to grind dry materials such as chemicals.

Balmer, Johann Jakob   Swiss chemist (1825-1898) noted for his work on spectral series and his 1885 formula for the wavelengths of hydrogen.

banana oil   Amyl acetate; used to fit-test respirators.

Banbury mixer   A temperature-controlled device used in the rubber industry.

bandwidth (BW)   A range of wavelengths, frequencies, or energies.

Bang's disease   A synonym for brucellosis, also known as undulant fever or Malta fever, a disease caused by the brucella bacteria now mainly confined to meatpacking industry workers, livestock producers, and veterinarians.

bar   A unit of measure of pressure equal to 10^5 pascals, 10^5 newtons/m^2, or 10^6 dynes/cm^2.

barbiturate   A habit-forming derivative of barbituric acid, available by prescription; used as a sedative and anesthetic because it depresses the central nervous system, e.g., pentobarbital.

Bardeen, John   American physicist (1908-1991) who shared two Nobel Prizes in Physics (1956 and 1972).

baritosis   A benign pneumoconiosis caused by inhalation of the dust or fumes of insoluble barium compounds.

barium (Ba)   A soft, silvery white, alkaline earth group IIA metallic element, atomic number 56, atomic mass 137.34.

barometer   A device used to measure barometric or atmospheric pressure.

barometric pressure   The pressure exerted by air at a particular point, referenced to sea level and expressed in millimeters or inches of mercury, inches of water, or pounds per square inch absolute, psia, or atmospheres, atm.

barricade   A substantial dividing wall; an intervening barrier, natural or artificial, of such type, size, and construction as to limit, in a prescribed manner, the effect of an explosion or an undesired incident or exposure on individuals, the environment, or nearby buildings. *See* exposed site, firewall, shielding.

barrier   A physical method of separating individuals or an environment from an exposure to an undesired area, event, or substance.

barrier cream (lotion)   A protective cream that can be applied to the skin to provide minimal protection against selected chemical agents. The protection is not considered as effective as that provided by impervious gloves.

Barton, Sir Derek Harold Richard   English chemist (1918–) who shared the 1969 Nobel Prize in Chemistry for developing the concept of confirmation and its application to chemistry.

basal cell carcinoma   A skin tumor that is usually not metastatic.

basal metabolism   A measure of the amount of energy required by the body at rest.

base   A compound that yields hydroxyl ions in aqueous solution; a compound that forms a solution with pH greater than neutral. Also, a place for storing field supplies.

baseline   A point or line serving as a known reference from which a measurement is determined.

baseline data   Information collected before a specific time for later use in describing conditions before the project or activity began.

baseline survey   The first time an in-depth health and safety evaluation is performed, to which subsequent surveys will be compared.

BASIC   Beginner's all-purpose symbolic instruction code (a computer language).

basic   *See* alkali.

bat   Battery.

BAT   Best available technology.

batch   A quantity of material made at the same time and in the same manner.

BATF   US Bureau of Alcohol, Tobacco, and Firearms (within the Treasury Department).

Battelle Memorial Institute   Headquarters, 505 King Ave., Columbus OH 43201-2693.

Battelle Pacific Northwest Laboratories   PO Box 999, 902 Battelle Blvd., Richland WA 99352.

battery   A device that converts chemical energy into electrical energy.

battery acid   The electrolyte in batteries, i.e., sulfuric acid.

Bauer, Georg   *See* Agricola.

Baumé, Antoine   French chemist (1728-1804) who invented the hydrometer named after him and devised two scales for measuring specific gravity using this device: one for substances heavier than water and one for substances lighter than water.

Baumé   Specific gravity scales used on hydrometers, both of which were invented by French chemist Antoine Baumé.

bauxite   The naturally occurring mineral in which aluminum is found and its most common ore.

bauxite pneumoconiosis   A synonym for Shaver's disease; found in workers exposed to fumes of aluminum oxide when bauxite is smelted.

BBL   Body burden level; the level of a toxicant absorbed into the body through skin contact, inhalation, or ingestion.

bbl   Barrel.

bcc   Body-centered cubic.

BCD   Bar code decal.

BCF   Bioconcentration factor.

BCHM   Board of Certified Hazard Control Management.

BCL   Battelle Columbus Laboratories.

BCM   Blood-clotting mechanism effects.

BCSP   Board of Certified Safety Professionals.

BCSPM   Board of Certified Product Safety Management.

bd ft   Board foot (feet).

BDAT   Best demonstrated available technology.

Be   Beryllium.

Bé   Baumé.

BEA   Building evacuation area.

beam divergence   The spread arising from a ray or shaft of light or radiation.

Beckmann, Ernst   German chemist (1853–1923) who invented a very sensitive type of mercury thermometer used to very accurately measure small changes in temperature.

Beckmann thermometer   A very sensitive type of mercury thermometer used to measure small changes in temperature. Named after the German chemist Ernst Beckmann.

Becquerel, Antoine Henri   French physicist (1851-1908) who shared the 1903 Nobel Prize in Physics for his studies of radioactivity.

becquerel (Bq)   A unit, in the International System of Units (SI), for the measurement of radioactivity equal to one transformation or atomic disintegration per second (dps). There are 3.7×10^{10} becquerels per curie of radioactivity. Named after the French Nobel laureate physicist Henri Becquerel.

Beer, August   German physicist (1825–1863) who was one of the founders of photometry and after whom Beer's Law is named.

Beer's Law   States that the absorbance of a specific substance in a beam of monochromatic light is dependent on the thickness of the material and the concentration of the substance, expressed by the formula $A = abc$, where A is the absorbance, a is the absorptivity or extinction coefficient, i.e., the ability of a specific chemical to absorb light at a specific wavelength, b is the path length, and c is the concentration. Also known as the Beer-Lambert law; named after the German physicists August Beer and Johann Lambert.

beginning inventory   The quantity of any material present at the start of an accounting period.

behavior   Response or action as a response to stimulation.

BEI®   Biological exposure index.

Beilstein, Friedrich Konrad   German chemist born in Russia (1838–1906) who wrote the multi-volume *Handbook of Organic Chemistry* on organic chemical reactions and properties. Developed a test to detect halogens in organic compounds.

Beilstein test   A test developed by the German chemist Friedrich Konrad Beilstein for the detection of halogens in organic compounds.

BEIR   Biological effects of ionizing radiation.

BEJ   Best expert judgment.

belladonna   A bush whose very toxic leaves are used to make atropine; also known as deadly nightshade and banewort.

benchmark   An established reference point or standard measurements, to which subsequent measurements will be compared.

benchmarking   A study conducted to compare one product or practice to another to document changes or improve performance.

benign   Nonthreatening to life or health, as opposed to malignant. A benign tumor is one that does not spread from its primary site to secondary sites.

benzene C_6H_6; a leukogenic, flammable, high-production, aromatic ring chemical found in coal tar; The most simple member of the aromatic hydrocarbons.

benzidine   A carcinogenic, aromatic amine used in the production of certain dyes.

benzidine dye   A highly toxic, carcinogenic dye with an azo group (−N=N−).

benzine   A misleading, archaic term generally denoting volatile, flammable petroleum distillates.

Berg, Paul   American chemist (1926–) who shared the 1980 Nobel Prize in Chemistry for studies of the biochemistry of nucleic acids, particularly recombinant DNA.

Bergius, Friedrich   German chemist (1884–1949) who shared the 1931 Nobel Prize in Chemistry for the invention and development of high-pressure chemical methods.

beri-beri   A disease caused by a deficiency of vitamin B (thiamine).

berkelium (Bk)   A transuranic element, atomic number 97, atomic weight 247.

Berthollet, Claude Louis   French chemist (1748–1822) who discovered the use of chlorine for bleaching and, together with Lavoisier, devised a system of chemical nomenclature on which today's system is based.

berylliosis pneumoconiosis   A disease caused by exposure to beryllium dust or fumes.

beryllium (Be)   A light, strong, nonradioactive group IIA metal element, atomic number 4, atomic mass 9.01; possesses toxic properties and is a known human carcinogen; Exposures are connected to the disease berylliosis in some individuals.

Berzelius, Jons Jakob   Swedish chemist (1779–1848) who discovered and isolated selenium, cerium, thorium, titanium, zirconium, silicon, and niobium; also introduced the present system of chemical symbols.

beta   Greek lower case letter β.

beta decay   A radioactive decay process that gives off a beta particle.

beta particle   An elementary charged particle emitted from a nucleus during radioactive decay, with a mass equal to 1/1837 that of a proton. A negatively charged beta particle is identical to an electron. A positively charged beta particle is called a positron. Large amounts of beta radiation may cause skin burns, and beta emitters are harmful if they enter the body. Beta particles are easily stopped by a thin sheet of metal or plastic.

beta radiation   Irradiation by beta particles, intermediate in penetrating ability, i.e., penetrates paper but is stopped by thin plastic. Has significantly less health impact internally than alpha particles. *See* beta particle.

betatron   A donut-shaped device that accelerates electrons to high speeds with energies of millions of electron volts, thereby enabling production of new species.

BeV   Billion electron volts.

Bi   Bismuth; biot.

BIA   Brick Institute of America; Bureau of Indian Affairs.

bias   The difference between the expected value of an estimator and the true value being estimated; a persistent or systematic error that remains constant over a series of replicated measurements (also known as deterministic error, fixed error, or systematic error).

biennial   Every other year; a time interval not to exceed 30 calendar months.

bilateral   Having two sides, e.g., two sides of the human body or an organ.

bile   Fluid secreted by the liver into the small intestine by bile ducts.

bile acid   Any of several steroid acids derived from bile.

bimetal   A device composed of two different metals that are often bonded together because they expand at different rates on temperature change and thus may be used to control temperature.

binary   A system made of two parts or components.

binder   A material used to hold several parts or components together as in polymers, paints, drugs, foods, or construction materials.

binding energy   The work required to remove an electron from an atom; the energy released in binding protons and neutrons together in an atomic nucleus.

bioaccumulation   The build up of concentration of a particular substance in an organism over a period of time.

bioassay/bio-assay   The collection and analysis of human hair, tissue, nasal smears, urine, or fecal samples to chemically determine the amount of material that might have been deposited in the body. Routes of possible entry are inhalation, ingestion, or injection; The study of the effect of a substance or mixture on a living animal, system, or cell.

biochemical oxygen demand (BOD)   The amount of oxygen required for biological and chemical oxidation of a waterborne substance to take place.

biochemistry   The study of chemicals related to their reactions in living organisms.

biocide   A substance that kills or is destructive to living organisms.

bioconversion   The production of energy from or chemical changes to organic matter by a process that involves living organisms, e.g., fermentation.

biodegradable   Able to be broken down by a process that involves living organisms, such as microbes.

biodegradation   Biological decomposition of a material that occurs naturally.

bioengineering   The application of engineering principles to biological systems; often used to mean genetic alteration.

biohazard   A biological organism or product capable of self-replication that may produce harmful effects on other biological organisms, particularly humans, e.g., bacteria, viruses, molds, fungi, toxins, allergens, recombinant DNA products, parasites, etc.

Biological exposure indices (BEI®)   Reference values set by the ACGIH and intended for use as a guideline for the evaluation of potential health hazards resulting from levels of determinants most likely to be observed in specific biological specimens collected from healthy workers exposed to a specific chemical to the same extent as a worker with inhalation exposure to the threshold limit value (TLV®).

biological half-life   The time required for a biological system, such as that of a human, to eliminate by natural processes one-half the amount of a substance that is present within it.

biological hazard   A living agent that presents a risk to humans.

biologically effective dose   The amount of administered substance that actually reaches the internal target.

biological monitoring   The measurement of changes in or the amount of a specific substance present in human tissue, fluid, or expired air after exposure to a specific agent.

biological safety cabinet (BSC)   A cabinet designed to safely contain biological agents.

biological safety level   One of four levels of biological risk assigned on the basis of severity of consequences, with "1" being low or no risk and "4" being dangerous.

biological shield   A mass of absorbing material placed around a reactor or radioactive source to reduce the radiation to a level safe for humans.

bioluminescence   When a living organism emits visible light, e.g., a firefly; cold light.

biomarker   A substance whose detection or quantification indicates some biological phenomenon.

biomass   An organic, renewable energy source, e.g., wood or agricultural wastes; all of the bio-organic matter in a specified mass.

biomechanics   The study of the effects of forces on an organism; applying mechanical forces to an organism.

biopsy   A medical diagnostic tool that examines tissues removed from the body both macro- and microscopically.

biosynthesis   Production of complex molecules by a living organism; usually associated with compounds of medicinal importance.

biota   The plant and animal life of a region or ecosystem.

biotechnology   The application of scientific and engineering principles to the processing of any organic or inorganic substance by biological agents to provide goods and services.

biotransformation   A chemical change to a substance that occurs in an organism.

bird excrement   *See* guano.

bis   Prefix meaning twice. In chemical nomenclature it indicates that the group occurs twice in the molecule.

bismuth (Bi)   A white, brittle group VA metallic element, atomic number 83, atomic weight 208.98; used in low-melting alloys for fire safety products.

bit   A binary digit or information unit equal to one binary decision or the designation of one of two possible values.

bit/s   Bits per second.

bituminous coal   A mineral form of coal; soft coal as opposed to anthracite.

Bk   Berkelium.

black body   A physics term for a theoretical object that absorbs all incident radiation.

Black Death   A form of plague that swept through Europe and Asia for years after 1353.

black liquor   The liquor obtained from the kraft papermaking process, which cooks pulpwood in an alkaline solution.

black lung disease   The chronic lung disease associated with coal mining. *See* pneumoconiosis.

black powder   A potassium nitrate, charcoal, and sulfur ($\sim$ 75, 15, and 10%, respectively) mixture; low-explosive gunpowder.

blanch   To remove color; to whiten or bleach.

blank   A control sample treated identically to samples of interest but without inclusion of or exposure to the substance being measured.

blast furnace   A furnace that aids combustion by forcing air through it; usually associated with the production of steel.

blast gate   A sliding damper used to control air flow in a system.

blasting agent, blasting powder   *See* black powder.

BLD   Blood effects.

bleach   To remove the color, to make white or clear.

bleed   To lose blood, sap, or fluid from an organism or system.

bleomycin   A water-soluble blycopeptide powder; an antineoplastic and diagnostic agent.

BLEVE   Boiling liquid expanding vapor explosion.

BLM   Bureau of Land Management.

blood   The fluid circulated by the heart, essential for life. Carries oxygen and nutrients to cells and transports waste material to excretory areas.

blood-borne pathogens   Pathogenic microorganisms that may be present in human blood, blood components or products, body fluids, tissues, or organs. These pathogens include, but are not limited to, hepatitis B and C viruses (HBV, HCV) and human immunodeficiency virus (HIV), which may be spread by blood-soiled bandages, linens, needles, needle sticks, towels, wastes, etc. and hence are a concern to health care workers.

blood-borne pathogen standard   29 CFR 1910.1030.

blood-brain barrier   A barrier between the brain and circulating blood; keeps harmful substances from the brain.

blood count   The number of red and white blood cells in a given volume of blood.

Bloom's syndrome   An autosomal disorder that results in photosensitivity, dwarfism, and other abnormalities.

BLS   Bureau of Labor Statistics.

BM   Bowel movement; building manager.

BNL   Brookhaven National Laboratory.

BNWL   Battelle Pacific Northwest Laboratories.

Board of Certified Hazard Control Management (BCHCM)   8009 Carita Ct., Bethesda MD 20817.

Board of Certified Product Safety Management (BCPSM)   8009 Carita Ct., Bethesda MD 20817.

Board of Certified Safety Professionals (BCSP)   208 Burwash Ave., Savoy IL 61874-9571.

BOCA   Building Officials & Code Administrators International; Building Officials Conference of America.

BOD   Biochemical oxygen demand; a term used to indicate biological activity in the sample.

BOD5   Biochemical oxygen demand, 5-day incubation period.

body burden   The amount of radioactive material that, if deposited in the total body, will produce the maximum permissible dose rate to the body organ considered the critical organ.

body counter   Instrument used to count radiation in the body, primarily used to determine lung burdens.

body dosimetry badge   Dosimetry badge normally worn on the front of the upper body. Used to measure the penetrating radiation dose to which the wearer may be exposed.

Bohr, Niels   Danish physicist (1895–1962) awarded the 1922 Nobel Prize in Physics for his studies of the structure of atoms and the radiations from them.

boiler   A (usually closed) container for heating (boiling) liquids.

boilerscale   The crust formed inside boilers and tubes using hard water; usually calcium carbonates or sulfates.

boiling point (BP)   The temperature at which the vapor pressure of a liquid equals atmospheric pressure. Comparing boiling points can provide a relative index of volatility in many cases. Boiling temperature may be expressed in Fahrenheit (F) or centigrade (C). A physical property of chemical compounds commonly used to characterize them.

Boltzmann, Ludwig   Austrian physical chemist (1844–1906) who developed statistical mechanics as applied to the kinetic theory of gases and summarized his findings on their viscosity and diffusion in the Stefan-Boltzmann law.

Boltzmann constant   The ratio of the ideal gas constant to Avogadro's number.

BOM   Bureau of Mines.

bomb   A metal container holding pressurized gases or liquids; an explosive weapon.

bomb calorimeter   An instrument consisting of a sealed chamber in which combustion occurs and the resultant temperature change enables calculation of heats of combustion, reaction, etc.

bond   The force holding atoms together in compounds. Various types have been designated, e.g., ionic, covalent, hydrogen, double, triple.

bond energy   The energy needed to break a given bond.

bonding   The connecting of two objects. In chemistry, generally thought of as connecting atoms via ionic or covalent bonds; A connection between containers that prevents spark ignition of flammables, particularly when transferring.

bone meal   Coarse, powdery state of bones; used as fertilizer and animal feed.

bone seeker   A radioisotope that tends to accumulate in the bones when introduced into the body, e.g., strontium-90, which behaves chemically like calcium.

boral   A composite of boron carbide crystals in an aluminum matrix with aluminum cladding; used in reactor shielding.

borax $Na_2B_4O_7 \cdot 10H_2O$; a mineral, hydrated sodium borate.

boric acid H_3BO_3; a white, crystalline powder; used as an antiseptic or preservative and in fireproofing.

boride   Any compound with the boron anion in a -3 oxidation state.

boron (B)   A soft, brown, solid nonmetallic group III element; atomic number 5, atomic weight 10.81; used in hard alloys, flares, and reactor control rods.

Bosch, Carl   German chemist (1874–1945) who shared the 1931 Nobel Prize in Chemistry for the invention and development of high-pressure chemical methods.

botsball   A thermometer that is a sensor within a hollow copper sphere painted black and covered with black cloth.

botulism   A potentially lethal food poisoning caused by botulin.

Bouguer, Pierre   French scientist (1698–1758) who was one of the founders of photometry, the measurement of light intensities.

Bouin's solution or liquid   Picric acid, formalin, and acetic acid solution used to preserve tissues.

Boyer, Paul Delos   American chemist (1918–) who shared the 1997 Nobel Prize in Chemistry for elucidation of the enzymatic mechanism underlying the synthesis of adenosine triphosphate.

Boyle, Robert   Irish scientist and theologian (1627–1691) who studied the properties of air and formulated Boyle's law.

Boyle's law   States that the volume of a gas varies inversely with the pressure applied if the temperature is held constant. Works for ideal gases.

B.P., bp, b.pt.   Boiling point.

Bq   Becquerel.

Br   Bromine.

brackish   Briny, salt containing, distasteful; usually describes a solution.

bradycardia   Abnormally slow pulse (< 60 beats/min).

Bragg, Sir William Henry   English physicist (1862–1942) who shared with his son the 1915 Nobel Prize in Physics for X-ray crystallography.

Bragg, Sir William Lawrence   Australian-born physicist (1890–1971) who shared with his father the 1915 Nobel Prize in Physics for work on crystallography.

Bragg angle   The angle between incident X rays and those diffracted by the crystal planes.

Bragg-Gray law   The amount of ionization produced in a gas-filled cavity is proportional to the energy absorbed by the surrounding solid.

Bragg's law $n\lambda = 2d \sin \theta$; where n is an integer, λ is the wavelength of incident X ray, d is the distance between lattice planes, and θ is the Bragg angle.

brainstorm   An idea "out of the blue"; usually used as in group freewheeling interactions to solve a problem.

brake fluid   The fluid used in a hydraulic brake cylinder.

brake linings   Formerly asbestos-containing material used to brake a rotating shaft, most commonly encountered in automobiles.

braking horsepower   The power delivered to a shaft by a brake, expressed in horsepower units.

branch   A subdivision of a larger entity, e.g., artery, nerve; also a structural description of some organic molecules.

brass   An alloy of copper and zinc with other metals in various proportions.

braze   To decorate with brass; to harden or solder with a high-melting-point solder.

breakthrough   The loss of a substance, usually a fluid, through insufficient or defective means of transport, containment, or collection.

breakthrough time   The time required for a substance to penetrate a surface or barrier; often used in glove tests.

breast   The mammary gland; superior central part of the body; neck to abdomen.

breast milk   The colloidal liquid expressed from a female breast.

breathing zone   The volume from which a subject obtains breathing air. For humans, a 10-inch radius sphere around the nose.

breathing zone sampling   Sampling in the breathing zone.

breeder   A type of reactor wherein the neutron flux from the primary fuel consumption is used to produce other fissionable fuel.

Bremsstrahlung   Secondary photon radiation produced by deceleration of charged particles passing through matter.

brewing   A complex series of enzymatic reactions by which starch is eventually converted to yeast, e.g., the production of beer; may refer in general to any concoction or mixing of substances to make another substance.

brick   A material, generally with regular angular dimensions, that can be used structurally as well as for heat-resistant or -containing structures. Usually made of silicon or metal oxides.

Brick Institute of America (BIA)   11490 Commerce Park Dr., Reston VA 20191.

brick oil   Synonym for creosote.

Briggs, Henry   English mathematician (1561–1631) who first published logarithm tables in1617.

brightener   A compound or mixture added to a plating bath to produce a shiny, bright finish.

brightening agent   A substance that absorbs ultraviolet radiation and produces a bluish tint complementary to the yellows of an off-white substance.

brine   A solution of chiefly sodium chloride in water, but which may contain other salts.

British thermal unit (Btu)   The quantity of heat necessary to raise the temperature of one pound of pure liquid water 1°F.

brittle point   The temperature at which a substance shatters under pressure.

broad band   A wide range of frequencies.

bromcresol green   An acid-base indicator in the pH range 3.8–5.4; yellow crystals and solution at low pH, blue at high pH.

bromide   A binary chemical compound with the bromine anion in the -1 oxidation state; a CNS depressant.

bromine (Br)   A nonmetallic, corrosive, volatile, reddish-brown liquid group VII (halogen) element; atomic number 35, atomic weight 79.90; Br_2; used in dyes, fumigants, antiknock compounds for gasoline additives, etc.

bromphenol blue $C_{19}H_{10}Br_4O_5S$; an indicator, yellow at pH 3.0; purple at pH 4.6.

bronchial tube   A bronchus or any of its branches.

bronchitis   Inflammation of the bronchial tubes.

bronchoconstriction   Constriction of the bronchial tubes; making the bronchia smaller in diameter.

bronchopneumonia   Inflammation of the lungs resulting from bronchial infection.

bronze   An alloy of copper and tin, sometimes containing other metals.

Brookhaven National Laboratory (BNL)   DOE, PO Box 5000, Upton NY 11973-5000.

broth   A clear, thin aqueous solution resulting from boiling meat or vegetables in water.

Brown, Herbert Charles   American chemist (1912–) who shared the 1979 Nobel Prize in Chemistry for work with boron containing compounds.

brownian movement   The movement of colloidal particles caused by molecular motion impacts of the fluid suspending them.

browning   The color changing of a treated (usually heated) surface. Complex chemical reactions are involved.

browning reaction   Any of the reactions that change the surface color of baked goods or cooked meat.

brucellosis   Undulant fever, sometimes called contagious abortion.

brucine $C_{23}H_{26}O_4N_2 \cdot 2H_2O$; a poisonous white alkaloid.

BSC   Biological safety cabinet.

BSI   British Standards Institute.

Btu   British thermal unit.

bubble tube   simple, accurate, and inexpensive device used with a soap and water solution to calibrate airflow pumps or to directly measure air flow rate; also known as a bubble meter.

bubonic plague   A usually fatal, contagious disease transmitted by fleas.

Büchner, Eduard   German chemist (1860–1917), brother of Hans Büchner, awarded the 1907 Nobel Prize in Chemistry for showing that alcoholic fermentation of sugars is caused by enzymes in yeast; studied under Bayer; killed during World War I.

Büchner, Hans   German hygienist and bacteriologist (1850–1902), brother of Eduard Büchner, who showed that there are substances in the blood that protect against infection.

Büchner, Johann Andreas   German pharmacist (1783–1852) who established the scientific basis of pharmacy.

Büchner funnel   A porcelain funnel used for separating particulates that contains a perforated porcelain plate on which filter paper can be placed.

Buckminsterfullerene C_{60}; a spherical aromatic molecule, hollow and resembling a soccer ball.

Bucky balls   *See* Buckminsterfullerene.

buffer   A substance that reduces the change in the hydrogen ion concentration (pH) that otherwise would be produced by adding acids or bases to the solution.

buffer solution   A solution of certain salts whose pH is not appreciably changed by the addition of acid or alkali.

buffer zone   An area between the central facility security fence and the boundary perimeter fence.

Building Officials and Code Administrators International (BOCA)   4051 W. Flossmoor Rd., Country Club Hills IL 60478.

building-related illness   Any of a variety of complaints or symptoms given by individuals that they believe to be related to something in the building they occupy.

bulk chemical   Generally, any chemical produced or transported in large quantities.

bulk sample   A sample of a building material or other material taken for content analysis.

bulla   A large blister or vesicle.

buna rubber   Rubber made by polymerization of butadiene and sodium.

Bunsen, Robert Wilhelm   German chemist (1811–1899) who was the co-discoverer of cesium and rubidium; a pioneer of spectral analysis; invented the Bunsen burner.

Bunsen burner   A small lab burner with adjustable air flow to mix with the gas for variable flame characteristics.

burden   The amount of radioactive material present in the body or an organ of humans or animals.

Bureau of Labor Statistics (BLS)   Department of Labor, 441 G St., NW, Washington DC 20212.

Bureau of Mines (BOM)   Department of the Interior, 2401 E St., NW, Washington DC 20241.

buret, burette   A vertical tube with an adjustable stopcock near the bottom and graduations on the tube to permit accurate determination of the amount of liquid dispensed. Usually used in titrations.

burlap   A coarsely woven cloth, usually used to make large bags.

burn   To oxidize rapidly. An injury to the skin caused by flame, heat ,or chemicals. First-degree burns show redness of the unbroken skin. Second-degree burns show skin blisters and some breaking of the skin. Third-degree burns show destruction of the skin and underlying tissues.

burnt lime   *See* calcium oxide.

bursa   A bodily cavity.

bursitis   Inflammation of bursa, generally the knee, shoulder, or elbow.

butane C_4H_{10}; a colorless, highly flammable gas, narcotic in high concentrations. Low condensing pressure has led to widespread use of portable and fixed storage tanks of the liquid.

Butenandt, Adolf Friederich Johann   German chemist (1903–1995) who shared the 1939 Nobel Prize in Chemistry for work on sex hormones. The authorities of his country forced him to decline the award, but he later received the diploma and the medal.

butyl rubber   A synthetic rubber made by copolymerization of butylene with isoprene or butadiene. Noted for gas impermeability hence, used in tires.

BW   Bandwidth.

bypass position   The condition of a valve or damper such that flow is bypassed around a given component of a system.

byssinosis   A chronic lung disease associated with cotton flax and hemp workers. *See* pneumoconiosis.

C

C   Carbon; Celsius; Centigrade; TLV® ceiling limit; coulomb.

C-scale   A scale used in sound level meters to simulate the response and sensitivity of the human ear.

Ca   Calcium.

CA   Corrective action.

CAA   US EPA Clean Air Act.

CAAA   US EPA Clean Air Act Amendments.

CAD/CAM   Computer-aided design/computer-aided modeling.

cadmium (Cd)   A white, flammable, toxic, carcinogenic, soft group IIB metallic element, atomic number 48, atomic weight 112.40.

CAER   Community Awareness and Emergency Response.

caffeine   A white, solid, naturally occurring alkaloid found in coffee beans, tea leaves, kola nuts, and many carbonated beverages, especially colas; a stimulant.

CAG   Carcinogen Assessment Group in the US EPA.

caged zeolite   Sodium aluminosilicates used as drying agents, for the removal of undesired gases, and as catalysts.

CAGI   Compressed Gas Association Incorporated.

CAIR   Comprehensive assessment information rule.

caisson   A water-tight, airtight, pressurized chamber used by workers performing underwater construction.

caisson disease   A human barometric disease caused by going too quickly from conditions under pressure to conditions at or near atmospheric pressure; also called the bends or decompression sickness.

calamine   A suspension of zinc oxide and a small amount of ferric oxide used in ointments and lotions, sometimes for the treatment of a rash such as that caused by poison ivy.

calcination   The heating of a solid to just below its melting point to cause thermal decomposition, reaction, or a phase transition.

calcine   To burn to ashes or powder; a process used for the disposal of hazardous waste by drying liquid waste at high temperatures.

calcium (Ca)   A silver-white group IIA alkaline earth metallic element, atomic number 20, atomic weight 40.08; an essential element that reacts with water to form calcium hydroxide and heat.

Calcium oxide (CaO)   a white-grayish-white, odorless chemical; also known as lime, quicklime, burnt lime, unslated lime.

calibrate   To standardize by comparison to a known, approved, and accepted product or value and to note the deviation from the standard.

calibration standard   A material, product, or value of known purity or value used as a reference compared to which variation is noted.

californium (Cf)   A radioactive transuranic element, atomic number 98, atomic weight 251.

calomel   Mercurous chloride, Hg_2Cl_2; used as a fungicide, an insecticide, and an electrode in electrochemistry.

calorie   The amount of heat needed to raise one gram of water one degree centigrade at one atmosphere pressure. *See* kilocalorie.

calorimeter   A device used to accurately measure the heat absorbed or emitted from a chemical reaction.

camphor   A white or colorless, sublimable, naturally occurring ketone from the wood of camphor trees; used in flavorings and fragrances, medicine, insecticides, and chemical synthesis.

Campus Safety Association (CSA)   c/o National Safety Council, 1121 Spring Lake Dr., Itasca Il 60143-3201.

cancer   A malignant, cellular tumor that is often invasive and metastatic and can be fatal.

candidiasis   A fungal disease characterized by skin and mucous membrane lesions. Caused by *Candida albicans*; found in restaurant workers, bakers, packinghouse workers, cannery workers, and poultry processors.

canicola fever   *See* leptospirosis.

canister   The part of a respirator containing a particulate or sorbant filter through which inhaled air passes for purification before it is breathed.

cannabis   A mild hallucinogen sometimes used in medicine, containing 9-tetrahydrocannabinol; made from dried flowering tops of *Cannabis sativa* or hemp; also known as marijuana and hashish. Illegal in the USA.

Cannizzaro, Stanislao   Italian chemist (1826–1910) who marched with Garibaldi's Thousand. He was the first to appreciate the importance of Amedeo Avogadro's work in connection with atomic weights and demonstrated how to estimate atomic weights and that molecules of elements may consist of more than one atom. He integrated organic and inorganic chemistry and discovered the reaction named after him.

canola oil   *See* rapeseed oil.

canopy hood   The part of a ventilation system designed as an overhead enclosure or hood and located to capture emissions rising from a heated surface or area.

canthaxanthin   A carotenoid colorant in many natural products; found in edible mushrooms; used as a food and drug colorant; may cause loss of night vision.

capillary   A tube of very small diameter; a small, thin-walled blood vessel that connects arteries and veins.

Caplan nodules   Nodules observed in the lungs of coal miners with rheumatoid arthritis.

capture velocity   The point in front of a hood at which contaminated air overcomes ambient air currents and is captured, causing it to flow into the hood.

CAR, CARC   Carcinogen.

carbamate   A chemical class of pesticides derived from carbamic acid, NH_2COOH. Used as insecticides, fungicides, and herbicides. Carbamates are reversible cholinesterase inhibitors.

carbide   A very hard, chemically resistant, thermally stable, refractory compound of carbon and another element; used for abrasives, furnace refractories, and high-temperature applications.

carbinol   Common name for methanol; also any compound with a COH radical.

carbohydrate   The most abundant chemical class of organic compounds comprised of carbon, hydrogen and oxygen, and containing a saccharose unit. Found in sugar, starch, and cellulose and formed by green plants; a major source of animal food.

carbolic acid   Common name for phenol.

carbon (C)   A stable, abundant, nonmetallic group IVA element, atomic number 6, atomic weight 12.01; the element of life because it has an unusual ability to catenate, thus permitting the formation of very large and complex molecules. Diamond, graphite, and soot are all carbon.

carbon bisulfide   *See* carbon disulfide.

carbon black   A high-volume, energy-absorbing chemical found in soot and used in paints, dyes, inks, pigments, rubber, and batteries; also known as furnace black, thermal black.

carbon dioxide CO_2; a colorless, odorless, toxic, nonflammable gas. It is a high-volume chemical used for carbonation, aerosol propellants, fire extinguishers, and lasers and as a refrigerant in its solid state, "dry ice." As a solid $(-79°C)$ it is damaging to the skin, tissues, and eyes. It is also released during respiration, produced during fermentation, and removed from the atmosphere by photosynthesis.

carbon disulfide CS_2; also known as carbon bisulfide. It is a clear, nearly colorless, highly flammable, toxic liquid with a strongly disagreeable odor unless highly purified that can ignite by friction.

carbon monoxide   CO; a colorless, odorless, highly flammable, toxic gas formed by the incomplete combustion of organic materials. It has a much higher affinity for red blood cells (hemoglobin) than oxygen-forming carboxyhemoglobin. Signs and symptoms of exposure are headache, dizziness, nausea, vomiting, coma, and death. Initially, victims are pale; later mucous membranes become cherry red.

carbon tetrachloride CCl_4; a colorless, nonflammable, toxic liquid chlorinated hydrocarbon with a sweetish odor. Used as a fire extinguishant called pyrene until it was found to degrade into highly toxic phosgene gas and HCl on heating. It is not permitted in products used in US homes.

carbonyl group   A functional group comprised of a carbon atom double-bonded to an oxygen atom, $C{=}O$ which is present in numerous organic compounds. It combines with metals such as nickel to form highly toxic coordination compounds, e.g., $Ni(CO)_4$. It possesses a characteristic infrared absorbance.

carborane   Compounds made of carbon, boron, and hydrogen.

Carborundum^{TM}   A very hard, acid-resistant abrasive and refractory made of silicon carbide; used for grinding, cutting, and polishing.

carboxyhemoglobin (COHb)   Hemoglobin in the blood that has bound carbon monoxide to the iron. The affinity of hemoglobin for carbon monoxide is about 250 times that for oxygen.

carboxylic acid   A broad group of organic acids that include fatty acids and amino acids. The functional group is usually in the terminal position, depicted as $-COOH$ or $-CO_2H$, and is comprised of a carbon double-bonded to an oxygen and single-bonded to a hydroxyl group (OH).

carboy   Usually a 5- to 10-gallon glass container used to store chemical solutions; sometimes also used to designate large containers made of materials other than glass.

Carcinogen (CAR, CARC)   A substance known to produce cancer in animals and humans. A substance is considered a carcinogen if it meets the criteria defined by the International Agency for Research on Cancer (IARC), the National Toxicology Program (NTP), or OSHA. There are many other substances besides those listed in the above sources that are considered by many knowledgeable authorities to be carcinogens.

carcinogenesis   The process of developing malignant tumors.

carcinoma   A malignant new growth comprised of epithelial cells, which tends to invade other tissues and metastasize.

cardiovascular system   The heart, blood vessels, and circulatory system.

carnauba wax   The hardest and most expensive wax derived from the Brazilian wax palm, *Copernica cerifera*.

Carnot, Nicolas Léonard Sadi   French physicist (1796–1832) who was a pioneer in the study of thermodynamics; studied steam engines, and died of cholera.

Carnot cycle   An ideal closed cycle illustrating the second law of thermodynamics; used to describe a reversible heat engine.

carotene   A member of the class of pigments called carotenoids present in plants, algae, bacteria, and some animals, e.g., shark liver oil; a precursor of vitamin A converted to vitamin A in the liver; occurs naturally in plants in three forms: α, β, γ, the β form is used as a food additive.

carpal tunnel   The bony, narrow passage between the hand and the wrist, through which the median nerve and several tendons pass.

carpal tunnel syndrome (CTS)   A disorder resulting in the compression of the medial nerve by swollen tendons in the carpal tunnel. Associated with work or activities involving repeated motion, force, twisting, or flexing, e.g., of loggers, butchers, typists, assembly line workers, tennis players. Characterized by tingling, pain, and numbness in the thumb and first three fingers.

carrier   A person or animal that has an infectious disease but shows no signs of the disease and is a potential source of infection to others, e.g., Typhoid Mary. The gas used in gas chromatography to convey the volatile mixture being analyzed. A stable, nonradioactive tracer added to the radioactive material being analyzed.

carrier gas   The gas, usually nitrogen or helium, used in gas chromatography to convey a volatile mixture over packing material used to separate the mixture into components for identification and analysis.

CAS   Chemical Abstract Service.

cascade impactor   A sampling device used to separate airborne particulates by size based on their mean aerodynamic diameter.

case   A diseased individual having specific epidemiological characteristics.

case control study   An epidemiological study of a group of diseased individuals or those with particular health effects of interest and their relationship to a nondiseased group.

case-hardened   A surface hardening process used in making steel by heating it and raising the carbon or nitrogen content of the steel.

casein   The primary protein in milk used in food, plasticizers, paints, and binders.

cashew oil   A strongly irritating, sometimes sensitizing, phenolic oil made from cashew nut shells. Used for plasticizers, lubricants, resins, varnishes, germicides, and insecticides.

cask A thick-walled container, often made of lead, used to transport radioactive material. Also called a coffin.

CAS Registry Number A system used by the Chemical Abstract Service of the American Chemical Society to identify chemicals by assignment of a number.

castor oil An almost colorless oil used in animal feeds, foods, drugs, cosmetics, plasticizers, polymers, and resins. Used internally as a laxative and externally as an emollient. Seeds of *Ricinus communis* (in the same genus, Rhus, as poison ivy, poison oak and poison sumac) are used to produce castor oil. Allergic contact dermatitis may be caused by contact with the plant. Hay fever, asthma, and urticaria are common in castor bean processors. Also known as ricinus oil.

catalase An oxidizing enzyme in plants and animals used in food preservation.

catalyst A substance used to speed up or otherwise change, e.g., reaction temperature or the rate of a chemical reaction without itself being consumed.

cataract Clouding of the lens of the eye.

catenate To link together.

cathode The electrode where reduction occurs. The negative electrode of an electrolytic cell to which positive ions are attracted or the positive electrode of a Galvanic cell.

cation A positively charged ion.

cat-scratch fever A viral infection common to laboratory animal handlers, animal handlers, and veterinarians.

caustic Any strongly alkaline or basic substance that is highly corrosive.

caustic lime Calcium hydroxide.

caustic potash Potassium hydroxide.

caustic soda Sodium hydroxide.

CAUTION A sign used to warn against potential hazards or to caution against unsafe practices. The standard color of these signs is yellow with black letters. The least severe of the Danger, Warning, Caution sequence.

Cavendish, Henry English chemist and physicist (1731–1810) who determined the composition of air and water.

cavitation   The formation of holes or cavities in tissues or organs as a result of a disease. Also, the formation and collapse of bubbles in liquids formed by mechanical forces, such as from propellers.

CBA   Cost-benefit analysis.

CBOD   Carbonaceous biochemical oxygen demand.

CBT   Computer-based training.

CBW   Chemical and biological weapons.

cc   Cubic centimeter; also carbon copy.

CCHO   Certified Chemical Hygiene Officer; certified by the National Registry of Certified Chemists.

cd   Candela.

CD   Drag coefficient.

C$_D$ Cadmium.

CDC   Centers for Disease Control and Prevention.

Ce   Cerium.

CEC   Council on European Communities.

Cech, Thomas Robert   American chemist (1947–) who shared the 1989 Nobel Prize in Chemistry for his discovery of catalytic properties of RNA.

ceiling   The maximum allowable human exposure concentration that should not be exceeded during any part of the workday according to the ACGIH®.

ceiling concentration   The maximum concentration of a toxic substance allowed at any time or during a specific sampling period.

ceiling limit   A term used by the ACGIH® and designated as "TLV®-C," meaning the concentration that should not be exceeded during any part of the working exposure. According to the ACGIH®, if instantaneous monitoring is not feasible, then the TLV®-C can be assessed by sampling over a 15-minute period, except for substances that may cause immediate irritation during short exposure.

cell   The smallest part of a biological organism able to function independently, consisting of a wall or membrane, cytoplasm, and nucleus.

Celsius, Andres   Swedish physicist and astronomer (1701–1744) who developed the centigrade or Celsius temperature scale.

Celsius temperature (C)   A temperature scale named after the Swedish physicist and astronomer Andres Celsius, in which 0° equals the freezing point of water and 100° equals the boiling point of water. Symbolized by the letter C. *See* centigrade temperature.

CEM   Cost-estimating manager; continuous emission monitoring.

cement   A general term for many types of construction adhesives that can be poured or molded before setting into a solid form. *See* Portland cement.

center of gravity   The point at which all the weight is concentrated, i.e., the point at which a body is considered to be at equilibrium in any position.

Centers for Disease Control and Prevention (CDC)   US DHHS, PHS, 1600 Clifton Rd., NE, Atlanta GA 30333.

centigrade temperature (C)   A temperature scale in which 0° equals the freezing point of water and 100° equals the boiling point of water. *See* Celsius temperature.

centipoise   The metric unit of viscosity with the value of dyne-seconds/cm^2 or grams/cm-second; 1/100 poise.

centistoke   The kinematic unit of viscosity equal to the viscosity divided by the density in grams/cm^3, both at the same temperature; 1/100 stoke.

central nervous system (CNS)   The brain, spinal cord, and nerves.

centrifuge   A device that utilizes centrifugal force by spinning to separate materials by density or to simulate gravitational force.

CEP   Capital equipment project.

CEQ   Council on Environmental Quality.

CER   Complete engineering release.

cercarial dermatitis   Swimmer's, swamp, or clam digger's itch; a result of the penetration of certain species of freshwater *Schistosome cercariae* through wetted skin.

CERCLA   Comprehensive Environmental Response, Compensation and Liability Act of the US EPA.

cerium (Ce)   A highly reactive lanthanide metallic element, atomic number 58, atomic weight 140.12.

cer-pneumoconiosis   A condition resulting from exposure to cerium; reported in graphic art workers who use carbon arc lights.

certificate of conformance   A signed, or otherwise authenticated, document by an authorized individual verifying the degree to which items or services meet specified requirements.

certification   An official endorsement by contractor management of an individual who has completed a qualification and satisfactorily completed other position requirements, such as a medical examination. The endorsement is given by an individual or group other than the individual or group providing the training or the candidate's immediate supervisor. The act of determining, verifying, and attesting in writing to the qualifications of personnel, processes, procedures, or items according to specified requirements.

certified calibration   Standardization of an instrument against a primary standard or standard reference material, e.g., a material of known, stable, and documented purity or physical property, a device traceable to the National Institute of Standards and Technology.

Certified Chemical Hygiene Officer (CCHO)   A professional chemical hygiene officer certified as a CCHO on the basis of experience, education, and demonstrated knowledge through an examination administered by the National Registry of Certified Chemists.

certified gauge/tool   A gauge or tool that allows a product to be fabricated or accepted either without dimensional inspection or with partial dimensional inspection.

Certified Hazardous Materials Manager (CHMM)   A professional certified as a CHMM on the basis of experience, education, and demonstrated knowledge through an examination administered by the Academy of Certified Hazardous Materials Managers.

Certified Health Physicist (CHP)   A professional certified as a CHP on the basis of experience, education, and demonstrated knowledge through an examination administered by the American Academy of Health Physics.

Certified Industrial Hygienist (CIH)   A professional industrial hygienist certified as a CIH on the basis of experience, education, and demonstrated knowledge through an examination administered by the American Board of Industrial Hygienists.

certified instructor   An individual selected to teach a course who meets specified educational and experiential requirements, including teacher certification training.

Certified Manufacturing Operation (CMO)   Used to reduce the amount of gauging and inspection; also uses certified machining tapes and certified tools/fixtures to manufacture and accept product without subsequent inspection or with only partial inspection.

Certified Occupational Hygiene Nurse (COHN)   A professional certified as a COHN on the basis of experience, education, and demonstrated knowledge through an examination administered by the American Board for Occupational Health Nurses.

certified reference material (CRM)   Documents whose value are certified by a technically valid procedure accompanied by or traceable to a certificate or other documentation issued by a certified body.

Certified Safety Professional (CSP)   A safety professional certified as a CSP on the basis of experience, education, and demonstrated knowledge through an examination administered by the Board of Safety Professionals.

certified tape   A numerically controlled machine tape that allows a product to be accepted with partial dimensional inspection.

cerumen   Earwax.

cesium (Cs)   A highly reactive, flammable, and explosive group IA alkali metal element, atomic number 55, atomic weight 132.91. It violently reacts with water.

CESQG   Conditionally exempt small quantity generator.

CET   Corrected effective temperature.

cf   Abbreviation for the Latin word *confer*; compare.

CFC   Control frequency converter; chlorofluorocarbon.

cfm, CFM   Cubic foot (feet) per minute.

CFR   US Code of Federal Regulations; also cooperative fuel research.

cfs   Cubic foot (feet) per second.

cg   Centigram.

CGA   Compressed Gas Association.

CGL   Comprehensive general liability.

cgs   Centimeter-gram-second system of units.

Chadwick, Sir James   English physicist (1891–1974) awarded the 1935 Nobel Prize in Physics for discovery of the neutron.

chain   A series of atoms of the same element connected to each other by chemical bonds.

chain of custody   A written, documented record stating the time and location of a sample at all times.

chain reaction   A chemical or nuclear process in which one effect is carried on and affects the other members, particles, or components of the products being formed.

chalk   Calcium carbonate; when naturally occurring, it is the remains of marine organisms.

change notice   A document issued by Purchasing directing the contractor to perform work in accordance with attached drawing, specifications, sketches, etc.

channel   The combination of sensor, line, amplifier, and output devices that are connected to measure the value of a parameter.

channel calibration   A channel test that includes an adjustment of the channel so that its output corresponds with acceptable accuracy to known values of the parameter that the channel measures.

channel check   A qualitative verification of expected performance by observation of channel status. This verification, where possible, includes comparison of the channel with other independent channels measuring the same variable.

channel test   The introduction of a signal into the channel for verification that it is operable.

characteristic   Any property or attribute of an item, process, or service that is distinctly describable and measurable.

charcoal   A highly porous form of carbon made by heating organic substances such as wood with little or no oxygen. Charcoal has excellent adsorbing and absorbing properties and is often used to filter and purify, deodorize, or decolorize water, air, etc.

charged particle   An ion; an elementary particle with a positive or negative electric charge.

charge number   A code used for actual cost charging, e.g., job number, part number, tool order.

charge-to-mass ratio   The ratio of the charge on an ionized particle to its mass; usually expressed as e/m.

Charles, Jacques Alexandre César   A French physicist, chemist, and inventor (1746–1823) who anticipated Guy-Lussac's studies of expanding gases.

Charles' law   States that at constant volume, the pressure of a gas is directly proportional to its absolute temperature. *See* Guy-Lussac's law.

checker   A competent employee assigned to assess completion of a product or activity; not the same individual who designed the product or performed the activity, although can be from the same organization. The checker may be the originator's supervisor, provided the supervisor did not specify a singular design approach or rule out certain design considerations and did not establish the design inputs used in the design.

checklist   A standard form developed for each unit, operation, or area of concern to be used by shift personnel to aid the turnover process. The checklist provides a convenient method of denoting equipment in service, limiting conditions of operation (LCO) status, and other documents that oncoming shift personnel should review to ensure a complete transfer of building status information.

checkpoint   A point at which material may either be measured or physically verified.

check source   A radioactive source, not necessarily calibrated, used to confirm the continuing operations of an instrument.

chelate   Coordination compound whose central metal is attached by coordination bonds to nonmetal atoms or organic compounds called ligands. These compounds derive their name from the Greek word chele for "claw." Hemoglobin and chlorophyll are examples of chelation compounds.

chelating agent   Any compound that bonds with a metallic ion forming a ring structure in the molecule incorporating the metal ion. These compounds can be useful and effective but can also be hazardous. They should be used only for treating metallic ion poisoning under the direction of a physician. An example is ethylenediaminetetraacetic acid (EDTA), used to treat lead poisoning and to reduce water hardness by removal of certain metals.

chemical   A substance made up of atoms or molecules; also relating to the study of chemistry and the properties of the substances studied in chemistry.

Chemical Abstracts   A weekly publication by the Chemical Abstracts Service of the American Chemical Society of condensed summaries of major international articles and patents pertaining to chemistry.

Chemical Abstracts Service (CAS)   2540 Olentangy River Rd., PO Box 3012, Columbus OH 43210-0012; the part of the American Chemical Society that indexes, abstracts, and categorizes chemicals and the international chemical literature. The service is based on a system of unique numbers used to identify specific chemicals; they also publish *Chemical Abstracts*.

Chemical Abstracts Service number   A number, assigned by the Chemical Abstracts Service of the American Chemical Society; used to uniquely identify specific chemicals.

chemical agent   A chemical substance that acts on the human system.

chemical asphyxiant   A substance that reacts chemically to interfere with the body's uptake and transport of oxygen.

chemical burn   A burn similar to those caused by heat. After emergency first aid, treatment is usually the same as for thermal burns. However, some chemicals, such as hydrofluoric acid and phenol, require treatment of the systemic effect of the chemical. *See* burn.

chemical cartridge   A device attached to the inhalation portion of an air-purifying respirator that contains a sorbant capable of removing low concentrations of airborne vapor or gaseous chemical contaminants from air that is breathed.

chemical change   The rearrangement of atoms or other portions of a substance that produce a new substance with new properties. The process is called a chemical reaction.

chemical compound   A substance composed of two or more kinds of elements combined in fixed proportions by weight.

chemical equivalent   The atomic weight of an element divided by its oxidation state; the molecular weight of a compound or ion divided by the number of that species required for the reaction being considered.

chemical fallout   Chemical pollution deposited on land or water from contaminated air.

chemical formula   A representation depicting the kinds and numbers of each element present in a compound.

chemical hazard   Harm or danger that may result from a chemical.

chemical hood   A device, enclosed on five sides and with a sash, that exhausts laboratory air through the enclosure to the outside to minimize the escape of contaminants in the hood from going back to the laboratory or from exposing individuals working at the hood. Also known as a laboratory hood or, improperly, as a fume hood or fume cupboard, because its use is not restricted to fumes. *See* laboratory hood.

Chemical hygiene officer (CHO)   The employee defined according to the OSHA standard on "Occupational Exposure to Hazardous Chemicals in the Laboratory" (1910.1450), and qualified by experience and training to provide technical guidance in the development and implementation of the chemical hygiene plan.

Chemical hygiene plan (CHP)   A written program, as defined by the OSHA standard on "Occupational Exposure to Hazardous Chemicals in the Laboratory" (1910.1450) that describes the employer's procedures, equipment, and work practices to protect employees from workplace health hazards.

Chemical Industry Institute of Toxicology (CIIT)   PO Box 12137, 6 Davis Dr., Research Triangle Park NC 27709.

Chemical Manufacturers Association (CMA)   Former name for the American Chemistry Council (ACC).

chemical oxygen demand (COD)   The amount of oxygen in ppm consumed in the oxidation of matter present in industrial wastewater corrected for the effect and influence of chlorine.

chemical pneumonitis   Inflammation of the lungs caused by inhalation of chemical vapors or gases.

chemical reaction   A change in arrangement of atoms or molecules of a substance resulting in a new substance with different properties.

chemical smoke   An aerosol produced by a chemical reaction; used to fit-test respirators for leakage, by the military, and in theater to produce fog or smoke.

Chemical Transportation Emergency Center (CHEMTREC)   Part of the Chemical Manufacturers Association, organized to provide emergency information about transportation accidents involving hazardous chemicals.

chemical warfare   The use of chemical agents by the military and police to cause casualties or evacuation, or to conceal operations, e.g., poisoning, burning, irritation, asphyxiation, fire, smoke, contamination.

chemical waste   Unusable products that result from chemicals or as products of chemical reactions or use.

Chemical Waste Transportation Council (CWTC)   c/o National Solid Waste Management Association, 1730 Rhode Island Ave., NW, Washington DC 20006.

Chemical Workers Union (CWU)   1655 W. Market St., Akron OH 4313.

chemiluminescence   The emission of absorbed energy caused by a chemical reaction, which includes categories called bioluminescence such as the light given off by lightning bugs.

chemisorb   To take up or bind a chemical to the surface of another substance.

chemistry   A physical science concerned with the study of the composition, structure, properties, and reactions of substances.

chemosorb   *See* chemisorb.

chemosterilant   A chemical agent or process used to sterilize insects.

chemotaxis   The motion of living organisms with respect to a chemical substance, such as when *E. coli* bacteria move away from toxic chemicals but toward nutrients if both substances are present.

chemotherapy   The use of chemicals to treat or prevent disease.

chemotropism   The movement or growth of living organisms in response to chemicals.

CHEMTREC   Chemical Transportation Emergency Center.

Cheyne, John   Scottish physician (1777–1836) who lived in Dublin, who, along with William Stokes, helped recognize an abnormal breathing pattern in sleeping and unconscious adults that is common in infants.

Cheyne-Stokes respiration   Abnormal breathing in sleeping or unconscious individuals characterized by periods when breathing appears to stop for several minutes before suddenly starting again; common in healthy infants. Named after the Scottish physician (in Dublin) John Cheyne and the Irish physician William Stokes.

χ Lower case Greek letter chi.

chilblain   An itchy irritation of the hands, feet, or ears caused by exposure to moist cold.

chill   A feeling of coldness, sometimes accompanied by a fever; also hardness of a metallic substance caused by rapid cooling.

chimney effect   The tendency for air in a vertical duct to rise naturally when heated.

chimney sweep   A person who cleans chimneys; first associated with cancer of the scrotum by Sir Percival Potts in the 1700s, because of exposure to coal combustion products.

chiral   A term used to describe the optical rotation of light by a chemical, based on its asymmetry. Called the optical activity of a molecule; when light passes through a chemical or a chemical solution it can rotate either to the right (dextro, d, $+$) or to the left (levo, l, $-$).

chloracne   A skin condition resembling acne caused by exposure to certain chlorinated hydrocarbons such as chlorinated biphenyls, chlorinated dioxins, and chlorinated furans.

chloral CCl_3CHO; a colorless, odorless, toxic liquid that irritates the lungs and is used to make chloral hydrate and DDT.

chloral hydrate $CCl_3CH(OH)_2$; a colorless, chemical, hazardous to the eyes, also known as knockout drops; used as a sedative and hypnotic drug and to make DDT.

chloramine NH_2Cl; a colorless, unstable, liquid, with a pungent odor; formed when amines come in contact with chlorine, e.g., urine in a swimming pool.

chloramine-T   A solid organic chemical, sodium *p*-toluenesulfochloramine, used to reduce microbial contamination, i.e., sanitize.

chlorate ClO_3^-; a very strong class of oxidizing agents that can be explosive when mixed with organic compounds.

chlordane $C_{10}H_6C_{18}$; a colorless, odorless, toxic liquid chemical used as an insecticide and fumigant.

chlorinated hydrocarbon   An organic compound composed only of carbon, hydrogen, and chlorine.

chlorinated solvents   Chemical agents, such as methylene chloride, often used for cleaning or for their broad solubility properties.

chlorine (Cl)   A greenish-yellow, irritating, highly toxic, nonflammable, dense, gaseous group VIIA halogen element, atomic number 17, atomic weight 35.45. Cl_2, widely used to disinfect water supplies, has a pungent, irritating odor.

chlorine number   The amount of chlorine absorbed over a specific time under specific conditions; used as a measure of bleach consumption.

Chlorofluorocarbon (CFC)   A class of chemicals composed of carbon, hydrogen, chlorine, and fluorine. Use of these chemicals was prohibited under the 1987 Montreal Agreement because they cause depletion of the ozone layer in the atmosphere.

chloroform $CHCl_3$; a colorless, toxic, carcinogenic liquid with a characteristic odor; once used as an anesthetic, still used as a solvent.

chloroprene   A colorless, flammable, toxic liquid used to make neoprene.

CHMM   Certified hazardous materials manager.

CHO   Chemical hygiene officer.

cholesterol $C_{27}H_{45}OH$; the most common animal steroid, which also occurs in the human blood system.

cholinesterase   An enzyme occurring in the human body, found in the brain, nerves, and blood; important in the mechanism of nerve action. Acts by hydrolyzing acetylcholine in the nervous system.

cholinesterase inhibitor   Chemical, such as organophosphate pesticides and carbamates, that deactivates the mechanism of the enzyme cholinesterase by retarding the hydrolysis of acetylcholine.

CHP   Certified health physicist; chemical hygiene plan.

CHRIS   Chemical Response Information System; a reference tabular listing of compatible chemicals.

chromatography   A separation technique used to selectively separate components of mixtures based on differences in the physical properties of the components.

chrome ulcer   Skin lesion caused by exposure to hexavalent chromium compounds.

chromium (Cr)   A hard, brittle, gray group VIB metallic element, atomic number 24, atomic weight 52.00; named after the Greek word for color. Hexavalent compounds cause dermatitis and skin ulcers and are carcinogenic.

chromophor   A chemical functional group, which when present in the compound, produces color in the visible spectral range.

chromosomal abortion   A change in the normal appearance of a chromosome, i.e., number, size, shape.

chromosome   A threadlike body in cells, derived from chromatin and found in the nucleus, that contains genes.

chronic   Occurring repeatedly over prolonged time.

chronic effect   A symptom or result that occurs repeatedly over prolonged time.

chronic exposure   A repeated, low-level exposure over prolonged time.

Ci   Curie.

cigarette tar   The condensed smoke from cigarettes containing several thousand chemicals including polycyclic aromatic hydrocarbons.

CIH   Certified industrial hygienist.

CIIT   Chemical Industry Institute of Toxicology.

cilia   Short, hairlike structures found on the surface of protozoa and in the bronchi of human lungs. Their movement helps to propel (protozoa) or remove particulate contaminants (in the respiratory system).

ciliatoxic   A substance that, when present, retards or stops cilia movement.

cinnabar   HgS; Naturally occurring mercuric sulfide.

circadian rhythm   Routine biological activities that recur at set time periods, e.g., sleep.

cirrhosis   A chronic, progressive liver disease that hardens, hinders and eventually stops liver function, and results in death.

cis-   A prefix used in chemistry to describe geometric isomers when there is a double bond between two carbon atoms restricting rotation and where two identical atoms are on the same side of the double bond. Derived from the Latin word meaning on the same side.

Cl   Chlorine.

clam digger's itch   *See* cercarial dermatitis.

Clapeyron, Benoit Paul Emile   French engineer (1799–1864) who developed a theoretical relationship between the temperature of a liquid and its vapor pressure showing that it is not a straight line.

clarification   The removal or reduction of suspended matter from waste or effluent water.

class   A collection of related data, information, topics, or properties; a specific type of chemical group.

class A fire   Paper, wood, rag, rubbish, and other solid matter fire that can be extinguished by water.

class B fire   Flammable and combustible liquid fire that must be extinguished by removing oxygen, i.e., smothering, by the use of foam or dry chemical extinguishers.

class C fire   Energized electrical fire that must be extinguished by first deenergizing the device involved and using extinguishants that are nonconductive.

class D fire   Fire of combustible metals, such as sodium or magnesium, that requires special fire extinguishers.

class 1 laser (OSHA, ANSI)   Extremely low-powered laser considered nonhazardous; no precautions needed.

class 2 laser (OSHA, ANSI)   Low-powered, visible laser that could be dangerous if viewed for longer than the 0.25-second eye response (blink reflex) time. Limited precautions needed.

class 2a laser (OSHA, ANSI)   Low-powered, visible laser that could cause damage if viewed for longer than 1000 seconds; used in supermarkets.

class 3a laser (OSHA, ANSI)   Intermediate-power laser that could be dangerous if viewed through an optical device, such as a telescope. Limited controls are recommended.

class 3b laser (OSHA, ANSI)   Moderately powerful laser that may cause eye damage during incidental exposure without optical devices. Specific controls are recommended.

class 4 laser (OSHA, ANSI)   High-powered laser that can harm the skin and eyes; diffuse reflection may also be dangerous. Rigorous control measures are necessary.

class I flammable liquid (NFPA)   Liquid with flash point below 100°F.

class IA flammable liquid (NFPA)   Liquid with flash point below 73°F and boiling point below 100°F.

class IB flammable liquid (NFPA)   Liquid with flash point below 73°F and boiling point at or above 100°F.

class IC flammable liquid (NFPA)   Liquid with flash point at or above 73°F and below 100°F.

class II and III combustible liquids (NFPA)   Liquids with flash points above 100°F.

class II combustible liquid (NFPA)   Liquid with flash point at or above 100°F and below 140°F.

class IIIA combustible liquid (NFPA)   Liquid with flash point at or above 140°F and below 200°F.

class IIIB combustible liquid (NFPA)   Liquid with flash point at or above 200°F.

classification   The process of dividing, separating, or categorizing chemical, physical, or biological materials or substances into groups based on related or similar properties. Used in chemistry, biology, engineering, and geology.

clastogen   A chromosome-breaking agent; a substance that can cause chromosomal abnormalities.

Clausius, Rudolf Julius Emmanuel   German mathematical physicist (1822–1888) who developed the second law of thermodynamics.

Clausius-Clapeyron equation   The relationship that states that the rate of change of vapor pressure of a liquid (ergs/cc) with respect to absolute temperature ($^{\circ}$K) equals the heat of vaporization (ergs/g) divided by the product of the absolute temperature and the increase of volume (cc), i.e., $dp/dT = \Delta H/T\Delta V$.

cleaning solution   A solution made by mixing concentrated sulfuric acid and sodium dichromate; used for cleaning glassware.

clean room   A specifically designed and constructed room whose purpose is to limit the amount of airborne particulates and to precisely control temperature, humidity, and air movement. Often used in the micro-electronic industry.

clearance   A certification of an individual's right to certain information or locations; elimination of a substance from the organism; can also be transport to a specific site within the organism.

cleaving   The process of breaking or separating a solid substance like a crystal or mineral or a chemical bond.

Cleveland Open Cup (COC)   A standardized device, open to the atmosphere, used to determine flash point.

clinical chemistry   The branch of chemistry that deals with properties, components, and toxicology of bodily fluids and tissues.

Clostridium botulinum   The human pathogen that causes the very toxic botulism.

Clostridium perfringens   One of the human pathogens that cause food poisoning.

Clostridium tetanii   The human pathogen that causes tetanus.

cluster   A group of individuals with similar health symptoms or diseases who share the same job, environment, exposure, etc.

Cm   Curium.

CMA   Chemical Manufacturers Association.

CMO   Certified manufacturing operation.

CNS   Central nervous system.

CNS depression   A reversible suppression of the central nervous system exhibited by drowsiness, dizziness, stupor, or unconsciousness.

CNS effects   Signs and symptoms such as drowsiness, dizziness, stupor, or unconsciousness that indicate that there is a toxic effect that is depressing the central nervous system.

CO   Carbon monoxide.

coagulant   A substance that coagulates a liquid dispersion; used to separate phases.

coagulation   The aggregation of finely divided solids into a manageable mass, e.g., blood clotting.

coagulation value   The concentration of coagulant necessary to achieve some standard change.

coal   A commercial mineral, essentially carbon; used as a fuel.

coal oil   The oil obtained from destructive distillation of bituminous coal.

coal tar   The black, viscous liquid obtained from destructive distillation of bituminous coal. It is a carcinogen, toxic by inhalation with widespread commercial uses.

coal tar distillate   The lower-boiling fraction of coal tar.

coal tar dye   Any of the dyes based on coal tar or its derivatives.

coal tar pitch   The higher-boiling part of coal tar; an amorphous residue with many commercial uses.

coalesce   To grow together, unite, or fuse.

cobalt (Co)   A shiny, gray, hard, ductile, malleable, magnetic group VIII metallic element, atomic number 27, atomic weight 58.93; used in magnetic and high-temperature alloys.

cobalt-60 (^{60}Co)   An important radioactive isotope of the element cobalt that emits beta and gamma radiation; readily available and cheaper than radium; used for radiation therapy, testing of welds, and irradiation of foods.

cobalt blue   A blue to green colorant, mixture of cobalt oxide and alumina; used in cosmetics.

COBOL   Common business-oriented language (a programming language).

COC   Cleveland open cup; a technique to determine flash points; continuity of combustibility.

cocaine $C_{17}H_{21}NO_4$; a white crystal or powder alkaloid; CNS stimulant used as a local anesthetic; available by prescription only.

coccidioidomycosis   A fungal disease endemic in arid and semiarid regions of the southwestern US, Mexico, and Argentina caused by inhalation of the spores of *Coccidioides immitis* present in dust. Workers at risk include laboratory workers handling the organism, cotton mill, construction, farm, and migrant workers, and military personnel.

coccus   A spherical bacteria.

cochineal   A red dye made from dried insect bodies.

cochlea   A coiled, cone-shaped tube that forms part of the inner ear.

COCO   Contractor-owned, contractor-operated.

COD   Chemical oxygen demand.

code(s)   Compilation(s) of laws, regulations, and requirements; computer program(s).

codeine $C_{18}H_{21}NO_3 \cdot H_2O$; a white crystal or powder alkaloid; medically used as an analgesic; habit forming, hence available only by prescription.

Code of Federal Regulations (CFR)   A compilation of all the federal regulations promulgated under US law.

cod liver oil   A fishy-smelling and -tasting yellowish liquid obtained from cod livers. Formerly used for home medicine.

COE   Cost of energy; (US Army) Corps of Engineers.

coefficient of variation (CV)   A measure of precision; the standard deviation divided by the average, usually given as a percentage.

coefficient of viscosity   A measure of the shear resistance of a fluid. The poise is the unit for viscosity.

coenzyme   The organic molecule that, when attached to a specific protein, produces an active enzyme.

COHb   Carboxyhemoglobin.

coherence   A state or condition of orderly array.

coherent light   Laser light, i.e., monochromatic radiation in phase.

COHN   Certified Occupational Hygiene Nurse.

cohort   A group or band of similar things. A term used in epidemiology. Sometimes an associate or companion.

coke   The solid, almost pure carbon material left after destructive distillation of bituminous coal; a cleaner-burning fuel than coal.

coke oven gas   The gas emitted from coking bituminous coal; mainly H_2 with CH_4, N_2, CO, and assorted hydrocarbons.

coking   Changing into coke, converting coal into coke.

cola   An extract of the kola nut used in medicine and in soft drinks.

cold trap   A device used to contain air contaminants by immersing the apparatus in a coolant such as dry ice, dry ice and acetone, liquid nitrogen, or a mechanical device with a circulating liquid coolant.

colic   A painful spasm in any hollow soft organ.

collagen   A ropelike protein, molecular weight $\sim 100,000$, in the connective tissues of animals.

collection trap   Any container (usually cooled) that is used to separate out or collect fractions from a mixture.

colloid   A stable dispersion; not a solution, with particle sizes of $\sim 1-100$ nm.

colorant   Anything that colors the matrix material to which it is added.

colorimetry   An analytical technique for quantifying a compound with a known absorption coefficient.

color index   A commercial dye listing that assigns a specific number to identify each dye.

colorless dye   *See* brightener.

columbium   Obsolete name for niobium.

combinatorial chemistry   Chemical studies that facilitate the synthesis of chemical libraries, as well as their subsequent characterization and screening for properties of interest; generally computerized for high-volume throughput.

coma   An unconscious state resistant to normal means of arousal. Generally caused by trauma, disease, or ingestion of a toxic substance.

combustible   Capable of being burned; generally used for solids difficult to ignite and liquids with flash points $> 100°F$.

combustible liquid   (ANSI) Any Class II or Class III liquid having a flash point at or above $100°F$ ($37.8°C$) but less than $200°F$ ($93.4°C$). This excludes liquid mixtures in which the combustible liquids make up 1% or less of the total volume of the mixture; (DOT) A liquid chemical waste having a flash point greater than 100°F and less than 140°F; (NFPA) Liquid that has a flash point at or above 100°F and will burn or support combustion under normal atmospheric conditions. These liquids are categorized as Class II and Class III combustible liquids. *See* flammable liquid.

combustion gas detector   A device used to detect or measure nonspecific-combustible gases and vapors that may be present in air; often used to determine whether concentrations are immediately dangerous to life and health (IDLH); concentrations are sometimes expressed as percentages of the lower explosion limit (LEL); not suitable for particulates.

combustion tube   The cylindrical container used in combustion analyses.

commission   To give a commission; to verify that an area is suitable for occupancy; to give authority to carry out some task or duty or to give certain powers.

committed dose equivalent   Predicted total dose equivalent to a given organ or tissue over a 50-year period after an intake of a radionuclide into the body.

common ion effect   The term used to describe an equilibrium shift in a dissociated compound by the addition of a compound that supplies one of the ions in the original compound.

compatibility   A condition in which substances exhibit no apparent reactivity toward one another; an important consideration in storage of chemicals by compatibility.

complement   A protein component of immune serum that interacts with antibodies to destroy foreign cells.

complex   An ion or compound containing a central metal ion bound to one or more other ions or compounds. Complexes may have a neutral, positive, or negative electrical charge.

complexing agent   Any substance that interacts with an ion to form a different moiety.

complex ion   An ion that consists of more than one element.

compliance   The act of yielding to a request or demand, often with legal implication.

composite   A solid mixture of two or more materials.

composting   The aerobic decomposition of organic (generally biochemical) compounds by bacteria.

compound   A chemical combination of two or more elements combined in a fixed and definite proportion by weight. Over ten million chemical compounds are known.

compressed gas   Any mixture or compound in a container whose absolute pressure exceeds ambient pressure by a certain amount (usually 40 psi).

Compressed Gas Association (CGA)   1725 Jefferson Davis Highway, Suite 1004, Arlington VA 22202-4102.

compressibility   The capacity to be compressed, i.e., made more compact or forced into a smaller space.

Compton, Arthur Holly   American physicist (1892–1962) who shared the 1927 Nobel Prize in Physics for the discovery of the Compton effect, of importance in the attenuation of X rays and gamma rays because it ejects an excited electron from the atom.

Compton effect   A process in which incident radiation ejects an electron from the absorbing material as the mechanism for energy transfer.

computational chemistry   Use of computers to simulate chemical reactions of interest, from syntheses to biological efficacies.

computer-based training (CBT)   Training programs that have been developed on, and are delivered by, computer.

conc   Concentrated or concentration.

concentration   The number of grams (or moles, etc.) of a substance present in a liter (or kilogram, etc.) of solution.

concrete   A widely used conglomerate of various solids (pebbles, rock chips, etc.) in cement or mortar.

condensation   The phenomenon of a vapor changing state to a liquid or solid; generally effected by cooling and often used to separate compounds.

condition adverse to quality   An all-inclusive term used in reference to any of the following failures, malfunctions, deficiencies, defective items, and nonconformances. A significant condition adverse to quality is one that, if uncorrected, could have a serious effect on safety or operability.

conductance   The capability of a material or solution to transmit electron flow.

conduction   The transfer of something (heat, electrons) through a material.

conductive hearing loss   A condition for which sound transmission to the cochlea has been interrupted.

conductivity   The degree to which a material or solution transmits heat or electron flow.

confidence coefficient   The probability of a value being found within a given confidence interval; generally expressed as a percentage, e.g., 95%, 99%.

confidence interval   The range of values that has a stated probability of including the value of interest.

confidence level   The degree of comfort with a given value or result; the probability that a given value will lie within the confidence interval.

confidence limit   Generally plural, used to indicate the range of values that are considered to be in accord with the given data; the boundaries of a confidence interval.

confidential   Level of classification for information or material that, in the event of an unauthorized disclosure, could reasonably be expected to cause identifiable damage to security.

configuration   The arrangement of atoms in a given molecule.

confined space   An enclosed area that has the following characteristics: primary function is something other than human occupancy; restricted entry and exit, (entry and exit restrictions will be determined on an individual basis); may contain potential or known hazards (e.g., toxic, radioactive, flammable, reactive, or corrosive liquids, solids, and/or vapors); may contain inert gases in sufficient quantity to displace the air; may contain physical hazards.

confirmed human carcinogen   Any material associated with industrial processes recognized to have carcinogenic potential for humans.

conformance   An affirmative indication or judgment that a product or service has met the requirements of the relevant specifications, contract, or regulation; the statement of meeting the requirements.

congener   A member of the same group, genus, or kind; related structures that are not identical.

Congo Red   Sodium diphenyl-bis-α-naphthylamine sulfonate; a brownish-red powder dye that can be used as an indicator or in diagnostic medicines.

coniferin $C_{16}H_{22}O_8$, a glucoside found in conifers; used to manufacture vanillin.

conjugate   Joined or paired.

conjugated fatty acid   A fatty acid that contains conjugated double bonds.

conjugate layers   Two different liquid mixtures in layered equilibrium.

conjunctiva   The mucous membrane under the eyelids and on the surface of the eyeball.

conjunctivitis   Inflammation of the membrane that lines the eyelids and the front of the eyeball.

consensus standard   A standard agreed on by a group of experts or knowledgeable groups or individuals or those impacted by the standard.

consent decree   A negotiable agreement ordered by court and enforceable.

conservation of energy   The first law of thermodynamics: energy is neither created nor destroyed.

consistency   The resistance to change; generally refers to the thickness or viscosity of a fluid.

consolute temperature   The critical solution temperature.

constant boiling point   A characteristic of an azeotrope or a pure compound.

constant-volume pump   A sampling pump that draws a given, fixed volume of gas with each full stroke.

Consumer Products Safety Commission (CPSC)   5401 Westbard Ave., Bethesda MD 20816.

contact dermatitis   The skin irritation caused by contact with a sensitizing chemical.

contact process   A process for producing H_2SO_4 by catalytic oxidation of SO_2; the chief commercial method.

container   Anything used to hold another substance, whether for reacting, storing or transporting, ranging in size from ampoule to tanker.

containment   A testable enclosure; preventing a spill or leak from spreading.

contamination   The presence of unwanted material (e.g., radioactive or chemical matter).

contamination, radioactive   The deposition of uncontained or unwanted radioactive material on the surface of structures, areas, objects, or personnel.

contamination survey   An evaluation by smear or swipe and/or direct measurement to determine the absence or level of any contamination on a surface or in a volume.

continuous distillation   A separation process using a continuous introduction of a mixture and continuous removal of the fractions.

continuous wave (CW)   Describes a laser that operates continuously.

contractor   An employee working under contract, not employed directly by the managing corporation.

control   The blank or reference sample, condition, etc.; in occupational hygiene, the last, most important of the four primary responsibilities. Controls can be engineered, established work practices, administrative, or use of personal protective equipment.

control area   An area posted with signs such as an asbestos work area. Entry into these areas is limited to those who must enter, and personal protection is required.

control chart   A graph of values, usually chronological, to permit easy visual determination of trends for a parameter of interest.

controlled release   A substance manufactured (encapsulated or coated) to furnish a constant amount over time rather than all-at-once release. Used in, e.g., drugs and fertilizers.

controlled substance   One that requires a prescription to purchase or is otherwise legally monitored.

control limits   The established values beyond which any variation is considered to indicate the presence of an assignable cause. Control limits established at the 95% confidence level are called warning limits. Those established at the 99% confidence level are called alarm limits.

control room   An area in a plant from which most of the power production and emergency safety equipment can be operated by remote control.

control velocity   *See* capture velocity.

convection   Transmitting or conveying as in heat transfer, generally via a fluid.

conversion   A change from some state or condition to another.

conversion ratio   The ratio of fissionable atoms produced over the fissionable atoms consumed in a reactor.

convulsion   A violent spasm or series of jerkings of the face, trunk, or extremities; seizure.

Coordinate bond   A bond in which a pair of electrons are donated by one of the two atoms forming the bond.

coordination number   The number of entities (ligands) attached to the central atom of concern.

copolymer   A polymer produced from two or more different monomers.

copper (Cu)   A reddish colored, ductile, conducting group IB transition metal element, atomic number 29, atomic weight 63.55; widely used in electrical wiring and applications. Symbol from the Latin cuprum.

cordite   Smokeless powder made from nitrocellulose and nitroglycerin plus petrolatum.

cork   The light outer bark of *Quercus suber*, which is light, water resistant, and insulating.

cornea   The clear transparent portion of the eye that serves to refract incident light.

corona   A crownlike structure or appearance.

Corps of Engineers (US Army COE)   20 Massachusetts Ave., NW, Washington DC 20314.

corrected effective temperature (CET)   The scale that uses the globe temperature rather than the dry bulb temperature; includes radiation effects.

corrective action   Action taken to conform to a standard or required condition; action taken to remedy an error or to remove a deviation. Often accompanied by recurrence control. Measures taken to rectify conditions adverse to quality and, where necessary, to preclude repetition.

Corrective action report (CAR)   Findings observed during an audit are documented in the body of the report and on the corrective action record (CAR). Describes each problem, analyzes the cause, and provides corrective actions, and recurrence controls to be implemented by responsible management. *See* follow-up.

corrosion   Physical change, usually deterioration or destruction, brought about through chemical or electrochemical action as contrasted with erosion caused by mechanical action.

corticoid hormone   Any hormonal steroid compound; obtained from the adrenal gland.

corticosteroid   Any steroid compound; obtained from the adrenal gland.

cortisone   An adrenal gland hormone important in metabolism of fats, proteins, etc.; an anti-inflammatory compound.

corundum Al_2O_3; An abrasive, used in polishing, grinding, etc. Also called emery.

cosmic radiation (cosmic rays)   Penetrating ionizing radiation, both particulate and electromagnetic, that originates in space. Secondary cosmic rays, formed by interactions in the earth's atmosphere, account for about 45 to 50 millirem of the 125-millirem background radiation that an average individual receives in a year.

cosmochemistry   Chemistry of the planets or all stellar matter.

Cottrell, Fredrick Gardner   American chemist (1877–1948) who invented electrostatic precipitation for the removal of suspended particulate matter from air and other gases.

cough   A forceful expiration preceded by an atypical inhalation. There are many causes.

coulomb (C)   Quantity of electricity, electrical charge.

Council on Environmental Quality (CEQ)   722 Jackson Pl., NW, Washington DC 20006.

counter   A general designation applied to radiation detection instruments or survey meters that detect and measure radiation. The signal that announces an ionization event is called a count. *See* Geiger-Müller counter.

covalent bond   The bond between two atoms formed by sharing electrons.

Coxiella burnetti   An organism that causes a fever.

cp   Chemically pure.

CP   Center of pressure.

CPAF   Cost plus award fee.

CPF   Carcinogenic potency factor.

CPFF   Cost plus fixed fee.

cpm   Counts per minute; cycles per minute.

CPM   Critical path method.

CPR   Cardiopulmonary resuscitation.

cps   Counts per second; characters per second.

CPSC   Consumer Products Safety Commission.

CPU   Central processing unit.

Cr   Chromium.

cream of tartar $KHC_4H_4O_6$; potassium bitartrate; component of baking powder.

creosote   A yellow-brown wood tar mixture of phenols; wood preservative and carcinogenic.

cresol   A coal tar liquid.

Crick, Francis Henry Compton   British biophysicist (1916–) who, with J. D. Watson, and with the help of X-ray diffraction photographs, constructed a molecular model of DNA and proposed that DNA determines the sequence of amino acids in a polypeptide. He shared the 1962 Nobel Prize in Physiology or Medicine with Watson and Wilkins.

criteria   The standards on which a decision is based.

criteria document   A document that contains the recommendations to OSHA for establishing a standard. It is based on NIOSH research and may contain a wide range of topics, from acquisition of samples to record keeping.

critical constant   A maximum or minimum value for a physical property of interest.

criticality   Self-sustained nuclear fission reaction. In radiation physics, a state in which the number of neutrons released by fission is exactly balanced by the neutrons being absorbed (by the fuel and poisons) and escaping the pile. A reaction is said to be "critical" when it achieves a self-sustaining nuclear chain reaction.

critical mass   The minimal mass of a fissionable substance that will permit an explosion.

critical organ   The body organ receiving a radionuclide or radiation dose that results in the greatest overall risk.

critical point   The temperature above which a vapor cannot be liquified via pressure.

critical pressure   The pressure at which the vapor phase of a substance coexists with the liquid phase at the critical temperature.

critical solution temperature   The temperature above and or below which two liquids become miscible in all proportions.

critical temperature   The temperature above which pressure alone cannot liquify the gas.

critical volume   The volume of a unit mass of a substance at the critical temperature and pressure.

CRM   Certified reference material.

cross-link   To fasten polymer chains together by bridges.

crossover   The unintentional blending of two or more quantities of material (e.g., containing the same element but containing different percentages of a specific isotope) to the extent that one or more of the original quantities may no longer be identified as being of the original material type.

cross section   The effective area presented by an atom for a nuclear reaction. Generally is expressed in units of barns.

croton oil   A strong cathartic substance, no longer used medically.

CRT   Cathode ray tube; cargo restraint transporters.

crucible   A cone-shaped refractory container used in thermal laboratory work.

crud   A colloquial term for corrosion and wear products (rust particles, etc.).

cryogenic   Pertaining to very low temperature.

cryogenics   The study of matter and phenomena at low temperatures.

crystal   A solid with characteristic shape.

crystallization   The formation of crystals from vapor or solution by accretion.

crystallography   The study of crystals to determine atomic locations (geometry and distances).

Cs   Cesium.

CS   Cold storage area; US Department of Commerce, Commerce Standard.

CSA   Canadian Standards Association.

CSL   Chemistry Standards Laboratory.

CSO/ACSSO   Computer security officer/alternate computer system security officer.

CSP   Certified Safety Professional.

CSV   Central storage vault.

C_T Thrust coefficient.

CTD Cumulative trauma disorder.

CTS Carpal tunnel syndrome.

Cu *Cuprum* (Latin); copper.

cumulative dose The total dose resulting from repeated exposures to the same region, or to the whole body, over a period of time.

cumulative trauma disorder (CTD) A repetitive strain injury.

cupola, cupula Dome shape.

curare A poisonous class of resins from South American trees.

Curie, Marie (née Manya Sklodowska) French physicist born in Warsaw (1867–1934) who worked with her French husband, Pierre Curie (1859–1906), on magnetism and radioactivity and discovered radium. Pierre and Marie Curie shared the 1903 Nobel Prize in Physics with Becquerel for the discovery of radioactivity. After her husband's death in a horse-drawn carriage accident, she isolated polonium and radium and was awarded the 1911 Nobel Prize in Chemistry.

Curie, Pierre French chemist (1859–1906) who shared the 1903 Nobel Prize in Physics with his wife, Marie Curie. He and Marie Curie discovered radium and polonium in their investigation of radioactivity.

curie (Ci) The basic unit used to describe the intensity of radioactivity in a sample of material; one curie equals 37 billion (3.7×10^{10}) disintegrations per second or approximately the radioactivity of 1 gram of radium. Commonly divided into smaller units. Named for Marie and Pierre Curie, who discovered radium in 1898.

curie point That temperature above which ferromagnetism does not exist.

curing The final step in converting some product to its most useful or attractive condition; generally done by heat, chemical treatment, or simply aging.

curium (Cm) A silvery transuranic metallic element, atomic number 96, atomic weight 247.

custodian The generally responsible foreman or manager of a building or department; employee ultimately responsible for a radioactive source, including its use and storage.

cutaneous Pertaining to the skin.

cutaneous lesions Skin lesions.

cutaneous sensitizers   A substance that does not cause changes in the skin on first contact but induces cellular changes that then manifest after an incubation period, usually of 5 or 7 days, so that further contact with the material results in an acute dermatologic response; examples include nickel, potassium dichromate, formaldehyde.

cutting oil   A petroleum product used in machining to wash away particles and cool or lubricate the material being machined. Exposure can result in "cutting oil acne."

CV   Curriculum vitae, a résumé; cost variance; coefficient of variation.

CW   Continuous wave; cold water; clockwise.

CWA   US EPA Clean Water Act.

CWTC   Chemical Waste Transportation Council.

CWU   Chemical Workers Union.

cyanogen C_2N_2; a gas used in organic syntheses; any compound derived from cyanogen.

cyanosis   A bluish gray to dark purple skin discoloration caused by abnormal amounts of reduced hemoglobin in the blood.

cyclamate   The general name for diet sweeteners derived from cyclohexylamine or cyclamic acid. Legal in Canada but not in the USA.

cyclic compound   Any compound that contains a ring structure; three types are named alicyclic, aromatic and heterocyclic.

cyclone   A dust-collecting apparatus utilizing air flow similar to that observed in natural cyclones.

cyclotron   An electromagnetic accelerator, usually circular, that imparts high energies to positively charged particles.

cytochrome   The group of colored iron porphyrin proteins found in most plants and animals.

cytoplasm   Organized colloidal mixture of organic and inorganic substances that is external to the nuclear membrane of a cell; generally an aqueous solution or suspension of protein protoplasm.

D

δ   Lower case Greek letter delta.

Δ   Upper case Greek letter delta.

d   Deuteron; day; density; prefix for dextrorotatory.

D   Deuterium.

2, 4-D   Dichlorophenoxyacetic acid.

d/m/l   Disintegrations per minute per liter.

D-test   Destruction testing.

D&D   Decontamination and decommission; decontamination and disposal.

DA   Disbursement authorization; double amplitude; destructive analysis.

DAC   Derived air concentration.

Dakin, Henry Drysdale   An English chemist (1880–1952) who co-authored the *Handbook of Chemical Antiseptics*, developed Dakin's solution for treating wounds during World War I, and was awarded the Davy Medal by the Royal Society in 1941.

Dakin's solution   An aqueous solution of 0.5% sodium hypochlorite used as an antiseptic for wounds.

Dalapon   A chlorinated organic herbicide that is a strong irritant.

Dalton   The unit of measure used for atomic weight equal to 1/12 the mass of ^{12}C.

Dalton, John   An English chemist and physicist (1766–1844) who discovered the law of partial pressures of gases, discovered the law of multiple proportions, maintained the electric origin of the aurora borealis, and gave the first detailed description of color blindness.

Dalton's law   States that in a mixture of gases, each component exerts its pressure independently as if the other components were not there and that the solubilities of mixed gases in a liquid are proportional to the partial pressure of each individual gas.

damper   A device for regulating airflow in a duct or plenum by increasing or decreasing the size of the air passage; often used to balance airflow.

dander   The material from the coats and skins of animals, e.g., dogs, cats, laboratory rodents, and horses, that may cause allergic responses in susceptible persons.

DANGER   A labeling term denoting the highest degree of hazard. Lesser degrees are Warning and Caution.

Darcy, Henri-Philibert-Gaspard   French hydraulic engineer (1803–1858) who first derived the equation that governs the laminar flow of fluids in homogeneous, porous media and thereby established the theoretical basis of groundwater hydrology.

Darcy's law   The volumetric rate of water flow through a sand bed is directly proportional to the cross-sectional area of the bed and the pressure difference across the bed and inversely proportional to the bed thickness.

Darling, Samuel Taylor   American physician (1872–1925) and chief of laboratories in the Panama Canal Zone, who discovered the pulmonary infection caused by histoplasma, which is a fungus grown in soil contaminated by bird excrement.

Darling's disease   A worldwide, opportunistic primary pulmonary infection caused by *Histoplasma capsulatum*, a fungus grown in soils enriched by bat, chicken, and other bird excrement. Problematic to farm workers and, occasionally, construction workers from exposure near pigeon roosts and demolished grounds where wild birds have nested.

Darwin, Charles Robert   English naturalist (1809–1882) who developed the theory of natural selection and the pangenesis hypothesis, i.e., that humans are derived from anthropoids.

DAS   Data acquisition system; document accountability system.

data   Information; the plural of datum.

database   A comprehensive file of information arranged to meet the needs of a number of applications as opposed to one file for each application; often computerized.

data validation   The systemic review of information to identify and possibly remove data that do not belong because they do not fit pre-established selection criteria, e.g., statistical tests.

dative bond   *See* coordinate bond.

datum   The singular form of data.

daughter element(s)   The nuclide(s) formed by the radioactive decay of another nuclide or parent.

daughter products   Isotopes that are formed by the radioactive decay of another radioisotope. For example, radium-226 decays to 10 successive daughter products, ending with the stable isotope lead-206.

Davy, Sir Humphry   English chemist (1778–1829) who discovered the effects of nitrous oxide when inhaled; invented the miners lamp; showed that chlorine was an element, and demonstrated that diamond is carbon.

dB   Decibel.

dBA   Sound level in decibels measured on the A scale, the frequency scale that approximates human response to sound and discriminates against low frequencies.

DBA   Design basis accident.

dBC   Sound level in decibels measured on the C scale, the frequency that does not discriminate against low frequencies like the human ear does.

DBCP   1,2-dibromo-3-chloropropane.

DBE   Design basis earthquake.

DBT   Design basis tornado.

DBW   Design basis wind.

D & C   The designation used by the US FDA to certify that ingredients can be used in drugs and cosmetics.

DC   Direct current; design criteria, duty cycle.

DCG   Derived concentration guide.

DCW   Domestic cold water.

DDE   Dichlorodiphenyldichloroethylene; degradation product of DDT and also a contaminant in DDT residues.

DDT   Dichlorodiphenyltricholorethane; a formerly widely used pesticide, currently banned because of implications of environmental damage.

DDVP   Dimethyl dichlorovinyl phosphate; also known as dichlorovos and Vapona.

DEA   Drug Enforcement Agency.

dead man control   A switchlike device on a power tool or moving vehicle such as a train or subway, which enables power to reach the tool or vehicle only when the switch is depressed. When pressure from the operator is released, power immediately stops.

debility   Abnormal weakness or lack of bodily strength.

deblooming agent   A chemical added to oils to stop or hinder efflorescence.

Debye, Peter Joseph Wilhelm   American physicist (1884–1966) born in the Netherlands, who lived in Germany until 1935 and was awarded the 1936 Nobel Prize in Chemistry for studies of molecular structure through dipole moments and X-ray diffraction; worked on the Manhattan Project in the USA.

Debye-Huckel theory   A theory of interionic attraction in solution and ionic strength that assumes that strong electrolytes are completely dissociated into ions.

deca   Prefix meaning 10^1.

decaborane $B_{10}H_{14}$; a colorless solid, stable at room temperature, that may explode on heating or on contact with halogenated or oxygenated solvents; used as a catalyst, fuel additive, and propellant.

decarboxylase   Type of enzyme in living cells that removes carbon dioxide from carboxylic acids without oxidation.

decay   Decomposition or destruction, usually biological; deterioration to a lower state of health, mental ability, or energy level.

decay, radioactive   Decrease in the amount of any radioactive material over time, caused by spontaneous emissions of alpha or beta particles or gamma radiation from the atomic nuclei.

decelerate   A decrease in velocity.

decertification   Loss of certification status.

Decibel (dB)   The unit of sound power level, intensity, or pressure. It is expressed as a logarithmic ratio with respect to an arbitrary reference value that is defined as the lower threshold of human hearing in air.

decision theory   A mathematical system based on probability, used to determine choices and interpret results.

decoction   The liquid produced by boiling one or more substances in water and filtering.

decolorizing agent   A substance that removes color by chemical or physical means, e.g., bleach, charcoal.

decommission   The official removal from service of a laboratory, area, or building after it is inspected by qualified personnel and indicated as free of contamination. The procedure necessary before such areas can be renovated or removed.

decomposition   The breakdown of a biological or chemical substance into smaller parts, units, or elements by physical, chemical, or biological means.

decompression   Loss or reduction of pressure from atmospheric pressure.

decompression sickness   Also known as caisson disease. The presence of nitrogen bubbles in blood and body tissues because of a sudden decrease in atmospheric pressure. Underwater divers and deep underground workers are at risk.

decontamination   The reduction or removal of contaminating radioactive, biological, or chemical material from a structure, area, object, or person. Decontamination may be accomplished by treating the surface to remove or decrease the contamination or letting the material stand so that the radioactivity is decreased as a result of natural decay.

decortication   The physical removal of the hard coating on vegetables, fruits, or nuts.

deenergize   Completely remove power, e.g., electrical or mechanical, to render the system safe for work.

deer fly fever   *See* tularemia.

DEET™   An insect repellent.

defatting   A process that removes fat or oil, as from skin.

deflagration   A rapid fire or low-level explosion at the surface of a substance.

deflocculation   The physical or mechanical separation of individual particles of a mixture. Also, the suspension of dispersed phase particles of a colloid into the dispersion medium.

defoaming agent   A substance used to reduce foaming caused by proteins or dissolved gases.

defoliant   An herbicide used to remove leaves from living trees and plants, e.g., Agent Orange, 2,4-D, 2,4,5-T.

degradation   Decomposition, or breaking down into smaller components, that may result from biological action, oxidation, heat, infection, sunlight, etc.

degreaser   A chemical solvent, such as trichlorethylene, used to clean or remove grease or oil.

degreaser's flush   Skin inflammation experienced by workers using degreasing chemical solvents.

degree(s)   Increments on temperature scales such as Kelvin ($°K$), Celsius ($°C$), and Fahrenheit ($°F$). Degrees Fahrenheit are used by the US public. Degrees Celsius are metric units commonly used in science and by most of the public outside the US.

degrees of freedom   The number of variables in a system that must be specified to define the system. In chemical engineering, the number of variables that can be changed without producing a phase change. Statistically, an index used to determine the distribution necessary to analyze a system.

dehumidification   The removal of moisture from air.

dehydration   The removal of water from a substance; chemically, the removal of one or more molecules of water from a compound.

dehydrogenase   An enzyme that catalyzes the removal of hydrogen.

dehydrogenation   The removal of hydrogen from a chemical; considered to be a form of oxidation.

deionizing   A method for purifying water by passing it through an ion exchange resin bed.

Deisenhofer, Johann   German chemist (1943–) who shared the 1988 Nobel Prize in Chemistry for the determination of the three-dimensional structure of a photosynthetic reaction center.

Delaney, Jim   Democratic congressman from Queens, NY, who introduced the amendment (which bears his name) to the 1958 legislation (Section 409) on food additives. It has since been repealed.

Delaney clause   Part of the 1958 Food Additives Act of the US FDA that states that no chemical shown to be a carcinogen in any animal species, at any level of exposure, can be used as food additive. It has since been repealed.

deleading   The removal of lead from the human body by use of a chelate, e.g., ethylenediaminetetraacetic acid (EDTA). The lead is excreted in urine. Also the removal of lead paint by physical means to prevent or minimize human exposure.

deleterious   A chemical, biological, or physical agent capable of causing harm, injury, or disease.

delineation   The marking of an area contaminated chemically or biologically.

deliquescent   The tendency of a solid substance to absorb atmospheric water vapor and become a liquid solution. *See* hygroscopic.

delisting   The formal process by which a substance is removed as a designated agent from a regulatory list.

deluge shower   An emergency or safety shower, capable of rapidly delivering a large volume of water to wash off or dilute hazardous chemicals that have been spilled on a person.

delustrant   A chemical substance, such as barium sulfate, used to dull the surface of textiles.

demand respirator   A supplied-air respirator that admits air when a negative pressure is created inside the facepiece by inhalation.

demineralization   A water purification process that removes dissolved minerals from water, often by distillation or ion exchange.

de minimis **violation**   A minor infraction or violation of OSHA regulations that presents no threat to an employee's health or safety.

Democritus   A fourth and fifth century Greek philosopher, known as the Abderite and laughing philosopher, who adopted and extended the atom theory of Leucippus and used the term "atoms" to describe the smallest indivisible parts of matter.

demulcent   An oily substance used to relieve pain in irritated mucosal surfaces.

demulsify   To chemically (addition of substances) or physically (e.g., centrifugation, heating) break or remove emulsions.

demurrage   A charge by shippers for the use and or storage of a product such as a compressed gas cylinder.

demyelination   Deterioration of the myelin sheath around neurons.

denaturing   The process of changing the structure of a protein by heat or addition of an acid or base.

denatured alcohol   Pure, 95% ethyl alcohol to which another substance has been added to make it unfit for drinking.

dendrite   The branched part of nerve cells that transmits impulses.

dendron   A single dendrite.

dengue fever   A mosquito-borne fever with associated joint pain.

denier   The unit used in textiles to indicate filament fineness expressed in mg/m.

densitometer   A device used to measure density.

Density (d)   The physical property of a material expressed as the ratio of mass to volume. The higher the ratio, the greater the density.

dental alloy   An amalgam of metal containing mercury used to fill former dental caries locations.

deodorant   A substance used to eliminate, reduce, or mask unpleasant odors.

deoxidation   The process of chemically removing oxygen from a chemical.

deoxy-   Prefix denoting the replacement of a hydroxyl group by a hydrogen in a chemical.

deoxyribonucleic acid (DNA)   Nucleic acid with a double helix, spiral type structure, found in chromosomes of animals and plants that contains the genetic information. Structure determined in 1953 by James Watson and Francis Crick.

Department of Agriculture (USDA)   14th & Independence, SW, Washington DC 20250.

Department of Commerce (DOC)   Constitution Ave. & E St., NW, Washington DC 20230.

Department of Defense (DOD)   The Pentagon, Washington DC 20301

Department of Energy (DOE)   1000 Independence Ave., SW, Washington DC 20585.

Department of Health and Human Services (DHHS)   200 Independence Ave., SW, Washington DC 20201.

Department of the Interior (DOI)   1849 C St., NW, Washington DC 20240.

Department of Justice (DOJ)   950 Pennsylvania Ave., NW, Washington DC 20530.

Department of Labor (DOL)   200 Constitution Ave., NW, Washington DC 20210.

Department of Transportation (DOT)   400 Seventh St., SW, Washington DC 20590; US agency responsible for regulations on the transportation of materials.

depilatory   A substance used to remove hair from humans and animals, usually containing sulfides.

depleted uranium   The substance remaining after part of the fissile uranium has been removed. It has a percentage of uranium-235 less than the 0.72% found in natural uranium. *See* mill tailings.

deposit   Material transported and relocated by water, wind, gravity, or human activity.

deposition   Testimony given by a witness outside of the courtroom, usually during the discovery process.

DeQuervain, Frits   Swiss physician (1868–1940) who discovered that certain types of repetitive hand use cause numbness and tingling in the fingers.

DeQuervain's disease   A type of tenosynovitis of the wrist and thumb caused by repetitive hand use (e.g., tools) or motions (e.g., wringing, opening and closing the hand) that cause inflammation of the tendons.

derived unit   A unit of measure derived from time, length, or mass, such as speed or density.

derma-, dermal   Pertaining to the skin.

dermatitis   Inflammation or irritation of the skin.

dermatophytoses   Diseases caused by fungi; found among farmers, animal handlers, athletes, and animal laboratory workers.

DES   Diethylstilbestrol.

desalination   The process of removing minerals including salts, e.g., sodium chloride, from seawater. Methods may include distillation, ion exchange or osmosis.

Descartes, René   French scientist, mathematician, and philosopher (1596–1650), resident of Holland, whose Latin name was Renatus Cartesius; developed the Cartesian system of coordinates and mathematical certainty and modern scientific thinking summarized by the words, *cogito, ergo sum* ("I think, therefore I am").

desiccant   A substance that absorbs and removes water, e.g., activated alumina, calcium chloride, or silica gel.

desiccator   A laboratory device, usually made of glass or metal and capable of being put under partial vacuum, in which chemicals or substances are placed to remove water or to dry.

designated area   An area used for work with select carcinogens, reproductive toxins, or substances of high or acute toxicity as specified by the OSHA Laboratory Standard. A designated area may be an entire laboratory, an area of a laboratory, or a device such as a laboratory hood.

design-basis phenomena   Earthquake, tornado, hurricane, flood, etc., phenomena that a facility may be designed and built to withstand, without loss to the systems, structures, and components necessary to assure public health and safety.

design feature (DF)   A product feature or specification, for which the design agency retains design responsibility. A statement about design or engineering conditions or features important to safety and to which alterations are not to be made before safety review. Passive, physical facility safety features that remain constant throughout the life of the plant.

design process   A technical and management process that begins with the identification of design input and that leads to and includes issuance of design documents.

design review   The formal review of existing or proposed design(s) to detect and remedy deficiencies that could affect fitness for use and environmental aspects of the product, process, or service; the identification of potential improvements necessary in performance, safety, and economic aspects.

desorption   The release of an adsorbed substance from a solid, usually by heat, reduced pressure, or replacement by another substance.

destructive distillation   The process by which coal, charcoal, or other carbonized material is subjected to high temperature in the absence of air.

detector   A material or device sensitive to a stimulus (e.g., specific chemical or chemical group, radiation), that produces a response signal suitable for measurement or analysis.

detergent   An agent used for cleaning that reduces the surface tension of water, often at an oil-water interface, by use of emulsifying substances.

determinate error   An identifiable error that can be removed or corrected.

detonation   A violent, supersonic chemical reaction within a chemical compound that evolves heat and pressure. The result is extremely high pressure on the surroundings, propagating a supersonic shock wave. Detonations usually produce craters if the explosion is on or near the ground.

detoxify   To reduce or eliminate the toxic properties of a substance; also to convert to a less toxic substance.

deuterium (D)   An isotope of hydrogen, atomic number 1, atomic weight 2, symbol D; also known as, heavy hydrogen.

deuteron   The nucleus of a deuterium atom, atomic mass 2, charge 1.

developer   A substance used in photography to produce visible images from film; a substance used in the dye industry, which when mixed with other substances produces a third color; a person or group who develops a hazardous waste treatment, storage, or disposal facility.

developmental toxicology   The toxicological study of the harmful effects of chemical, biological, or physical substances on the development of an embryo or fetus after conception.

dewater   Remove water from a suspension by filtration., centrifugation, or settling.

dew point   The temperature at which a gas condenses to a liquid; also, the temperature at which air is saturated with moisture.

dextran   Also known as macrose. A high-molecular-weight polymer used as a blood plasma substitute, food additive, and filtration gel.

dextrin   A gum formed from hydrolysis of starches.

Dextrorotatory (d)   A chemical, designated by d or +, in which light is rotated to the right as it passes through the substance.

dextrose   *See* glucose.

dextrose equivalent   The total amount of reducing sugars present.

DHHS   Department of Health and Human Services.

DHW   Domestic hot water.

di-   Prefix meaning two.

diagenesis   Physical and chemical changes that take place in sediments after burial.

diagnosis   The identification of a disease by examination.

dialysis   The selective passage of a solute going from higher to lower concentrations through a membrane to separate smaller molecules from larger molecules based on differences in rates of diffusion.

diamond   A very hard, stable, crystalline form of carbon. Can be made synthetically by heating carbon at 3000°F under high pressure.

diaphoresis   Perspiration, especially profuse.

diaphragm cell   An electrolytic cell used to produce sodium hydroxide and hydrogen from sodium chloride.

diarrhea   Excessive elimination of watery fecal matter.

diastereoisomer   An optically active isomer with two asymmetric carbons, which is part of a pair of enantiomers (mirror images) and part of a set of four total isomers.

diatomaceous earth   The prehistoric remains of algae, used for filtration, decolorizing, pigments, chromatography, ceramics, etc.

diatomic   A gas whose molecules are formed by two identical atoms, e.g., oxygen, O_2, nitrogen, N_2, chlorine, Cl_2.

diazepam   An addictive tranquilizer used as a sedative because it depresses the central nervous system.

diazo   A chemical in which the structure $Ar—N{=}N—X$ is present, where Ar is an aromatic ring and X is an inorganic group.

diazotize   To react a primary aromatic amine with nitrous acid in the presence of excess acid. Widely used to make dyes.

dibasic   An acid having two reactive protons.

dibenzofuran $C_{12}H_8O$; diphenylene oxide; a solid obtained from coal tar and used as an insecticide.

diborane B_2H_6; a colorless, foul -smelling reactive gas. It is flammable and toxic and reacts violently with oxidizers. Used in syntheses of boron compounds.

dibromochloropropane (DBCP)   A colorless, combustible liquid. A carcinogen used to fumigate soil and in pesticides and nematocides.

dicamba   The generic name for 2-methoxy-3,6-dichlorobenzoic acid. A solid used for pest control and in herbicides.

dichlorobenzene $C_6H_4Cl_2$; the *ortho*, *meta*, and *para* isomers are all used as insecticides.

Dichlorodiphenyldichloroethylene (DDE)   Degradation product of DDT and also a contaminate in DDT residues.

Dichlorodiphenyltricholorethane (DDT)   A formerly widely used pesticide, currently banned because of implications of environmental damage.

dichloromethane   *See* methylene chloride.

dichlorophenoxyacetic acid (2,4-D)   A defoliant used in Agent Orange in the Vietnam war. Formerly often used commercially as a weed killer.

dichlorovos   A toxic liquid organophosphorus insecticide and fumigant; toxic via dermal absorption.

dichroism   The property of a crystal that refracts light in two directions, thus appearing to be a different color when viewed from different angles.

dichromate   The anion $Cr_2O_7^=$.

die   A device used for forming, usually a metal; death of an organism.

dieldrin $C_{12}H_{10}OCl_6$; a tan solid used as an insecticide. It is toxic by ingestion, inhalation, and dermal absorption and is carcinogenic.

dielectric   A substance having very low electrical conductivity, e.g., insulators such as glass, rubber, and wood.

dielectric constant   A quantitative indication of the electrical conductivity of a substance compared to a vacuum, which has a value of 1. Lower values are better insulators.

diene   *See* diolefin.

diesel oil   fuel oil #2; a petroleum distillation cut used for diesel engines. *See* fuel oil.

dietary food supplement   Vitamins or minerals added to food in a quantity that would deliver more than 50% of the FDA Recommended Daily Allowance in a single serving.

diethyleneglycol   A colorless, odorless, sweet, noncorrosive, hazardous liquid commonly used in antifreeze.

diethyl ether $(C_2H_5)_2O$; a colorless, volatile, hygroscopic liquid with distinctive odor; very flammable and forms explosive peroxides; CNS depressant via inhalation or dermal absorption.

diethylstilbestrol (DES)   A carcinogenic compound; a synthetic estrogen formerly used to prevent spontaneous abortion.

differential gravimetric analysis (DGA)   A differential thermal analysis technique utilizing the rate of change in weight with heating.

differential pressure   The difference in pressure between two points of a system, such as between the inlet and outlet of a pump.

differential thermal analysis (DTA)   An analytical technique that measures the temperature and rate of change in temperature for the material being analyzed as heat is added or removed.

diffraction   Modification of light or a beam of radiation, e.g., X rays, by passage through or around something.

diffuser   A device used to distribute gas flow evenly.

diffusion layer   The liquid layer immediately around an electrode with a concentration gradient.

diffusion rate   The tendency of one gas or vapor to disperse and mix with another gas or vapor. The diffusion rate depends on the density of the vapor or gas as compared with that of air. Sampling methods that use organic vapor monitors are based on this principle; also applicable to species in the liquid or solid state.

digester   Any container or apparatus used to digest or put into solution a material of interest.

digitalis   A drug used to treat cardiac diseases made from the leaves of the foxglove plant.

digitoxin $C_{41}H_{64}O_{13}$; a white, odorless powder with bitter taste. Used in cardiac disease treatment; can be fatal if given as overdose.

dike   An embankment used to contain a spill or unwanted spread of a liquid.

diluent   Anything added to reduce the concentration of a substance to the desired level. May be in any state of matter.

dilute   To lower the concentration of a solute or weaken the potency, strength, or purity.

dilution ventilation   Exposure control via dilution, i.e., mixing a contaminant with large volumes of air sweep through the general workplace.

dimensional analysis   A technique to verify the validity of a calculation by utilizing the units for each component. Thus, if calculating the density of something, one must have mass units divided by volume units.

dimer   The combination of two identical entities, often used with respect to large molecules.

dimethicone   A silicone oil used to prevent irritation of the skin by water-soluble species.

dimethyl dichlorovinyl phosphate   Known as (see) dichlorovos and Vapona.

dimethyl sulfide $(CH_3)_2S$; a colorless, flammable, volatile liquid with a foul odor. Used as a gas odorant.

dimethyl sulfoxide (DMSO) $(CH_3)_2SO$; a colorless, hygroscopic liquid; common aprotic solvent. It readily penetrates skin and is used in human and veterinary medicine.

DIN   Do it now.

dineric   A solution of two immiscible solvents achieved via a solute soluble in each.

dinitrophenylhydrazine   Used as a test for aldehydes and ketones; explosive.

diol   See glycol.

diolefan   An aliplatic compound with two double bonds. See diene.

dioxane $(CH_2)_4O_2$; a colorless, flammable liquid with an etherlikeodor. May form explosive peroxides; toxic by inhalation; a carcinogen.

dioxin   Commonly used name for 2,3,7,8-tetrachlorodibenzo-p-dioxin (TCDD). A white solid carcinogen, teratogen, and mutagen whose presence as a contaminant of 2,4,5-T led the FDA to ban that herbicide.

dipole   A charge separation within a molecule, which overall is neutral. The difference in electronegativity between H and Cl atoms, e.g., leads to a dipole in HCl.

dipole moment (μ)   The distance between the charge centers in a molecule multiplied by the charge in electrostatic units.

diquat $(C_5H_4NCH_2^-)_2Br_2$; A toxic yellow solid, soluble in water and used as a herbicide, often to defoliate marijuana.

Dirac, Paul Adrien Maurice   English physicist (1902–1984) who shared the 1933 Nobel Prize in Physics for his work in quantum mechanics, notably the book, *The Principles of Quantum Mechanics*, published in 1930.

direct production/process engineering operating costs   Includes process and industrial engineering, tool and equipment design engineering, and product engineering direct support (salaries and wages, fringes, other manpower costs, and supplies and services) that can be identified with a specific system.

direct reading instruments   Real-time monitors that read or measure in actual time of the observation.

disability   The inability to function in what is considered the normal fashion for humans. A term used in the DOJ Americans with Disabilities Act.

disassembly   The process of breaking a unit into all of its components.

disaster   An occurrence with widespread destruction, death, and/or distress; usually occurring suddenly.

discharge   To let go, or terminate, e.g., releasing of a pollutant into a waste stream.

disease   An abnormal, deleterious condition of an organism, usually associated with an infection.

disease cluster   An atypical incidence of some abnormal condition linked by time, location, or exposure.

disinfectant   An agent that eliminates the likelihood of infection by killing or inactivating harmful microorganisms.

disintegration   *See* decay, radioactive.

dispersing agent   A substance added to a suspension to maximize uniformity of the solid distribution.

dispersion   A system with one component finely dispersed within the other. Suspensions or colloids are examples, but solutions refer to a different condition.

displacement   Generally the substitution of some entity (e.g., an ion in a molecule) with another different entity.

dispose   To discard, throw away, get rid of; to arrange.

disproportionate   The condition in a chemical reaction in which atoms within the same molecule serve as oxidizers and reducers; something being of a markedly different size from what is considered normal.

dissociate   To separate a substance into its components. Often used to describe polar compound status in aqueous solution, i.e., positive and negative ions.

dissolve   To put into solution.

dissolved oxygen (DO)   A self-explanatory term used to indicate the environmental condition of waters. Fish and living organisms require a certain concentration to survive.

distal   Away from the center.

distill   To vaporize a liquid and condense the vapor elsewhere to purify the liquid or separate its components.

distribution   The set of values collected for some measurement; the range of an organism; the dispersal of something.

diuresis   The elimination of an abnormally large amount of urine.

diuretic   A substance that increases urination to remove water from the organism.

diurnal   Occurring daily.

DL   Detection limit.

dl   Equal rotation.

DMSA   2,3 Dimercaptosuccinic acid.

DMSO   Dimethyl sulfoxide.

DNA   Deoxyribonucleic acid.

DNAPL   Dense non-aqueous phase liquid.

DNI   Do not incorporate. An indication appearing when instructions are for immediate and temporary use rather than permanent incorporation.

DO   Dissolved oxygen.

Döbereiner, Johann Wolfgang   German chemist (1780–1849) who discovered the catalytic properties of platinum and palladium; anticipated the periodic table by observing "triads of elements", e.g., chlorine, bromine, and iodine.

DOC   Department of Commerce.

documented evidence   Written proof that an action has occurred. *See* quality evidence and objective evidence.

DOD   Department of Defense.

DODES   Division of Disaster Emergency Services.

DOE   Department of Energy.

DOI   Department of the Interior.

DOJ   Department of Justice.

DOL   Department of Labor.

dollar of reactivity   The amount of reactivity increase needed for a reactor to go from critical (k = 1.000) to prompt critical. One one-hundredth of a dollar is one cent of reactivity.

Dollo, Louis (Antoine Marie Joseph)   French paleontologist (1857–1931) who worked in Brussels; stated the law of irreversibility in evolution.

Dollo's law   The principle that radical evolutionary changes are irreversible.

dolomite $CaMg(CO_3)_2$; a mineral with widespread use in fertilizers, ceramics, etc. to remove acid stack gases and as a source of magnesium.

don   To put on, such as a respirator.

donor   An atom that contributes both electrons in a covalent bond; an individual who donates an organ or part thereof to another individual.

DOP   Dioctylphthalate.

L dopa   L-Dihydroxyphenylalanine.

dopant   Something added to another substance, e.g., a semiconductor, usually in small quantities, to alter its properties.

dope   A narcotic; a lubricant.

Doppler, Christian Johann   Austrian physicist (1803–1853) known for his work with sound, especially the effect named for him.

Doppler effect   The phenomenon of wavelength shift when the emitter is moving relative to the observer. When approaching, the shift is to shorter wavelength, hence higher frequency.

DOS   Department of State; disk operating system; dosimetry-internal and external.

dose   A quantity (total or accumulated) of ionizing radiation received. Often used in the sense of the exposure, expressed in roentgens, which is a measure of the total amount of ionization that the quantity of X-ray or gamma radiation could produce in air. Distinguished from absorbed dose, given in rads, which represents the energy absorbed from any radiation in a gram of any material. The biological damage to living tissue from the radiation exposure. The amount of any substance absorbed by an organ or individual.

dose equivalent   Term used to express the amount of biologically effective radiation when modifying factors have been considered. The product of absorbed dose multiplied by a quality factor multiplied by a distribution factor. Expressed numerically in rem.

dose rate   The radiation dose delivered per unit time and measured (e.g., in rems per hour).

dose-response relationship   The relationship between the concentration and duration of exposure to an agent or chemical and the incidence and severity of the resultant health effects. Used to help establish safe exposure levels. Based on the Paracelsus quote, "All substances are poisons: there is none which is not a poison. The right dose differentiates a poison and a remedy."

dosimeter   A combination of absorber(s) and radiation-sensitive elements that is used to provide a cumulative record of absorbed radiation dose or dose equivalent received.

dosimetry   The theory and application of the principles and techniques involved in the measurement and recording of radiation doses. The practical aspect is concerned with the use of various types of radiation instruments with which measurements are made. *See* survey meter.

DOT   Department of Transportation.

DP   Defense programs; data processing; delta pressure; differential pressure.

dpm   Disintegrations per minute.

dps   Disintegrations per second.

Dragendorf's reagent   Potassium iodobismuthate; a reagent to test for alkaloids. Sometimes called Kraut's reagent.

Draize test   The animal (rabbit) test for corrosivity of chemicals.

dram   A unit of weight in pharmacology equal to 3.888 grams, 0.125 ounce, or 60 grains.

drier   Any substance used to catalytically expedite the drying of paints, inks, etc.

DRIERITE™   Trademark-protected name for a special form of anhydrous calcium sulfate drying agent.

dross   *See* slag.

drug   A generic term for a substance that kills or disables pathogens or enhances the activity of an organ or bodily function. The FDA definition is a substance "intended for use in the diagnosis, cure, instigation, treatment or prevention of disease, or to affect the structure or function of the body."

dry bulb thermometer   The standard thermometer for measuring ambient air temperatures.

dry cell   A primary battery used in flashlights, most commonly zinc anode and carbon cathode in a solid paste, but variations exist.

dry chemical extinguisher   A fire extinguisher containing a chemical agent that extinguishes fire by interrupting the chain reactions involved in a fire. The chemicals used prevent the reaction of free radicals in the combustion process so that combustion does not continue.

dry ice   Solid carbon dioxide that sublimes at $-78.5°C$. Widely used coolant that avoids the necessity of dealing with a liquid phase.

dry powder extinguisher   A fire extinguisher designed for use on combustible metal fires, e.g., sodium, potassium.

DTA   Differential thermal analysis.

duct   A passageway through which something, usually a fluid, is conveyed; an organism passageway for secretions.

duct velocity   The fluid flow rate through a duct cross section.

Duhring's rule   The postulate that the vapor pressures of similar liquids will behave in a similar fashion, therefore enabling one to estimate the vapor pressure of a compound whose vapor pressure has not been determined by analogy to one whose vapor pressure has been well characterized.

Dulong, Pierre Louis   French chemist (1785–1838) noted for Dulong and Petit's law, which relates the specific heat capacity of a solid element to its atomic mass, which for over a century was the way to approximate atomic weights.

Dulong and Petit's law   States that the atomic heat capacity of elementary substances is a constant of 6.2 cal/g K at room temperature. Atomic heat capacity is atomic weight times the specific heat.

Dursban™   Chlorpyrifos; a formerly widely used insecticide, withdrawn from commercial availability.

dust   Solid particles generated by handling, crushing, grinding, rapid impact, detonation, or decrepitation of organic or inorganic material. Dusts disperse readily in air and settle slowly. Dusts are normally characterized by an average particle size smaller than 74 microns (200 mesh). *See* nuisance dust.

duty cycle (DC)   The ratio of on-time RF emissions to the total time of operation.

DVM   Doctor of Veterinary Medicine; digital voltmeter.

DWS   Drinking water standard.

Dy   Dysprosium.

dye   A substance used to color something.

dynamic   Changing.

dynamite   An explosive consisting of nitroglycerin or ammonium nitrate in a solid matrix.

dyne   A unit of force, one centimeter-gram-second.

dys-   Painful, bad, or difficult.

dysbarism   Symptoms resulting from less than atmospheric pressure exposure. In the extreme sometimes called decompression sickness.

dysphagia   Difficulty in swallowing or the inability to do so.

dysphasia   Impairment of speech.

dyspnea   Difficult or labored breathing; also shortness of breath.

dysprosium (Dy)   A lanthanide element, atomic number 86, atomic weight 162.50.

dystrophy   A disorder caused by nutritional deficiency.

dysuria   Painful urination.

E

ε   Lower case Greek letter epsilon.

E   Antenna; armature; arrester, lightning; binding post; brush, electrical contact; exposure level; contact, electrical; east; energy; uppercase Greek letter epsilon; exa- (prefix $= 10^{10}$); environment/environmental (category).

e⁻   Electron.

E⁰   Redox or oxidation-reduction potential at standard conditions.

EA   Environmental assessment.

EA&C   Environmental analysis and control.

EACT   Emergency action and coordination team.

EAP   Employee assistance program; emergency action plan.

E&OH   Environmental and occupational health.

ear   The organ for hearing; composed of three parts: the external ear, the middle ear, which includes the ossicles, and the inner ear, which includes the semicircular canals, vestibule, and cochlea.

ear muff   A hearing protection device that covers the external ear; also a device worn over the external ear to protect against cold.

ear plug   A hearing protection device that is inserted into the ear.

ear protection   Devices either worn over the external ear or inserted into the external ear to protect against hearing loss; also worn to protect against cold.

Earth   The third planet from the sun in this solar system.

earth   The land surfaces of the world; dirt or soil.

Ebola virus   An exceptionally deadly human pathogen believed to originate in Africa; symptoms include excessive internal bleeding. Exposure results in a high rate of death.

ebullator   A solid substance or surface that helps prevent superheating of liquids above their boiling point.

EBW   Electron beam welding.

EC   Eddy current; electron capture; European Community.

EC$_{50}$ The effective concentration of a substance that causes an observable effect in 50% of the test animals.

ECAD   Electronic computer-aided design.

ECC   Estimated cost at completion.

eccentric   Deviating from the normal; a type of contraction in which the muscles lengthen while tensing.

eccrine   Sweat from the eccrine glands, which consists primarily of diluted salt water; a response to aid in cooling the body.

eccrine gland   A sweat gland that occurs in almost all parts of the skin.

ECD   Electron capture detector.

ECG   Electrocardiogram. *See* EKG.

ECL   Exposure control limit.

e/cm^2   Electrons per square centimeter.

E. coli   *Escherichia coli.*

ecology   The science or the study of the relationship between organisms and their environment.

ecosystem   A unit comprised of the living system and its physical environment.

ECS   Emergency control station; engineering computer system; engineering control system.

ecto-   Prefix meaning exterior.

ectoparasite   A parasite living on the exterior of another organism, e.g., a flea.

eczema   A generic inflammatory condition of the skin, often characterized by scaling, itching, and burning.

ED   Effective dose.

ED$_{50}$ Effective dose that affects 50% of the population dosed.

EDB   Ethylene dibromide; a chemical implicated in male sterility.

eddy   A current or flow in air or water that moves contrary to the main current or flow.

edema   Abnormal accumulation of fluid in cells, tissues, or cavities of the body, resulting in swelling.

EDF   Environmental Defense Fund.

Edison Electric Institute (EEI)   701 Pennsylvania Ave., NW, Washington DC 20004.

EDL   Economic discard limit.

EDP   Electronic data processing (also referred to as ADP, automatic data processing).

EDS   Energy-dispersive spectroscopy.

EDTA   Ethylenediaminetetraacetic acid.

EEG   Electroencephalogram or electroencephalography.

EEGL   Emergency response guidance level.

EEI   Edison Electric Institute.

EEL   Environmental exposure limit.

EEOC   Equal Employment Opportunity Commission.

EF   Emission factor.

effective charge   The charge that, when multiplied by the actual distance between two atoms in a diatomic molecule, gives the dipole moment.

effective dose (ED)   The amount of chemical necessary to produce an observable change in a healthy animal.

effective half-life   The time necessary for the amount of radioactivity to diminish by 50%.

effective temperature (ET)   The actual temperature felt by a human.

effervescence   The rapid escape of a gas from a liquid.

efficiency   The fraction or percentage obtained by dividing the useful output by the total input; a factor used to calculate the actual counting rate of a detector used to measure radioactive decay or disintegration.

efflorescence   The loss of waters of hydration when a chemical is exposed to air, which results in partial decomposition.

effluent   A fluid emitted from an area; often as in a waste stream.

e.g.   Abbreviation for the Latin phrase *exempli gratia* ("for example").

egress   To exit, leave, or go out.

ehp   Effective horsepower.

Ehrlich, Paul   German bacteriologist (1854–1915) awarded the 1908 Nobel Prize in Medicine or Physiology; a pioneer in immunology and chemotherapy; developed methods to stain tubercle bacillus, a remedy for syphilis, and a diphtheria antitoxin.

EHS   Extremely hazardous substance.

EI   Emissions inventory.

EI&C   Electrical instrumentation and control.

Eigen, Manfred   German chemist (1927–) awarded the 1967 Nobel Prize in Chemistry for his work on extremely rapid chemical reactions, hydrogen ions, and enzyme control.

Einstein, Albert   German-born (1879–1955) naturalized Swiss and American theoretical physicist awarded the 1922 Nobel Prize in Physics for developing the general and specialized theories of relativity and unified field theory of electromagnetism and gravitation, explaining Brownian movement, and developing the law of photoelectric effect.

einstein   A unit that expresses the energy acquired by a gram molecular weight of a molecule when it absorbs a quantum of excitation energy.

einsteinium (Es)   A radioactive transuranic element with atomic number 99 and atomic weight 253.

EIS   Environmental impact statement; effluent information system.

ejector   An air-moving device used for sampling air contaminants.

EKG   *See* ECG.

EL   Explosive limit; exposure level.

elastic   Capable of returning to the original state; capable of adapting to circumstances; showing no loss of energy on collision.

elastomer   A synthetic thermosetting polymer that has the properties of vulcanized natural rubber, e.g., nitrile, butyl, silicone rubber.

electric field   The area of electric charge surrounding or emitted from a substance or material possessing electromagnetic energy or conducting electricity.

electric field strength   The intensity of the field in the area surrounding the body of the substance with electromagnetic energy.

electric furnace   An enclosed chamber powered by electricity, used to heat materials, such as diamonds, steel, salts, and graphite, to extremely high temperatures (3000°C).

Electric Power Research Institute (EPRI)   PO Box 10412, Palo Alto CA 94303.

electrochemistry   The study of the relationship between electrical forces and chemical reactions.

electrocoat   A primer paint covering applied to a metal utilizing an electrical process; used for appliances and automobiles.

electrode   One of two substances having differing electromotive activity that allows electricity to flow when an electrolyte is present; often one component is termed positive and one negative or one an anode and one a cathode.

electrodeposition   The process in which a material is deposited on one electrode when electricity flows through a solution or a suspension of particles in air.

electrodialysis   A process by which ions in solution are separated by an electric current, e.g., the desalination of seawater.

electrolysis   Chemical decomposition or change produced by an electric current in a solution containing an electrolyte; also destruction of living tissue, e.g., hair, by an electric current.

electrolyte   A substance that conducts an electric current by forming ions.

electrolytic cell   A system through which an electric current is passed to produce electrolysis; electroplating is an example.

electromagnetic field   The force field surrounding a body or substance having an electric charge in motion that consists of an electric and a magnetic component.

electromagnetic radiation (EMR)   A traveling wave motion resulting from changing electric and magnetic fields. Familiar electromagnetic radiation ranges from gamma rays of short wavelength, through visible regions, to radiowaves of long wavelength. All electromagnetic radiation travels in a vacuum with the velocity of light. *See* photon.

electromagnetic separation   The process of separating isotopes by first accelerating them through an electrostatic field and then passing them through a magnetic field.

electromagnetic spectrum   The all-inclusive range of radiations, from frequencies of 0 to 10^{23} cycles per second; high-energy cosmic radiation to radio waves; *see* electromagnetic radiation.

electrometer   A device used to measure the potential difference between two points.

electromotive force (emf)   The potential difference between two electrodes that produces a current flow.

electromotive series   The arrangement of metals based on their ability to react with acid or water.

Electromyogram (EMG)   A graph of electric signals produced by muscle contraction.

electron (e⁻)   A fundamental particle of matter with a unit negative electrical charge and a mass 1/1837 that of a proton. In a classical representation, electrons surround the positively charged nucleus of an atom and determine the chemical properties.

electron capture (EC)   Radioactive decay in which orbital electrons merge with a proton in the nucleus, followed by emission of an electron or photon.

electron capture detector (ECD)   An extremely sensitive detector used in gas chromatographs particularly for the analysis of halogen-containing compounds.

electron microscope (EM)   A type of microscope that uses electrons emitted from a cathode in vacuum to illuminate the object. This is particularly useful when viewing objects with a wavelength shorter than visible light.

electron paramagnetic resonance spectroscopy (EPR; ESR)   Spectroscopic analysis using microwave radiation and a high magnetic field; used to study free radicals and materials with unpaired electrons.

electron volt (eV)   A very small unit equal to the energy acquired by an electron as it passes potential difference of 1 volt; equal to about 1.6×10^{-19} joules, $1.6^{\mathrm{V}}10^{-12}$ ergs; often expressed as KeV, MeV, or BeV.

electronegativity   The possession of a negative charge; the relative capability of attracting electrons.

electronic mail   The electronic transmission of messages, data, and information via computers. *See* e-mail.

electrophile   A chemical that takes on or accepts electrons in a chemical reaction.

electrophoresis   The motion of charged particles through a stationary liquid toward an electrode, under an electric field.

electroplating   The process of depositing a thin layer of material onto the surface of a metal by passing an electric current through a solution of the material being deposited.

electropolishing   Polishing by reversing the electroplating process.

electrostatic   A stationary electric charge.

electrostatic bond   An ionic attraction between two oppositely charged ions. *See* electrovalent or ionic bond.

electrostatic precipitator (ESP)   A device designed by Fredrick Cottrell that removes particles suspended in a gas by creating an ionized atmosphere in which undesired charged particles are deposited on an electrode; useful for removing dust and particulate matter from air.

electrovalent bond   *See* electrostatic bond or ionic bond.

element   One of the basic chemical building blocks (e.g., hydrogen, lead, uranium) that make up all matter and that cannot be chemically divided into simpler chemical substances.

elephantiasis   Hypertrophy and fibrosis of the skin and subcutaneous tissue, especially in the lower extremities and genitalia, caused by chronic lymphatic obstruction; common after chronic infection by filarial worms.

ELF   Extremely low frequency.

ELF-EMF   Extremely low-frequency electromagnetic field.

elimination   Clearance of a substance from the body; a chemical reaction in which part of the reactants is removed, e.g., water.

Ellenbog, Ulrich   Austrian physician who described lead and mercury poisoning in 1473.

eluent   A liquid used to separate or extract one material from another.

elutriator   A device that separates respirable and nonrespirable particles.

eluviate   To separate, clean, or wash out by settling, washing, or decanting.

EM   Equipment management; environmental management; electron microscope.

e-mail   Electronic mail.

embalmer   A person who treats a corpse with preservatives, such as formaldehyde, to prevent or prolong decay.

embalming fluid   A fluid, such as formaldehyde, used to prevent or postpone the decay of a corpse.

embolism   An obstruction of a blood vessel by, e.g., a transported clot or a mass of bacteria.

embryo   The early stage of development of an organism; In mammals, the stage between ovum and the fetus. In humans, the embryo stage is generally from one week after conception to the end of the eighth week.

embryogenesis   The early cellular stage in the development of a human embryo.

embryotoxin   A substance that is harmful to a developing embryo.

emergency   An unexpected, serious situation requiring immediate response or action.

emergency action plan (EAP)   A document that describes the responsibilities, training, action, and communication to be taken in the event of a potentially hazardous situation, condition, or circumstance that requires timely response in a manner that has been carefully anticipated, thought out, and communicated to all individuals potentially involved both on-site and external to the organization, e.g., police, fire, medical, departments.

emergency contingency plan   *See* emergency action plan.

emergency egress   The path to be taken by personnel to exit a facility in the event of an unplanned, unscheduled potentially hazardous situation, condition, or circumstance.

emergency eyewash   A device used to drench and flush eyes with water when they have been contaminated with or exposed to a potentially hazardous substance.

Emergency Planning and Community Right-to-Know Act   Part of the US EPA Superfund Act that requires states and local municipalities to develop and implement emergency response plans for emergencies involving hazardous substances. Similarly, certain industries are required to report unanticipated environmental releases to designated governmental agencies.

Emergency Response Guidance Level   Various levels of response described in the Emergency Response Planning Guideline established for the Department of Defense to provide guidance for emergency conditions in which personnel may be exposed to unusual hazardous airborne substances.

Emergency Response Planning Guideline(s) (ERPG)   A multistage document designed for the military in the event of an unintended exposure to unusual hazardous airborne substances.

emergency shower   A device used to drench and wash the entire body if it is contaminated with or exposed to a potentially hazardous substance. *See* deluge shower.

emergent beam diameter   The diameter of the exiting beam of a laser.

emery   *See* corundum.

emetic   An agent that induces vomiting.

emf   Electromotive force.

EMG   Electromyogram.

emi   Electromotive interference.

emission   Material(s) discharged into waterways or the atmosphere by industry, residences, or vehicles.

emission rate   The amount per unit time of a specific substance being discharged into the environment.

emission spectroscopy   The study of the specific wavelength of light emitted from molecules after they have been excited by an external energy source and are allowed to relax to their normal energy states; excitation is usually accomplished by a flame, arc, or spark.

emission standard   The maximum amount of a substance allowed by law to be discharged into the environment.

EML   (DOE) Environmental Measurements Laboratory.

emphysema   An irreversible lung disease in which the alveolar walls lose their resilience, resulting in an excessive reduction in lung capacity.

empirical   Relying on observed and experimental data and information, rather than on theoretical data.

employee   A person who works for another individual or organization for money.

employee assistance program (EAP)   An employee benefit program designed to provide short-term assistance to individuals with family, relationship, or physical or mental health conditions and circumstances.

employee safety escort   A trained and qualified employee who is continually with and accepts responsibility for a nontrained individual.

employer   A person or organization who pays money to another to perform work for them.

EMR   Electromagnetic radiation.

EMS   Emergency Medical Service.

EMT   Emergency Medical Technician.

emu   Electromagnetic units.

emulsion   A mixture of two or more immiscible substances suspended in a liquid, which will not settle out, separate, or precipitate on standing.

enamel   The outer surface of a tooth; an opaque paint containing pigments dissolved in a resin with a hard, glossy surface.

enantiomer   One of a pair of optical isomers with one or more asymmetric carbon atoms, which will rotate light either to the left (levo) or to the right (dextro).

encapsulate   To enclose in a protective film, coating, or layer.

encephalitis   Inflammation of the brain.

enclosure   A contained space; an area enclosed on all sides and kept under negative pressure relative to the outside.

ENCOP   Energy conservation project.

endangered species   A species of plant or animal threatened by extinction.

endemic   A disease or symptom present in a community, group, or region of people.

endo-   Prefix meaning interior.

endocarditis   Inflammation of the endocardium, or inner part of the heart.

endocrine glands   Glands with no ducts whose secretions are absorbed directly into the blood.

endoergic   The absorption of heat.

endogenous   Originating within an organ or part of an organ.

endomorphin   Any polypeptide found in brain tissue believed to control transmission of signals to nerves.

endosmosis   An old term for osmosis in a direction toward the interior of a cell or a cavity.

endosperm   Nutritional tissue of a plant seed that surrounds and nourishes the embryo.

endothermic   A process that absorbs heat.

endotoxin   A toxic substance produced by a microorganism, such as Gram-negative bacteria, that can produce undesirable effects in humans, such as fever.

endotrophic   An organism that receives nourishment from other organisms, such as fungi.

end point   In chemistry, the point in an analytical titration marked by an indication, such as a color change, that means that the amount of reagent added is chemically equivalent to the solution titrated; also, a term meaning the object or goal desired.

endrin   A stereoisomer of the insecticide dieldrin that is toxic by inhalation and skin absorption.

-ene   A suffix meaning an open-chained, unsaturated hydrocarbon with one double bond.

Energy (E)   The capacity to do work; $E = h\nu$; also expressed by Einstein's theory of relativity as $E = mc^2$, where E is energy, m equals mass, and c is the speed of light.

energy balance   The amount of energy necessary to sustain a chemical reaction or process conditions.

energy band   A general term that refers to all or part of the electromagnetic spectrum characteristic of the electrons of a particular substance.

energy level   A stationary energy state in the quantum mechanics of an atom or compound.

enfleurage   The extraction of odorous components of flowers in the production of perfumes and essential oils.

enforcement   The action of a regulatory agency or power, such as EPA or OSHA, to obtain compliance with a standard, regulation, or requirement.

engineering controls   The changes taken or modifications made, such as ventilation or isolation, to eliminate or decrease to acceptable levels, undesired human exposure to potentially harmful physical, chemical, or biological agents.

enhancer   A food additive, such as monosodium glutamate, that brings out the taste of the original food without imparting its own taste.

enol   An organic chemical group with a double bond and a hydroxyl group.

enology   The study of wine; the study of wine making.

enrichment   Increasing the amount of a specific radioactive isotope; addition of oxygen to air; addition of a nutrient, such as a vitamin or mineral, to a food.

ensilage   Livestock feed produced by long-term storage in an air-tight silo in which anaerobic fermentation occurs.

ensure   To make sure, certain, or safe.

Entamoeba   A parasite that infects humans and animals

enteric   Related to the intestines.

enteritis   Inflammation of the intestines, particularly the small intestines.

entero-   Prefix meaning pertaining to the intestines.

enthalpy (H)   A thermodynamic term used to express an increase in heat content of a substance or system as a result of a change in going from one state to another at constant pressure; also a measure of the internal energy plus the product of the volume and pressure.

entomology   The study of insects.

entrain   To capture, as in the capture of an effluent fluid.

entropy (S)   A measure of the disorder of a system; a thermodynamic term used to denote the capacity of a system to undergo spontaneous change, expressed as $dS = dQ/T$, where dS is the entropy change in the system, dQ is the amount of heat absorbed at a specific absolute temperature (T).

entry   An opening or passageway; an item in a register or list.

entry loss   Loss of static pressure when air flows into a hood or duct.

entry permit   Under a permitting system, e.g., confined space, a document that authorizes an individual to enter or use the system in question under specific written controls.

entry requirement sign   Sign posted at entries to "controlled areas" specifying requirements for entry to those areas.

environment   Surroundings; the external physical conditions that influence an organism.

environmental audit   A system defined by the US EPA for reviewing a facility to ensure that it is in compliance with regulations and requirements.

environmental chemistry   The branch of chemistry concerned with pollution or changes to the environment. Studies apply to air, land, and water as well as to plants or animals.

environmental engineer   An engineer who has received specific training in the methods and techniques necessary for the prevention, control, and elimination of biological, chemical, and physical agents that may be potentially hazardous to the environment; sometimes called a sanitary engineer.

environmental health   The study of the situations, conditions, and circumstances associated with diseases and disorders that result from exposure to biological, chemical, and physical agents present in the air, food, and water.

environmental impact statement (EIS)   A document prepared to indicate what influence or change to the environment will result from a proposed action.

environmental monitoring   A system for the sampling and analysis of biological, chemical, and physical agents present in the air and water that may impact on the well-being of the environment and its occupants.

Environmental Protection Agency (EPA)   401 M St., SW, Washington DC 20460.

environmental quality   Values of exposure to biological, chemical, and physical agents present in the air and water that are defined as the appropriate limits of exposure sufficient to ensure a state of well-being for the occupants of the environment studied.

environmental sampling   A method of sampling of biological, chemical, and physical agents present in the air and water that ensures the values obtained on analysis represent the values present when the samples originally existed in their environment.

environmental spill or release   A chemical spill or release that has a real potential of causing pollution of the air, land, or water.

enzyme   Biochemical catalyst; entity necessary for the proper functioning of living organisms.

enzyme induction   The increase in an enzyme concentration resulting from an increase in its rate of synthesis or decrease in its rate of destruction.

E⁰   Standard potential.

EOC   Emergency operations center.

eolian   Pertaining to the wind.

eosin $C_{20}H_8Br_4O_5$; a red crystalline powder used to color dyes, inks, etc.

eosinophil   A granular leukocyte; a cell stained by eosin.

eosinophilia   The condition of an unusual number of eosinophils in the blood.

EP   Emergency preparedness; extreme pressure.

EPA   Environmental Protection Agency.

EPA identification number   A specific code number assigned by the US or specific state EPA to individual generators and transporters and storage, treatment, and disposal facilities.

EPC   Emergency planning commission.

epi-   Prefix indicating on or upon, around, or toward.

epicondylitis   The inflammation of the eminence at the articular end of the humerus and tissues around it.

epidemic   An incidence of disease or health problems affecting a large number of people at the same time within a given geographic area.

epidemiology   The study of the incidence and distribution of disease in a population.

epidermis   The outer protective layer of the organism (skin) or plant.

epilation   Hair extraction; loss of hair caused by some condition or exposure.

epimer   An isomer of a compound differing in the relative positions of hydrogen and the hydroxyl group. Commonly used by sugar chemists.

epinephrine $C_9H_{13}NO_3$; a white or tan crystalline compound used to stimulate the heart and to treat asthma; also known as adrenaline; an adrenal hormone stimulant.

epiphora   Excessive flow of tears.

epistaxis   Nosebleed; nasal hemorrhage.

epithelioma   A tumor in the epithelium; a carcinoma.

epithelium   The epidermal cells of skin and surfaces of membranes.

epoxide   Any organic compound that contains the reactive oxygen atom, usually bonded in cyclic fashion to adjoining carbon atoms by single bonds, e.g., R———R.

$$\underset{O.}{\diagdown\diagup}$$

epoxy glue   Any glue utilizing an epoxide.

EPR   Electron paramagnetic resonance; ethylene propylene rubber.

EPRC   Emergency planning review committee.

EPRI   Electric Power Research Institute.

epsom salts   *See* magnesium sulfate.

eq   Gram equivalent weight.

equation of state $PV = nRT$; the equation relating pressure, volume, and temperature for a substance.

equilibrium   The condition of constant concentrations of the species in a given reaction or transformation. Forward and reverse reactions are occurring at the same rate.

equilibrium constant (K)   The numerical value that expresses the relationship between concentrations of species in a chemical equilibrium. It varies with temperature.

equipment specification review   An evaluation activity that ensures that all characteristics of items listed in a formal specification document will satisfy the mandatory operational requirements.

equivalent weight   The weight of an acid that furnishes one mole of hydrogen ions; the weight of any compound that reacts with a specified weight of another compound.

Er   Erbium.

ER   Emergency room; electrorefining.

erbium (Er)   A lanthanide element, atomic number 68; atomic weight 167.26.

ERDA   Energy Research and Development Administration; now known as the Department of Energy (DOE).

erg   The unit amount of work when a 1-dyne force acts through a distance of 1 cm.

ergonomics   Derived from the Greek, meaning literally "rules for work," it generally means the study of human characteristics and corresponding stresses in the work and living environment. A multidisciplinary field that includes the study of human capabilities, limitations, psychology, the entire work environment, and the equipment used in that environment. *See* human factors.

ergot   A toxic fungal growth, used in medicine.

Erlenmeyer, Emil   German chemist (1825-1909) who proposed the formula for naphthalene and designed the flat-bottomed, cone-shaped glass flask named after him.

Erlenmeyer flask   A flat-bottomed conical type of glassware with a cylindrical neck.

Ernst, Richard Robert   Swiss chemist (1933–) awarded the 1991 Nobel Prize in Chemistry for developing Fourier transform NMR, which enabled NMR to analyze small quantities of material.

erosion   Destruction of a surface.

ERP   Emergency response plan.

ERPG   Emergency response planning guidelines.

error   A mistake; false knowledge; a deviation from truth; the difference between the actual value and the experimentally obtained, calculated, or derived value.

ERT   Emergency response team.

erysipeloid   An infective dermatitis, usually of the hands, as a result of handling meat, poultry, or shellfish.

erythema   An abnormal redness of the skin caused by distension of the capillaries with blood. Can be caused by different agents (e.g., heat, drugs, ultraviolet rays, and ionizing radiation).

erythr-, erythro-   A prefix meaning red.

erythrocyte   A red blood cell.

erythroleukemia   A type of acute myelogenous leukemia.

Erythromycin   An antibiotic, most effective against Gram-positive bacteria.

Es   Einsteinium.

eschar   A dry scab.

Escherichia coli   A common bacteria of the alimentary canal. They are short, mobile Gram-negative, non-spore-forming bacilli. Normally non-pathogenic, but outside the intestines under certain conditions can cause infection.

escort   An employee who is fully indoctrinated and knowledgeable of safety or security requirements of areas entered by the visitor under the employee's care.

ESD   Electron-stimulated desorption.

ESH   Environmental safety and health.

-esis   A suffix indicating a condition, process, or state.

ESP   Electrostatic precipitator.

especially hazardous   Refers to situations involving extremely toxic or highly toxic substances, strong carcinogens, mutagens, Class I flammable liquids, explosives, or materials in tanks and other sealed containers in which high or low pressures are present or could develop.

ESR   Electron spin resonance. *See* electron paramagnetic resonance spectroscopy.

essential   Indispensable; an adjective often applied to those amino acids necessary for nutrition but not produced by the organism; eight are identified for humans.

essential oil   A volatile substance usually associated with the flavor or odor of a plant, hence, a substance used to flavor food or as a perfume.

ester   A class of organic compounds containing the $O{=}C{-}O{-}$ group. Formed by reaction between organic acids and alcohols.

esterase   Any enzyme that catalyzes the hydrolysis (breakdown) of an ester.

esterification   The formation of an ester; a reaction that produces an ester.

ester number   The weight in milligrams of alkali needed to saponify glyceryl esters in a fat or oil.

estradiol $C_{18}H_{24}O_2$; a hormone found in the ovaries; commercial product used to treat estrogen deficiency.

estrogen   A group of steroid hormones that produce estrus and the secondary female sex characteristics.

esu   Electrostatic unit(s).

et   Shorthand notation for the ethyl group; Latin word for "and," as in *et seq.*, and "the following."

ET   Effective temperature; eddy current testing; emission testing.

-et   A suffix indicating smallness.

ETA   Event tree analysis.

etch   To wear away a surface, generally to produce some design or pattern and often accomplished by an acid, e.g., HF on glass.

ethanol C_2H_5OH; ethyl alcohol; the generic term used to denote the alcohol most often consumed by humans.

ether   R—O—R; The class of organic compounds that connect two organic groups via an oxygen atom. The generic term for diethyl ether.

ethical drug   A prescription drug.

ethyl alcohol   *See* ethanol.

ethylene glycol $C_2H_6O_2$; A colorless, somewhat viscous liquid used as an antifreeze. O ethylene oxide CH_2—CH_2; epoxyethane; a colorless flammable gas, listed as a known human carcinogen; used as a sterilant, fumigant, etc.

ethylenediaminetetraacetic acid (EDTA)   A common chelating agent. Colorless crystals used in detergents, plating solutions, etc.

ethyl ether $(C_2H_5)_2O$; diethyl ether; a volatile, flammable liquid.

etiological agents   Substances known to contain viable agents (bacteria, viruses, microbes), or their toxins, infectious to humans. Requires additional handling, packaging, and labeling standards. *See* infectious waste, waste.

etiology   All of the factors that contribute to the cause of a disease or an abnormal condition. A branch of medical science concerned with the causes and origins of disease.

EtOH   Ethyl alcohol.

ETS   Environmental tobacco smoke.

ETU   Environmental test unit.

Eu   Europium.

eu   Entropy unit.

eukaryote   An organism in which a membrane surrounds the cell nucleus.

euphoria   The feeling of exaggerated well-being.

europium (Eu)   A rare earth element, atomic number 63, atomic weight 151.96.

Eustachian tube   A tube connecting the tympanic cavity with the pharynx.

eustress   A coined word indicating good or agreeable stress, as contrasted with the usual meaning given to the word "stress" in the safety and health professions.

eutectic   A mixture (usually of metals) that has a defined minimum melting point at a certain relative composition.

eutectic point   The melting point of a eutectic (mixture).

eutrophication   The condition where increasing levels of various substances decrease the dissolved oxygen, hence eliminating animal life but still enabling plant life.

eV   Electron volt(s).

evacuation   To empty or exhaust some container or volume. To create a vacuum.

evaluation   The judgment, often quantitative, of relative value or merit of something.

evaporation   The transformation of a liquid to a vapor, usually considered as occurring under ambient or natural conditions.

evaporation rate   The rate at which a material will vaporize from the liquid or solid state when compared to the rate of vaporization of a known material (usually *n*-butyl acetate). With *n*-butyl acetate rated at 1.0, an evaporation rate of 3.0 or greater is considered a fast evaporation rate, between 0.8 and 3.0 is considered medium, and less than 0.8 is considered slow.

evolution   The gradual change of anything, generally to a better state.

excimer   An excited state dimer formed between solute and solvent with short lifetime.

excipient   An inert diluent or carrier for a drug.

excitation   Adding energy to a system to produce unstable energy states above its ground state.

excited state   An energy level above the ground state; the unstable state, caused by energy transfer to the entity; one can characterize the species by the radiation wavelength emitted returning to ground state.

exciton   A bound state of an electron and a hole within a crystal.

exclusion   Removal of something from a group; a legally allowed omission, exemption, or deviation.

exclusion area   The area around a facility to which access is controlled.

excrement   Waste expelled from an organism; generally, fecal matter.

excursion   Deviation from the normal.

exemption   A legally allowed exception.

exfoliate   To remove or shed, e.g., skin or tree leaves.

exhalation valve   The valve on a respirator through which expired air is emitted into the area outside of the device.

exhale   To breathe out.

exhaust air   Air discharged from another space or enclosure, usually into the atmosphere.

exhaust emission control   A form of minimizing or eliminating air pollution.

exo-   Prefix meaning external or outside.

exosmosis   Fluid flow through a permeable membrane into a lower-density fluid.

exothermic   Characterized by, or formed with, the evolution of heat.

exothermic reaction   A reaction that releases heat.

exp   Exponential; expanded.

expectoration   Spitting or coughing up phlegm.

experiment   A test conducted to demonstrate the validity of a hypothesis.

experimental design   A selection of conditions/settings/materials for specific process variables to be used in a series of tests for the purpose of identifying their individual and combined influences on the results of those tests; The arrangement in which an experimental program is to be conducted and the selection of one or more factors or factor combinations to be included in the experiment.

expert witness   Any individual qualified by a judge to offer testimony in a specialized area of consideration.

expiration date   The date on medication indicating when the substance can no longer be regarded as having efficacy.

expire   To terminate or end; to be no longer effective; to exhale.

explosion   A chemical reaction of any chemical compound or mechanical mixture that, when initiated, undergoes rapid combustion or decomposition, releasing large volumes of highly heated gases that exert pressure on the surrounding medium; A mechanical reaction in which failure of the container causes the sudden release of pressure from within a pressure vessel, e.g., when the pressure ruptures a steam boiler. Depending on the rate of energy release, an explosion can be categorized as a deflagration, a detonation, or a pressure rupture.

explosive   A term applied to chemical compounds or mechanical mixtures that, when subjected to heat, impact, friction, detonation or other suitable initiation, undergo rapid chemical change, evolving large volumes of highly heated gases that exert pressures on the surrounding medium; A term applied to materials that either detonate or deflagrate.

explosive limit (EL)   *See* flammable limits.

explosive range   The region between the lower explosive limit (LEL) and the upper explosive limit (UEL); also known as the lower and upper flammability ranges (LFL, UFL).

exposed site (ES)   A location exposed to the potentially hazardous effects (fragments, debris, and/or heat) of an occurrence.

exposure   The act or condition of being subject to the effect or risk of a chemical substance or a field of radiation or dispersion of radioactive material. Acute exposure is generally accepted to be a large exposure received over a short period of time. Chronic exposure is exposure received over a long time. *See* dose.

exposure assessment   The determination of the extent of exposure to an undesired item or effect.

exposure limit (EL)   That value of exposure agreed upon below which there is no hazardous effect to ordinary healthy people under ordinary circumstances.

exposure path   The means by which a person is exposed to the source of an undesired agent.

exposure point   The place at which a hazardous agent may gain access to the body.

exposure rating   An estimate of exposure.

exposure route   How a hazardous agent enters the body: inhalation, ingestion, injection, skin, eyes, etc.

expression   The transfer of a fluid from a solid matrix by application of pressure.

extension   The spreading or separating of components; straightening a limb as opposed to flexion.

extensor muscle, tendon   The muscle or tendon that extends some part.

exterminator   Anyone who completely destroys something, generally insects, rodents, or pests.

external audit   An audit of those portions of another organization's quality assurance program, not under the direct control of or within the organizational structure of the auditing organization.

external radiation dose equivalent   Radiation dose received from sources of radiation outside the body, usually expressed in rem or millirem (mrem).

extinguishing media   The type of fire extinguisher or extinguishing method appropriate for use on a specific material.

extraction   Pulling out, e.g., a tooth; Chemically separating the component of interest from the bulk matrix.

extrapolation   To extend or predict data from existing data that do not bracket that point of interest.

extremely hazardous substance (EHS)   Substances listed under US EPA Superfund Amendments and Reauthorization Act, (SARA), Title III.

extremely low frequency (ELF)   The part of the electromagnetic spectrum usually designated between 30 and 300 Hz; the range of nonionizing radiation arising from household appliances and power lines.

extremities   Outermost points; bodily appendages (e.g., hands, forearms, feet, and ankles).

extremity dosimeter   Also known as the "wrist badge"; measures the radiation dose received by the hands and forearms.

extrinsically safe   Term describing a system that requires external controls and procedures to ensure safety.

extrusion   A process of forcing a substance through a small space, e.g., die or rollers, to shape; may be done either in the hot or cold state with or without additional lubricants or other additives.

exude   To emit gradually, e.g., sap oozing or any substance emitted from pores.

eye piece   The part of a protective device that covers the eyes or through which one looks.

eye strain   Fatigue of eye muscles; may be caused by improperly fitting protective eyewear or computer monitors, etc.

F

f   Femto (10^{-15}).

F   Fluorine; Fahrenheit; farad (C/V).

FAA   Federal Aviation Administration.

fabric filter   A device, sometimes pleated, to collect airborne particulates, often in bags.

face velocity   Average air velocity entering into an exhaust system which is measured at the opening of the hood or booth. *See* capture velocity, slot velocity.

facepiece   The part of a respirator covering the user's nose and mouth in half-mask respirators or nose, mouth, and eyes in full-face respirators. Facepieces are designed to form a gas-tight and/or dust-tight seal with the face.

Factory Mutual Engineering Research Organization (FMER)   1151 Boston-Providence Turnpike, PO Box 9102, Newton MA 02062.

Fahrenheit, Gabriel Daniel   German physicist (1686–1736) who lived in Holland and England; used mercury instead of alcohol in the thermometer and introduced the Fahrenheit scale of measurement.

Fahrenheit (F)   A temperature scale in which 212° is the boiling point at 760 mmHg and 32° is the freezing point of water.

fail safe   A systems engineering term meaning in the event of malfunction a control will not be lost and disaster will not happen; may work by redundancy.

failure mode and effects analysis (FMEA)   A systems safety technique used in process safety to determine everything that could possibly fail and every possible effect and to then make recommendations for prevention of failure.

fallout   Material that resettles to earth after an explosion and is distributed by the wind.

false negative   A test result reported as negative, but which is actually positive.

false positive   A test result reported as positive, but which is actually negative.

fan   A device to cause air movement.

fan laws   Laws and equations used to relate fan speed, breaking horsepower, volume, pressure, and size.

fan rating curve   The output of a fan at different static pressures.

fan static pressure   The pressure added to a ventilation system by a fan.

farad (F)   A unit of capacitance equal to the capacitance of a capacitor between whose plates there is a potential of one volt when charged by one coulomb of electricity.

Faraday, Michael   British chemist, physicist, and apprenticed bookbinder (1791–1867) who worked with Sir Humphry Davy; made many contributions to science such as work on the liquefaction of gases, the conservation of force, electrolysis, polarized light, and the relationship between electricity and magnetism.

Faraday   The quantity of electricity transferred in electrolysis per equivalent weight of an element or ion equal to about 96,500 coulombs. Named after Michael Faraday.

farmer's lung   A disease caused by inhalation of actinomycetes spores or other fungus producing hypersensitivity. Other examples are bagassosis, maple-bark disease, mushroom worker's lung, suberosis; extrinsic allergic alveolitis.

fasciculation   Small, spontaneous local contractions of muscles, visible through the skin; muscular twitching.

fast   Permanent, as in dye; rapid, as in the speed of light; to go without food.

fat   A glyceryl ester of high-molecular-weight fatty acids, e.g., tallow, lard; usually solids.

fatigue   Weakening due to stress or constant use, as in metal fatigue.

fat rendering   To separate fat from other animal tissue.

fatty acid   A carboxylic acid derived from animal or vegetable oil or fat; may be saturated or unsaturated.

fault tree analysis (FTA)   A systems engineering technique used in process safety to determine the cause of an undesirable event.

fauna   Animals.

FBF   Fluid bed fluorination.

FC   Fail closed; a system or component that when it fails is left in the closed mode.

FCC   US Federal Communications Commission; US Federal Construction Council.

FD   Fire department.

FDA   US Food and Drug Administration; part of the US Department of Health and Human Services and Public Health Services.

FD&C color   FDA colorants approved for use in foods and drugs and as colorants.

FDCA   Food, Drug and Cosmetics Act.

FDR   Final design review.

FE   Facilities engineering.

Fe   *ferrum* (Latin); iron.

FEA   Finite element analysis.

Federal Aviation Administration (FAA)   600 Independence Ave., SW, Washington DC 20591.

Federal Communications Commission (FCC)   445 12th St., SW, Washington DC 20554.

Federal Emergency Management Agency (FEMA)   500 C St., SW, Washington DC 20472.

Federal Energy Regulatory Commission (FERC)   825 North Capitol St., NE, Washington DC 20426.

Federal Highway Administration (FHWA)   400 7th St., Washington DC 20590.

Federal Maritime Commission (FMC)   1100 L St., NW, Washington DC 20573.

Federal Mine Safety and Health Review Commission (FMSHRC)   1730 K St., NW, Washington DC 20006.

Federal Railroad Administration (FRA)   400 7th St., SW, Washington DC 20590.

Federal Register (FR)   The daily publication of the Code of Federal Regulations (CFR) that constitutes the US government documents promulgated under law.

Federal Tort Claims Act   The law that permits claims to be brought against the US government.

Federal Trade Commission (FTC)   CRC-240, Washington DC 20580.

Federal Water Pollution Control Act (FWPCA)   The EPA law known as the Clean Water Act.

Fehling, Hermann von   German chemist (1821–1885) who introduced the important oxidizing solution named after him.

Fehling's solution   A reagent used to test for the presence of sugars and aldehydes; made from copper sulfate and alkaline tartrate.

FEIS   Final environmental impact statement.

felt   Compressed, nonwoven woolen fabric; may cause dermatitis in felt hat makers.

FEMA   US Federal Emergency Management Agency.

femto- (f)   Prefix meaning 10^{-15}.

fenthion   An organophosphate insecticide and acaricide.

FEP   Field evaluation program.

feral animal   A wild animal or a domesticated animal that has reverted to the wild.

ferbam   A carbamate fungicide.

FERC   Federal Energy Regulatory Commission.

fermentation   A chemical reaction induced by living organisms that divide complex organic compounds into smaller units with the evolution of carbon dioxide and alcohol.

Fermi, Enrico   Italian-born American physicist (1901–1954) awarded the 1938 Nobel Prize in Physics; studied statistical mechanics, worked on the Manhattan Project to develop the first US atomic bomb, and was the first to achieve a controlled nuclear reaction.

fermium (Fm)   A transuranic, radioactive rare-earth element with atomic number 100 and atomic weight 254.

ferredoxin   An iron-containing protein in plants believed to be responsible, in part, for photosynthesis.

ferrum   The Latin name for iron.

fertilizer   A substance containing plant nutrients such as nitrogen- and phosphorus-containing chemicals.

fetotoxicity   A substance that induces harm or death in an unborn offspring or fetus. *See* also embryotoxin, reproductive toxicity.

fetus   Unborn offspring (human and other animals) in the postembryonic stage from the point when major structures are outlined (7–8 weeks after fertilization in humans) until the time of birth. *See* embryo, zygote.

FEV   Forced expiratory volume.

fever   Elevation of the normal body temperature.

feverfew   A common Southern US weed that may cause dermatitis in sensitive people.

FHWA   Federal Highway Administration.

FI   Facilities inspection.

fiber   A fundamental form of a slender, elongated solid with high tensile strength, flexibility, and elasticity. Characterized by a high ratio of length to diameter (i.e., aspect ratio), fibers may be derived from animal (wool), vegetable (cotton), mineral (asbestos), natural (metal) or synthetic (polyester) sources. Also used to denote a tissue or group of tissues such as muscle or nerves.

Fiberglas™   A patented product used for insulation, fibers, yarns, and plastics.

fibrillation   Very rapid and irregular contractions of the heart muscles resulting in an irregular heartbeat.

fibrin   An insoluble blood protein.

fibrosis   The formation of, or increase in, interstitial fibrous tissue. Also means fibrous degeneration.

Fick, Adolph Eugen   German physiologist (1829–1901) who studied and developed rules for diffusion.

Fick's law   Governs diffusion and states that the mass of solute diffusing through a unit area per second is proportional to the concentration gradient and dependent on temperature and thickness of the absorber.

FID   Flame ionization detector.

Fieser, Louis   American chemist (1899–1977) who synthesized vitamin K1; did research on aromatic carcinogens and steroids such as cortisone.

FIFO   First-in first-out.

FIFRA   US EPA Federal Insecticide, Fungicide and Rodenticide Act.

filler   A nonreactive, "inert" substance used in drugs, plastics, and other materials to add body.

film badge   A device containing photographic film used by workers to detect exposure to ionizing radiation. The extent of darkening or clouding of the film indicates exposure.

film dosimeter   A device used to measure the presence or exposure to ionizing radiation.

filter   A device used to separate components of a chemical, biological, or physical mixture. In various forms can be used to separate solids, liquids, ionizing and nonionizing radiation, pathogens, etc.

filth   Refuse; animal contamination; in general, dirty matter.

finding   A statement of fact regarding noncompliance with established policy, procedures, instructions, drawings, specifications, and other applicable documents, regulations, standards, or specifications.

fines   Finely crushed or powdered material or fibers; especially those smaller than the average in a mix. Small particles resulting from the operation of certain equipment, e.g., reciprocating saws, abrasive wheels, grinders, sanders, hones, or polishers; particles less than a specified size.

finishing compound   A chemical used to give color, softness, texture, fire resistance, water resistance, and/or flexibility.

fire brick   Bricks made of specific fire-resistant material, within required limits.

fire brigade   An emergency response team trained to respond in the event of a fire; especially useful if a facility is remote.

fire damp   A term coined by miners, generally refers to methane, but could apply to the accumulation of any flammable gas in a mine.

fire extinguisher   A device used to extinguish and control fires, divided into four classes: A, B, C, D and combinations thereof; the designation refers to the type of fire that fire extinguisher will control.

fire point   The lowest temperature at which a material produces sufficient vapors to induce continuous combustion. The fire point is always a few degrees higher than the flash point.

fire pyramid   The four interdependent components necessary to sustain a fire, i.e., fuel, oxygen, ignition, and a chain reaction.

fire retardant   A coating or component applied to combustible materials such as clothing to decrease or eliminate their ability to burn. *See* flame retardant, fire resistive.

fire resistive   Describes a structural component able to resist combustion for a specific time period under conditions of specified heat intensity, without burning or failing. *See* flame retardant.

fire suppression system   A system used to extinguish, control, and limit the spread of a fire and to protect equipment and the facility, e.g., water sprinkler system, carbon dioxide, dry chemical, foam, etc. Designed to be specifically applicable and compatible with the type of material present, e.g., water would not be appropriate for combustible metal, electrical, or computer equipment fires.

fire triangle   A pictogram showing the three components essential for a fire, i.e., fuel, oxygen, and ignition; now replaced by the fire pyramid, which includes the chain reaction mechanistic aspect of fires.

fire wall   A wall of fire-resistive construction designed to prevent the spread of fire from one side to the other for a designated time period. Also used to indicate screening computer access to and from off-site locations. *See* barricade, substantial dividing wall.

fire watch   A requirement that a person(s) be assigned for the sole purpose of looking for a fire in a designated location (called a hot area) where work is being performed that has a high potential fire risk. The person must be instructed regarding the specific risks and hazards and appropriate response and must remain on guard in the location for a specified time after the work is completed.

first aid   Immediate measures taken to aid individuals exposed to hazardous substances, conditions, or circumstances, using materials immediately or generally available to eliminate or reduce adverse health effects. Aimed principally at terminating exposure and providing comfort and relief.

first responder   According to HAZWOPER regulations, a trained person likely to either first detect or initiate a response to an emergency situation.

Fischer, Edmond Henri   German-born American chemist (1920–) who worked with Krebs on plant enzymes, growth regulators, hormonal mechanisms and metabolism; shared the 1992 Nobel Prize in Physiology or Medicine.

Fischer, Emil Hermann   German chemist (1852–1919) awarded the 1902 Nobel Prize in Chemistry for studies of the chemistry of carbohydrates and proteins.

Fischer, Ernst Otto   German chemist (1918–) who shared the 1973 Nobel Prize in Chemistry for his work with organometallic sandwich compounds.

Fischer, Hans   German chemist (1881–1945) awarded the 1930 Nobel Prize in Chemistry for studies of animal and vegetable pigments, hemoglobin, chlorophylls, porphyrins, and bilirubin.

Fischer-Tropsh process   The synthetic method used to make hydrocarbons by passing steam over hot coal; used by Germany in the 1930s to make synthetic fuels.

fish-liver oil   An oil from specific types of fish livers, high in vitamin A, used to lower human cholesterol levels; different than fish oil.

fish oil   A drying oil obtained from specific types of fish.

fissile   The term describing material capable of fracture into lighter elements, thereby releasing tremendous energy.

fission   The splitting of a heavy nucleus into two roughly equal parts (which are nuclei of lighter elements), accompanied by the release of a relatively large amount of energy and frequently one or more neutrons. Fission may occur spontaneously, but usually it is caused by the absorption of neutrons.

fissionable   Material capable of undergoing fission with neutrons but which will not do so spontaneously.

fit check   The process by which the wearer of a respirator determines whether there is any apparent leakage from the outside into the respirator and that the respirator fits properly and comfortably.

fit factor   A measure of the ability of a specific respirator to fit a specific wearer's face by using a specific test chemical and determining the amount present on the outside and on the inside while the subject is wearing the respirator.

fit test   A qualitative or quantitative test to determine whether a respirator fits the face of a wearer and does not allow inward leakage at the face seal.

fixative   A chemical such as formaldehyde used to preserve tissues.

fixed surface contamination   Material tightly adhered to or imbedded in a surface. The contamination cannot be removed easily.

Fl. Pt., fl. pt.   Flash point.

FL   Fail last; the last item or component in a system to fail.

flame ionization detector (FID)   A type of gas chromatograph detector based on conductivity used for analysis of organic materials.

flame photometric detector (FPD)   A type of gas chromatograph detector used for the analytical determination of phosphorus- and sulfur-containing organic compounds.

flame proof   Describes a condition of reduced combustibility, e.g., when a substance, such as boric acid, is added to or applied to a product.

flame propagation   The spread of a flame during combustion, outward from the point of origin.

flame retardant   *See* fire retardant, fire resistive.

flammable   A material capable of supporting combustion. Also known as inflammable.

flammable aerosol   According to the US Federal Hazardous Substances Act (15 USC 1261), an aerosol producing a flame projection of 18 inches at the full valve opening or a flashback at any degree of valve opening when tested. Substances meeting such conditions must be labeled flammable.

flammable chemical   *See* flammable gas, flammable liquid, flammable solid.

flammable gas   A gas that forms a flammable mixture with air at a concentration of 13% or less (by volume) at atmospheric temperature and pressure or that forms a range of flammable mixtures with air at concentrations greater than 12%, regardless of the lower limit; a gas that projects a flame of more than 18 inches beyond the ignition source with the valve opened fully or that projects a flame that flashes back and burns at the valve with any degree of valve opening, when tested in the Association of American Railroads' Bureau of Explosives flame projection apparatus; (DOT) A compound gas that satisfies the criteria for flame projection, lower flammability limit, and flammability range, according to the federal regulations code, Section 173.300(b).

flammable limits   The stoichiometric minimum and maximum concentrations, expressed in percentage of gas or vapor in air by volume, of an ignitable substance in air below or above which propagation or spread of a flame does not occur on contact with a source of ignition. The minimum concentration is called the lower flammable or explosive limit (LFL or LEL). The maximum concentration is called the upper flammable or explosive limit (UFL or UEL). The explosive range lies between the LFL and the UFL flammable liquids (1) (ANSI) Any Class I liquid having a flash point below 100°F (37.8°C). This excludes liquid mixtures in which flammable liquids make up 1% or less of the total volume of the mixture. (2) (DOT) Liquid chemicals (including chemical waste) having a flash point below 100°F. (3) (NFPA) Any liquid (including chemical waste) with a flash point below 100°F. Flammable liquids are divided into classes, according to National Fire Protection Association Standard No. 30.

flammable liquids   (NFPA) Any liquid (including chemical waste) with a flash point below 100°F and having a vapor pressure not exceeding 40 pounds per square inch absolute. Flammable liquids are divided into three classes, IA, IB, and IC, by the National Fire Protection Association; (ANSI) Any Class I liquid having a flash point below 100°F (37.8°C). This excludes liquid mixtures in which flammable liquids make up 1% or less of the total volume of the mixture; (DOT) Liquid chemicals (including chemical waste) having a flash point below 100°F. *See* combustible liquids.

flammable solid   A solid, other than an explosive, that can cause fire through friction, absorption of moisture, spontaneous chemical change, or heat retained from manufacturing or processing or that can be ignited readily and, when ignited, burns so vigorously and persistently as to create a hazard; (DOT) A nonexplosive material that can cause fire as a result of friction or heat retained from production or that, ignited, produces a serious transportation hazard, as defined by the Code of Federal Regulations, section 173.150. *See* combustible liquids.

flange   A rim, edge, or lip.

flanged hood   A hood with a front edge or lip added to reduce air from entering from behind the hood.

flare   To spread out; a pyrotechnic device containing a flammable substance that is lighted in emergency conditions to attract attention and summon assistance.

flash blindness   Temporary vision loss caused by the appearance of a sudden, bright light.

flash burn   Inflammation of the eye lens caused by excess exposure to UV light from welding.

flash point (fp)   The lowest temperature of a flammable liquid at which it gives off sufficient vapor to form an ignitable mixture with the air and produce a flame when a source of ignition is present near the surface of the liquid or within the vessel used. Common testing methods are the Tagliabue open or closed cup (TOC, TCC) and the Cleveland open or closed cup (COC, CCC). Open-cup methods generally approximate conditions representing situations open to the air, whereas closed-cup conditions represent closed systems.

flax   Combustible fibers from the linseed plant.

Fleming, Sir Alexander   A Scottish biochemist and bacteriologist (1881–1955) who developed penicillin from mold, which led to the development of many antibiotics.

flexion   Bending as opposed to extension.

flint   A naturally occurring form of silica; used as a quartz abrasive.

flocculation   The formation of a fluffy mass of material held together by weak forces. A term used in environmental engineering.

flood plain   A plain or area bordering a river or water source subject to flooding.

flora   Plants; a term often used for microscopic organisms that are found in a specific environment, e.g., the intestines.

Flory, Paul   American chemist (1910–1986) awarded the 1974 Nobel Prize in Chemistry for his work with polymers.

flotation   A technique used to separate solids by agitating the powdered solids in a water, oil, or some wetting agent liquid. Air sends nonwetted mixtures to the surface for easy physical separation.

flow coefficient   A correction factor used in computing volume flow rates of fluids through an orifice. The factor includes the effects of contraction, turbulence, compressibility, and upstream velocity. When the compressibility and upstream velocity are small, the flow coefficient is assumed equal to the coefficient of discharge. Also known as coefficient of discharge.

flow diagram   A chart that indicates sequential steps in a process.

flow indicator   A mechanical or electrical device by which flow is verified.

flow meter   A mechanical or electrical device with which the flow of fluids is measured.

FLSA   US Fair Labor Standards Act of 1938.

flue   A pipe or channel that conveys hot air, gases, smoke, etc., such as a chimney.

flue gas   Any gas passed through a chimney or other such tube.

fluid   A substance, usually a liquid or gas, with a tendency to flow and conform to the outline of its container.

fluidization   Any technique that makes a powder or finely divided solid behave like a fluid. Usually a suspension in liquid or gas and often done with a catalyst.

flume   A channel, generally narrow, used to transport water.

fluorescein $C_{20}H_{12}O_5$; an orange-red compound that is intensely fluorescent in alkaline solution; used as a tracing agent.

fluorescence   The emission of light from a molecule or atom with a very short time between energy absorption and photon emission; decay from an excited state to a lower one.

fluoridation   Adding fluorides to a drinking water system to prevent or minimize dental cavities.

fluorine (F)   The lightest halogen group VII element, atomic number 9, atomic weight 19.00; the most electronegative element. F_2 is a yellow, diatomic, reactive, corrosive gas.

fluorocarbon   Any hydrocarbon in which one or more H atoms have been replaced by an F atom.

fluoroscope   A fluorescent screen that gives a visual image of an object that has been placed between the screen and a radiant energy source.

fluorspar CaF_2; a mineral source for fluorine.

flush   Irrigation of something; a reddening of the skin.

flux   Solder, welding; continuous flow, movement, or rate of transfer of fluids, particles, or energy, as in luminous flux. Also a substance used to promote joining or fusion of two surfaces as in soldering.

fly ash   Finely divided alumina, metal oxides, and carbon resulting from combustion of powdered coal; usually swept into the air with flue gases, but precipitators can recover it.

FM   Frequency modulation; finished material; facilities manager.

Fm   Fermium.

FMC   US Federal Maritime Commission.

FMEA   Failure mode and effects analysis.

FMER   Factory Mutual Engineering Research Organization.

FMS   Flexible manufacturing system.

FMSHRC   US Federal Mine Safety and Health Review Commission.

FO   Fail open; a system or component of a system that when it fails is left in the open mode.

foam   A general name for the type of fire-fighting material consisting of small bubbles of air, water, and concentrating agents.

FOD   File or destroy.

fog   A visible suspension of fine liquid droplets in a gas such as air; an aerosol. *See* mist.

FOG   Fat, oil, and grease.

FOI   Freedom of information.

FOIA   Freedom of Information Act.

folliculitis   Inflammation of a follicle(s).

follow-up   The written response on the corrective action report (CAR) describing the action to be taken with an estimated completion date. Each CAR remains "open" until verification of corrective actions is completed. Audit results are summarized monthly and reported to involved management.

FOM   Figure of merit.

FONSI   Finding of no significant impact.

food additive   The National Research Council defines as "a substance or mixture other than a basic foodstuff that is present in food as a result of any aspect of production, processing, storage or packaging." May be intentional or unintentional.

Food and Drug Administration (US FDA)   5600 Fishers Ln., Rockville MD 20857.

food chain   A sequence of events in which smaller or more elementary organisms are consumed by larger and larger (or of increasing complexity) organisms.

food poisoning   Gastroenteritis caused by bacteria or toxins ingested with food. Symptoms occur from 30 minutes to 12 hours after eating.

food preservatives   Substances that inhibit, retard, or arrest the deterioration of food.

Food Safety and Inspection Service (FSIS)   14th St. & Independence Ave., SW, Washington DC 20250.

fool's gold   Iron pyrite; any mineral located in gold-colored veins or nuggets.

foot candle   A unit of illumination. The illumination produced by a uniform point source of one candle when measured perpendicularly on the surface at a distance of one foot from the source.

foot-pound   A unit of energy or stress measurement. The product of the force of one pound exerted over a distance of one foot.

force   The energy needed to change the state of rest or motion of a body; for units. *See* newton.

forced expiratory volume (FEV)   The maximum volume of air that can be expired in a specific time (usually seconds) from a maximum inspiration. Often expressed as a percentage of forced vital capacity.

forced vital capacity (FVC)   The total amount of air that can be exhaled as quickly as possible.

form   The dimensions or physical shape of the part/component; a document containing preprinted constant data or information with spaces for entry of written variable data; also includes certain printed matter such as tags, labels, and report covers.

formaldehyde   HCHO; a carcinogen; a gas with a strong odor, toxic by inhalation and also explosive.

formalin   An aqueous solution of formaldehyde.

formic acid   HCOOH; a colorless, fuming liquid with a piercing odor; dermal corrosive with various industrial uses.

formula   A combination of chemical symbols and subscripts, showing the components of a molecule or ion and their proportions by weight; a mathematical expression.

formulation   The choice of product components made to provide the best properties chosen or to meet specifications.

formula weight   The sum of the atomic weights represented in a chemical formula.

Fourier, Baron Jean Baptiste Joseph   French mathematician (1768–1830) who applied mathematics to heat flow, discovered the equation bearing his name, and demonstrated other applications by which a single variable can be expanded in a series.

Fourier transform   The computerized application of a mathematical series expansion technique named after Fourier and adapted by Ernst in 1991 to NMR that enhances the sensitivity of infrared and nuclear magnetic resonance spectrometry by 10 to 100 times.

fp   Flash point; freezing point.

FPA   Federal Pesticide Act.

FPD   Flame Phosphorus Detector.

FPM   Fine particle mass.

Fr   Francium.

FR   Federal Register.

FRA   Federal Railroad Administration.

fraction   A part of a mixture with similar properties; often used to group distillates.

fractional distillation   A distillation with the distillates collected by fractions rather than individual compounds.

fractionation   The separation of a mixture into components with similar properties.

fragmentation   The breaking up of the confining material of a chemical compound or mechanical mixture when an explosion takes place.

fragments   Pieces; may be complete items, subassemblies, pieces thereof, or pieces of equipment or buildings containing the items.

Francisella tularensis   The non-spore-forming Gram-negative organism that causes tularemia; rabbit fever.

francium (Fr)   A radioactive metallic element, atomic number 87, atomic weight 223.

frangible disk   A membrane, sometimes used as a safety device for compressed gases, which ruptures when pressure exceeds a pre-set limit.

Frasch process   A process, named for Herman Frasch; used to recover sulfur by pumping in superheated steam under pressure that melts the sulfur. Compressed air then forces the liquid sulfur to the surface.

Freedom of Information Act (FOIA)   A federal statute (5 US Code, USC, 552) that gives any person the right to request records from Federal agencies.

free radical   An entity with an unpaired electron(s), formed by scission of a normal bond; generally very reactive and often used to initiate chain reactions.

freeze dry   Lyophilization; a low-temperature dehydration effected by vacuum that leaves the solid (generally a biochemical) undamaged.

freezing point (fp)   The temperature at which a substance changes from liquid to solid at standard pressure.

Freon™   DuPont series of fluorocarbons (some containing chlorine also) used in cooling equipment. They are clear, waterlike liquids, practically inert.

frequency (v)   The number of complete vibrations or oscillations per unit time, as in cycles per second; for units. *See* hertz.

friable   Easily broken up or shattered.

friable asbestos   Any material containing asbestos that can be crumbled in the hand, easily creating a hazardous airborne dust. Chalky, hard pipe insulation that has been crushed or water damaged is considered friable asbestos unless proven otherwise.

friction   Rubbing one substance or surface against another; a force tangential to the interface of two bodies that resists the motion between them.

friction loss   A pressure loss caused by friction between a fluid and the surface it is passing over.

frit   Finely ground glass; used in glazes or to form specialty glasses.

fritted bubbler   A device in which sampled air is passed through a porous glass plate to form small bubbles that then pass through a collection liquid.

FRMAP   US Federal Radiological Monitoring and Assessment Plan.

frostbite   Tissue injury caused by exposure to extreme cold, e.g., liquid nitrogen.

FRP   Fiberglass-reinforced plastic (or polyester).

fructose $C_6H_{12}O_6$; called fruit sugar because it is found naturally in many fruits; sweet-tasting white crystals in common usage.

FSAR   A term used in systems safety hazard analysis meaning the final safety analysis report.

FSIS   Food safety and inspection service.

FTA   Fault tree analysis.

FTC   US Federal Trade Commission.

FTIR   Fourier transform infrared.

fuel   Any substance consumed to produce energy, e.g., wood, oil, gas.

fuel cell   Any electrochemical cell in which the energy of a reaction is converted directly into direct electric current.

fuel oil   Diesel oil; any petroleum product used in an engine or to heat.

fugitive emission   Waste emissions that are not contained or trapped by a capture system and hence are emitted into the environment.

Fukui, Kenichi   Japanese professor at Kyoto University (1918–) who shared the 1981 Nobel Prize in Chemistry for his work in quantum mechanics and reactivity.

fullerines   Closed spherical aromatic molecules with an even number of carbon atoms.

fuller's earth   Colloidal aluminum silicate used as absorbent, decolorant, carrier, etc.

full protective clothing   Protective gear that keeps gases, vapors, liquids, and solids from any contact with the skin, face, or hair and prevents them from being inhaled or ingested.

full work cycle   The amount of time necessary to complete an assignment or process, which may not be an 8-hour shift.

fulminate   Sensitive explosive(s); containing carbon, nitrogen, and oxygen groups, e.g., mercury fulminate; to cause to explode.

fumaric acid $C_4H_4O_4$; used in resins.

fume   Extremely small, solid particulate matter generated from heating a solid body (such as welding rods, lead, or cigarettes). The small particles are generated by condensation from the gaseous state. Fumes may flocculate (form several fluffy masses) and sometimes coalesce (form one mass). Odorous gases or vapors are not fumes, although they are often incorrectly referred to as such.

fume fever   An acute condition caused by a brief high exposure to the fumes of a metal or its oxides.

fume hood   This is a misnomer because this type of hood is used for protection against exposure to other materials in addition to fumes. Although this wording is still used, the preferred term is laboratory or chemical hood. Meaning a five-sided, boxlike structure with one side open intended for placement on a table or bench that may have provisions for lighting and utilities. Laboratory hoods are equipped with a vertically, horizontally, or vertically and horizontally movable sash and adjustable baffles through which air is exhausted from the top and back.

fumigant   A gas pesticide used to kill insects and undesirable organisms; may be toxic to humans.

fumigation   The use of fumes or toxic gases to eliminate insects or other undesirable organisms.

fuming (acid)   Most commonly, fuming nitric and sulfuric acids that are impure and emit small, visible concentrations of nitrogen dioxide and sulfur trioxide in the form of smoke when exposed to air. Sometimes a characteristic of highly active liquids, e.g., hydrofluoric acid, when they come in contact with air.

functional group   Any group of atoms capable of functioning chemically, e.g., —COOH, —O—O—, —OH; generally an organic chemistry term.

fungi   plural of fungus.

fungicide   Any substance that kills fungi.

fungus   A plantlike group of organisms that does not produce chlorophyll. Fungi may live on live or dead plants and animals or by parasitic attachment, often causing disease and infection. Examples are molds, mildew, mushrooms, rusts, and smuts. The highly toxic material produced by fungi are called mycotoxins. *See* mycotoxin.

furan C_4H_4O; C—C; Colorless flammable heterocyclic liquid.

HC———CH
$\;\;\|\qquad\quad\|$
HC$\qquad$CH
$\quad\searrow\quad\swarrow$
$\qquad$O

furfural   Colorless liquid, irritant.

furnace   An enclosed, heated volume or one in which heat is produced by combustion.

furnace black   Carbon black.

fuse   A circuit-breaking device, usually a metal strip that melts when too much current is passed; also to force together as in fusion.

fusel oil   A mixture of amyl alcohols; toxic via ingestion and inhalation.

fusion   Formation of a heavier nucleus from two lighter ones with the attendant release of energy, as in a hydrogen bomb.

FWPCA   Federal Water Pollution Control Act.

G

γ Lower case Greek letter gamma; gamma ray.

g   Gram; acceleration due to gravity.

G   Amplifier; gauss; giga (prefix $= 10^9$).

Ga   Gallium.

ga   Gauge, gage (obsolete).

GACT   Generally available control technology.

gadolinium (Gd)   A rare-earth, combustible, magnetic, lanthanide group IIIB metallic element, atomic number 64, atomic weight 157.25.

gage pressure   obsolete term. *See* gauge pressure.

Galen of Pergamum   Greek physician (129-c. 216); a vivisectionist who regarded anatomy as the foundation of medical knowledge; continued Hippocratic ideas; his physiology concepts influenced medicine for 1400 years.

gallium (Ga)   A silvery, liquid, group IIIA metallic element, atomic number 31, atomic weight 69.72.

gallium arsenide   GaAs, a toxic solid used in the semiconductor industry.

155

galvanic cell   A cell in which a chemical reaction produces electricity.

galvanize   To coat ferrous metals with zinc to reduce rust and corrosion.

galvanometer   A device to detect and measure electric current.

gamma radiation   High-energy, short-wavelength electromagnetic radiation emitted by the nucleus. Often accompanies alpha and beta emissions and always accompanies fission.

gamma ray (γ)   A nonparticulate photon ray capable of penetrating paper, plastic, or flesh. Gamma rays are very penetrating and are best stopped by dense materials such as lead or uranium. Gamma rays are identical to X rays and have the same energy (0.010 and 10 MEV). Lead provides an effective shield against gamma radiation. Adverse health effects are the same as those from X rays of the same energy.

ganglion   A group of nerve cells in the peripheral nervous system; a cyst usually attached to a tendon in the hand, wrist, or foot or connected to a joint.

gangrene   Death of tissue accompanied by infection and putrefaction.

garbage   Food or household waste, as opposed to laboratory, scientific, or industrial waste.

gardona   An organophosphate pesticide.

gas   A normally formless fluid that occupies and takes the shape and volume of its container. Gases can be changed to liquids or solids by increasing pressure and/or decreasing temperature.

gas chromatography (GC)   An analytical device or technique used to separate the components of a mixture by injecting the unknown into a gas that is heated and swept through an absorbing or adsorbing substrate that selectively retains the components of the mixture. Several different techniques are used for detection, e.g., flame ionization, flame photometry, electron capture.

gas constant (R)   The value of the term R in the ideal gas law, $PV = nRT$, equal to 0.0821 liter-atmospheres/mole-degree (K); units change depending on the units used in the equation.

gaseous diffusion   A method of isotopic separation based on the fact that gas atoms or molecules with different masses diffuse through a porous barrier or membrane at different rates. This method is used to separate uranium-235 from uranium-238 and requires large gaseous diffusion plants and huge quantities of electric power.

gasification   The production of gaseous hydrocarbons from coal for use as fuel.

gas mask   A type of full-face, air-purifying respirator used for protection against specific types of gases. A term more often used in the military than in scientific practice. *See* respirator.

gasohol   Gasoline to which 10% methanol or ethanol is added.

gasolene   *See* gasoline.

gasoline   A highly flammable, explosive, volatile mixture of hydrocarbons derived from petroleum and used as fuel; the most common fuel for motor vehicles.

gas pressure   The force exerted by a gas in all directions, usually expressed in pounds per square inch or pounds per square foot.

gastric   Pertaining to the stomach.

gastro-   Prefix meaning pertaining to the stomach.

gastroenteritis   Inflammation of the stomach and intestines.

gauge pressure   Pressure measured on a gauge relative to atmospheric pressure, as opposed to absolute pressure; often expressed as inches of pressure with respect to water (32 inches = 1 atmosphere) or inches, water gauge, or inches, water glass.

Gauss (G)   The cgs unit of magnetic flux density, equal to one maxwell per cm^2.

Gauss, Karl Friedrich   German mathematician and astronomer (1777–1855) who developed the first mathematical theory of electricity, participated in geodetic surveys, and developed an absolute system of magnetic units.

gavage   Feeding by means of a tube inserted via the mouth and throat into the stomach.

Gay-Lussac, Joseph Louis   French chemist and physicist (1778–1850) who made balloon ascents to study earth magnetism and composition of air; described the law of volumes named after him; invented the hydrometer; with von Humboldt, studied the composition of water.

Gay-Lussac law   States that at constant pressure the volume of a confined gas is proportional to its absolute temperature; a modification of Charles' law.

GC   Gas chromatography.

GC/MS   Gas Chromatography/Mass Spectrometry.

Gd   Gadolinium.

Ge   Germanium.

Ge (Li)   Abbreviation for lithium-drifted germanium detector.

Geiger, Hans   German physicist (1882–1945) who investigated beta-ray radioactivity and worked with Walther Müller to develop a counter that measured it.

Geiger counter   A radiation detection and measuring device consisting of a gas-filled chamber or tube containing electrodes between which there is an electrical voltage but no current flowing. When ionizing radiation interacts in the chamber, a short, intense pulse of current passes from the negative to the positive electrode and is measured or counted. The number of pulses per second is the intensity of radiation. Named for Hans Geiger.

Geiger-Müller counter, G-M counter   Amplifies the detection of a Geiger counter, named for Geiger and Müller.

gel   A colloid in which the dispersed phase has combined with the continuous phase to become a semisolid, jellylike material; also short for gelatin.

gelatin   A protein extract from collagen.

gene   A hereditary unit present in all organisms, composed mainly of DNA, with a fixed location arranged in linear order on a chromosome that transmits genetic information and controls the biochemical reactions in cells.

General duty clause   The part of the OSHA Act, section 5(a)1, that requires employers to provide a place of employment free from recognized hazards that may cause death or serous physical harm to employees.

general ventilation   The use of either natural or mechanically induced fresh air movement to mix with and dilute workplace contaminants. Not the recommended method of choice for contamination control. Also called dilution ventilation.

Generally recognized as safe (GRAS)   A term used by the US FDA for food additives accepted as safe.

generation   A group of organisms with common parents with a single line of descent; contemporary individuals; the average time between birth of parents and birth of offspring; the act or process of producing.

generator   Specifically defined waste producers, in various categories, according to the US EPA RCRA regulations; an apparatus, site, or location used for the production of energy or power.

generic   A general term used to apply to an entire group.

genetic code   The information stored in genes.

genetic effects   Changes in inheritable traits that can be passed on by cells during reproduction. *See* mutagen.

genetic engineering   A process of transference of genetic material from genes of one species or individual to those of another. Performed by uniting part of the DNA from one organism with that of another with the resultant molecule replicating in the same way as normal DNA molecules and being called a recombinant molecule. Also called gene splicing.

genome   A complete set of chromosomes.

genotoxic   Describes a substance that causes damage to, or is toxic to, genetic material.

gentian violet   A purple dye used as a biological stain and bactericide; the USP name for methyl violet.

genus   A taxonomic ranking below a family and above a species.

geometric isomer   A type of stereoisomer in which the chemical groups occupy different spatial positions with respect to a double bond. When the double bond is between carbon atoms, they are called *cis* and *trans*.

geometric mean   The nth root of the product of n factors.

geometric standard deviation (GSD)   In logarithmic distribution, a measure of dispersion; often used to describe particle size distribution.

geometry   The study of measurements, properties, and relationships of points, angles, lines, and surfaces.

germanium (Ge)   A brittle, gray group IVA nonmetallic element, atomic number 32, atomic weight 72.59; extensively used in the semiconductor industry.

germ cell   Cell that produces organisms.

germicide   A general term used to describe any agent able to destroy bacteria, fungi, viruses, and other similar harmful pathogens.

gestation   The period of time for carrying an offspring from conception to birth.

GFCI   Ground fault circuit interrupter.

GFI   Ground fault interrupter.

GI   Gastrointestinal.

Gibbs, Josiah Willard   American physicist (1839–1903) who studied thermodynamics, entropy, and the electrical properties of light; established the phase rule and the field of physical chemistry.

Giga (G)   Prefix = 10^9.

Gilbert, Walter   American biochemist (1932–) who shared the 1980 Nobel Prize in Chemistry for the study of the chemical structure of nucleic acids.

gingivitis   A disease causing inflammation of the gums.

GL   General Laboratories.

glacial   A term applied to acids when their freezing point is slightly below room temperature, e.g., acetic and phosphoric acid.

glacial acetic acid   So called because its freezing point is just below room temperature i.e., 16.6°C.

glanders   A chronic disease of horses and some members of the cat family, caused by *Pseudomonas mallei* and transmissible to humans.

glare   An intense shine or brilliance.

glass   A large class of solid materials that solidify from the molten state without crystallization; possessing optical properties.

glass blower's cataracts   Cataracts produced by long-term exposure to UV radiation; also found in foundry workers.

glass etcher   A substance which selectively frosts glass; glass etching is often performed by use of the very hazardous and corrosive hydrogen fluoride.

Glauber, Johann Rudolf   German chemist and physician (1604–1668) who lived in Holland; discovered hydrochloric and nitric acid; also studied the decomposition of salt by acids and bases.

Glauber's salts   Sodium sulfate, $Na_2SO_4 \cdot 10H_2O$; an ingredient of many natural laxative waters; also used as a cathartic.

glaucoma   A disease of the eye characterized by high ocular pressure, hardening of the eye, and loss of vision.

glazing   A thin, transparent, uniform, glossy coating; may contain lead or other toxic substances.

GLC   Gas liquid chromatography.

globe temperature (t_g)   A measure of radiant heat.

globe thermometer   A device used to measure radiant heat in which a thermometer is placed inside the center of a black metal sphere.

globulin   A type of protein made by the body when attached by infective organisms.

glovebag   A single-use control device used in small-scale, short-duration experiments, maintenance, and renovation operations; often used in the removal of asbestos.

glove box   A sealed enclosure with viewing windows and openings fitted with gas-tight gloves by which certain highly hazardous substances such as chemicals, radioactive materials. and/or biological agents may be safely handled.

GLP   Good laboratory practice(s).

glucose   A colorless, sweet, widely used type of sugar produced in the blood and by plants.

gluon   A massless, neutral elementary particle used to hold quarks together.

glutaraldehyde   An irritating liquid used as a tissue fixative and in tanning leathers.

glycerol   A clear, odorless, syrupy, colorless liquid alcohol used as a lubricant, in soap manufacturing, food, cosmetics, and many other applications.

glycol   A general term for compounds with two alcohol groups like glycerol.

GMP   Good manufacturing practice(s).

goggle   A tight-fitting shield worn over the eyes for protection; it may be ventilated or gas tight.

GOGO   Government-owned, government-operated.

gold (Au)   A yellow, ductile, soft group IB metallic element, atomic number 79, atomic weight 196.97; symbol derived from the Latin *aurum* meaning gold; valuable metal used in jewelry.

gonad   Reproductive organ(s).

Gooch, Frank Austin   American chemist (1852–1929) who developed analytical methods and a particular type of filter and crucible named after him.

Gooch crucible   A cone-shaped container made of refractory material with a perforated base to permit suction filtration.

Good manufacturing processes (GMP)   Written standards established by the US FDA for the food and pharmaceutical industries.

Goodyear, Charles   American inventor (1800–1860) who experimented with rubber and developed vulcanization.

googol   A number equal to 10^{100}.

GOPO   Government-owned, privately-operated.

Government Printing Office (GPO)   North Capitol & H St., NW, Washington DC 20402.

GPO   Government Printing Office.

grab sample   A sample, taken during a short or instantaneous time period, to determine constituents present during that precise period.

grade   Any of a number of standards set to classify purity, e.g., ACS, USP, NF, Tech., CP grade.

Graham, Thomas   Scottish chemist (1805–1869) who pioneered the study of colloids; established the law named after him; discovered dialysis and polybasic acids.

Graham's law   States that the rate of diffusion of gases through a membrane varies inversely with the square root of their densities.

grain   0.0648 gram; the smallest unit of the avoirdupois system for mass; a type of cereal, e.g., wheat.

grain alcohol   Ethyl alcohol.

grain elevator   A tall structure, e.g., a silo, in which grains, like wheat, are stored. Hazards may result from explosions of aerosolized fine particulates or the generation of toxic substances such as methane or carbon dioxide.

grain fever   An elevated temperature in humans as a result of bites by mites that are present in grain or grain elevators.

grain itch   Bites and inflammation and other symptoms in humans that result from bites by mites that are present in grain.

grain oil   *See* fusel oil.

gram (g)   The mass of one cm³ of water at 4°C.

Gram, Hans Christian Joachim   Danish physician (1853–1938) who developed the method and types of stains to identify specific bacteria.

gram atomic weight   The weight of an element expressed in grams, e.g., hydrogen = 1.008 grams.

gram molecular weight   The weight of a molecule expressed in grams, e.g., carbon monoxide = 28 grams.

Gram-negative staining   Certain bacteria appear red under the microscope when stained according to a specified procedure in which alcohol penetrates the cell wall.

Gram-positive staining   Certain bacteria appear blue under the microscope when stained according to a specified procedure because they have thick walls that become decolorized by alcohol.

Gram stain   The four-step procedure developed by Gram used to identify certain types of bacteria on the basis of the type of cell wall.

granule   A grainlike particle; a minute, discrete mass; a colony of the bacterium or fungus causing a disease; nonporous particulate matter.

granuloma   A tumorlike mass or nodule with actively growing fibrous tissue; characterized by inflammation due to infection.

granulomatous dermatoses   Chronic inflammations that heal with a scar; may be caused by bacteria, fungi, virus, asbestos, beryllium, silica.

granulomatous pneumonitis   An extrinsic allergic lung response caused by inhalation.

graphite   A planar form of carbon; widely used as a lubricant and the "lead" used in pencils; used as a moderator in some nuclear reactors.

graphium   An allergic pulmonary disease caused by inhalation of redwood sawdust; also known as sequoiosis.

GRAS   Generally recognized as safe.

gravimetric   Describes a measurement by weight; may apply to density or specific gravity.

gravimetric analysis   A type of analytical method using the production of a precipitate that can be weighed and measured after drying.

gray (Gy)   1 joule of absorbed energy per kilogram of matter; A unit, in the International System of Units (SI), of absorbed dose; 1 Gy = 100 rad.

Gray, George William   Scottish chemist (1926–) who made the first stable liquid crystal.

grease trap   A device used to separate and contain grease in a waste stream; often used in restaurants.

Greenberg-Smith impinger   An older air sampling device that bubbles air being collected through a liquid to ensure uniformity.

greenhouse effect   The gradual increase of environmental temperature because of the build-up of carbon dioxide and other gases that decrease the removal of heat from the Earth's surface.

greenockite   The only mineral in which cadmium is found.

green tobacco sickness   A disease contracted by tobacco harvesting workers believed to result from the presence of nicotine in the sticky surface of the leaves of the plant being collected.

griseofulvin   An antifungal feed additive used in medicine.

Grignard, Franois Auguste Victor   French chemist (1871–1935) awarded the 1912 Nobel Prize in Chemistry for introducing the use of organo-magnesium compounds (Grignard reagents) that form the basis of many organic synthetic reactions.

grit   Texture or degree of roughness of a surface of a stone, sandpaper, or polishing media.

groin   The area between the thigh and the body trunk

gross anatomy   That portion of anatomy that deals with disorders that can be seen by the unaided human eye.

ground   A conducting body that has a zero potential; one that is connected to the Earth.

ground fault interrupter, ground fault circuit interrupter (GFI, GFCI)   A device used for personal protection against injury caused by electricity by determining when electricity is passing to ground by a path other than the proper one and shutting down the circuit to the user.

grounding   The connection of one or more objects to ground (i.e., a large conducting body, e.g., the Earth) to neutralize the electrical charge between object(s) and ground.

ground itch   A general term for the infection caused when larvae, such as hookworm, invade feet.

ground water   Natural water derived from an underground source.

group   A class into which elements are separated such as the periodic table based on similar properties and reactions; a class separated by chemical functional similarities, e.g., carbonyls, alcohols, oxidizers; members, resident, or individuals of a class or area with similarities.

growth   Cell division and replication; increase in size or maturity.

growth hormone   A chemical substance that regulates growth.

GSD   Geometric standard deviation.

guano   A substance composed of mainly sea bird or bat dung; used as a fertilizer.

guard   In safety, a device on a machine such as a vacuum pump with a pulley/belt system that minimizes or prevents entanglement in moving parts.

guideline   A suggested practice that is not mandatory in programs intended to comply with a standard. The word *should* denotes a guideline; the words *shall* and *will* denote a requirement.

Gulf Coast tick   A tick capable of transmitting disease to humans by its bite; field laboratory workers are at risk.

gum   A natural or synthetic material that is insoluble in alcohol or organic solvents.

gum arabic   A gum from the African tree *Acacia senegal* used in pharmaceuticals, glues, and candies; it has been reported that printers became sensitized.

gum spirit   Turpentine.

guncotton   Nitrocellulose, used in explosives.

guthion   An organophosphorus cholinesterase inhibitor pesticide.

Gutzeit, Max   German chemist (1847–1915) who developed the tests for arsenic and antimony.

Gutzeit test   A test for arsenic.

Gy   Gray.

H

h   Hecto (prefix = 10^2); height.

H   Hydrogen, the lightest element; henry, a unit of measure of electrical induction; enthalpy.

ℏ   Planck's constant.

^{3}H   A naturally occurring radioactive isotope of the chemical element hydrogen (most abundant atomic weight 1) with an atomic weight of 3 amu. Also called tritium.

HA   Hazard analysis.

Ha   Hahnium.

Haber, Fritz   German chemist (1868–1934) awarded the 1918 Nobel Prize in Chemistry for his work in nitrogen fixation.

haemo-, hema-, hemo-   Prefixes pertaining to blood.

hafnium (Hf)   A silvery metallic element, atomic number 72, atomic weight 178.49.

Hahn, Otto   German physical chemist (1879–1968) awarded the Nobel Prize for Chemistry in 1944 for his pioneering work on atomic fission.

hahnium (Ha)   A transuranic element, atomic number 105, atomic weight 262.

half-life, biological   Time required for a biological system to eliminate by natural processes half the amount of a substance that has entered it.

half-life, effective   Time required for a radionuclide present in a biological system to be reduced by half as a combined result of radioactive decay and biological elimination.

half-life, physical   Time in which half the atoms in a radioactive substance disintegrate; varies from millionths of a second to billions of years.

half-thickness   The thickness of any given absorber that will reduce the intensity of a beam of radiation to one-half its initial value. This value varies with radiation energy and beam size and location of shielding. *See* attenuation, shielding.

halides   Binary compounds that include one of the halogens.

hallucinogen   Any chemical that induces false perceptions or delusions; e.g., LSD, mescaline.

halocarbon   Any hydrocarbon that contains one or more halogen atoms.

halogen   The name for the group VII nonmetallic elements.

halogenated hydrocarbon   *See* halocarbon.

Halon™   Various types of fire extinguishants composed of a combination of bromo-, chloro-, and/or fluorohydrocarbons. Signatories of the 1987 Montreal Protocol agreed to ban their use because of the damage they cause to the Earth's ozone layer.

Hamilton, Alice   American physician (1869–1970) regarded as the mother of American occupational medicine; first female faculty member of Harvard; studied lead poisoning, phossy jaw, and many other occupational disorders.

hammer mill   A device used to crush and grind stone, sugar cane, wood, and other products inside a steel cylinder.

HAP   Hazardous air pollutant.

hapten   A compound that combines with an antibody in vitro.

Harden, Sir Arthur   English chemist (1861–1940) who shared the 1929 Nobel Prize in Chemistry for studies on fermentation.

hardhat   A protective device worn on the head that conforms to ANSI Standard Z 89.1.

hardness   The resistance of a material to scratching or deformity, measured by the Mohs scale.

hard water   A term used for water that contains metallic salts, e.g., calcium and magnesium, that decreases the effectiveness of soap and adds to the scaling of boilers.

hashish   An extract of the *Cannabis sativa product*; marijuana that is more potent because it is more concentrated.

Hassel, Odd   Norwegian chemist (1897–1981) who shared the 1969 Nobel Prize in Chemistry for the development of the concept of confirmation and its application in chemistry.

Hauptman, Herbert Aaron   American biophysicist (1917–) who shared the 1985 Nobel Prize in Chemistry for work with X-ray crystallography used to determine the three-dimensional structure of various biochemicals.

Haworth, Sir Walter Norman   English chemist (1893–1950) who shared the 1937 Nobel Prize in Chemistry for determining the chemical structure of vitamin C.

hay fever   An allergy that affects the mucous membranes; caused by pollen(s).

HAZ   hazardous, as in HAZ-MAT (hazardous materials) team; heat-affected zone.

hazard   The danger, risk, chance, or probability of an unwanted event occurring. A hazard can be defined by the degree of probability that a certain material, activity, or condition will produce an unwanted event.

Hazard analysis (HA)   A term used in system safety engineering for the logical, systematic examination of an item, process, condition, facility, or system to identify and analyze the source, causes, and consequences of potential or real unexpected events occurring.

Hazard Communication Standard (HCS)   A 1985 OSHA Standard, 29CFR 1910.1200, often referred to as the right-to-know act because it requires employers specified in the standard to notify, train, and inform potentially exposed employees of hazardous chemicals and to provide material safety data sheets (MSDS), maintain a chemical inventory, label containers, and develop a written hazard communication plan.

Hazard and operability study (HAZOP)   A method of investigating elements of a system to determine those factors that may contribute to an accident, and to what extent, and to then redesign the system and eliminate these potential problems.

hazard rating system (HRS)   A system used according to NFPA 704, Standard System for the Identification of Fire Hazards of Materials.

hazardous chemical   A chemical that may present a biological, chemical, environmental, health, or physical hazard. A chemical or chemical mixture that is toxic, irritant, corrosive, oxidizer, sensitizer, combustible, flammable, or highly reactive or that otherwise may cause substantial injury or illness during, or as a direct result of, customary or reasonable handling or use. Also specifically defined in the OSHA Hazard Communication Standard 29 CFR 1910.1200. For hazard ratings, *See* NFPA 704, Standard System for the Identification of Fire Hazards of Materials.

hazardous location   A specific NFPA classification system that specifies the electrical and spark-producing equipment needed for locations by the type of chemical in use and the nature of the area. The system is divided into Classes I (flammable gases and liquids), II (combustible dusts), and III (ignitable fibers); each of which is further divided into two divisions.

hazardous material(s) (HAZMAT)   According to the US DOT and US EPA, a substance or material determined by the Secretary of Transportation to be capable of posing an unreasonable risk to health, safety, and property when transported and that is designated as such in 49 CFR 172.101, or the appendix to 172.101, or is subject to the Hazardous Waste Manifest Requirements of the EPA specified in 40 CFR 262.

Hazardous Materials Identification System   The four-part diamond system developed by the NFPA for the identification and gradation of flammability, health, reactivity, and special hazards of chemicals.

Hazardous materials response team (HZTM)   A group organized and trained in accordance with 29 CFR 1910.120 (HAZWOPER) to respond to and handle and control hazardous substance spills.

hazardous substance   Defined by the US EPA under CERCLA as material that, when discharged to the environment in sufficient quantities, poses an unreasonable risk to people or the environment. A substance is hazardous when, in a single container, an EPA-listed chemical is present in an amount (by weight) exceeding the listed reportable quantity (RQ).

Hazardous Substances Act   The Consumer Product Safety Commission (CPSC) regulation, 16 CFR 1500, that regulates the interstate commerce of defined hazardous substances.

hazardous waste   Defined by the EPA under RCRA, 40 CFR 261 as waste materials that, when transported and/or disposed of improperly, pose an unreasonable risk to health, safety, and the environment. Defined by the DOT as material that exhibits any of the characteristics of a hazardous waste or that is listed in the regulations as such.

Hazardous Waste Operations and Emergency Response (HAZWOPER)   An OSHA Standard, 29 CRF 1910.120, that regulates the safety and health of employees and specifies four levels of protection, A, B, C, and D, during emergency response to hazardous substances and clean-up of specified waste sites.

Hazardous Waste Treatment Council (HWTC)   1919 Pennsylvania Ave., NW, Washington DC 20006.

hazardous work   Performance of tasks involving the following substances and/or conditions   unguarded elevated surfaces greater than 4 feet; live unguarded electrical circuits, systems or battery banks; high-pressure systems greater than 15 lbs per sq. in.; nonradioactive hazardous biological, chemical, or physical agents; radioactive agents in quantities that could result in radiation exposures in excess of established requirements and guidelines; ergonomically stressful situations.

HAZCOM   The OSHA Hazard Communication Standard; 29CFR.1910.1200.

HAZMAT   Hazardous material(s).

HAZOP   Hazard and operability study.

HAZWOPER   Hazardous Materials Waste Operations and Emergency Response.

HBV   Hepatitis B virus.

HC   Hydrocarbon(s).

HCN   Hydrocyamic acid.

HCS   Hazard Communication Standard.

Hct   Hematocrit.

HDPE   High-density polyethylene.

He   Helium.

head space   The area above a substance in a closed container.

health   The overall condition of an organism.

health hazard   An agent or condition having the ability to impair the health of a person exposed to it. Included are biological, chemical, ergonomic, and physical agents and/or conditions such as acutely and chronically toxic (poisonous) agents, carcinogens, corrosives, irritants, mutagens, sensitizers, and agents with specific target-organ toxicity (i.e., damaging or impairing liver, kidneys, central nervous system, blood, skin, eyes, etc., and reproductive and developmental organs).

health physics (HP)   A discipline whose practice encompasses radiation physics and is concerned with human problems of radiation damage and protection. The science is concerned with the recognition, evaluation, and control of health hazards related to ionizing and nonionizing radiation.

Health Physics Society (HPS)   1313 Dolly Madison Blvd., McLean VA 22101-3926.

healthy worker effect   The epidemiologically important phenomenon that working people usually are healthier than the general population. Those with severe health problems generally are not in the work force.

hearing conservation program   A formal, written program required by OSHA in 29 CFR 1910.95 that includes medical surveillance and testing, provision of protective equipment, and workplace monitoring.

hearing impairment   A loss in ability to hear.

hearing level   The ability of an individual to hear different frequencies and intensities.

hearing loss   The deviation of an individual's hearing ability from that of the average person with no hearing loss.

heat   The energy associated with the motion of atoms or molecules that is transmitted by conduction, convection, and/or radiation.

heat balance   The difference in a system between heat input and heat output.

heat capacity   The amount of heat needed to raise the temperature of a specified mass of substance one degree C.

heat cataract   Cataract resulting from exposure to visible and infrared radiation, e.g., glass blower's cataracts.

heat cramps   Painful spasms in voluntary muscles caused by inadequate fluid and salt intake while doing strenuous activity in a hot environment.

heat exchanger (HX)   A device that transfers heat from one fluid (liquid or gas) to another fluid or to the environment.

heat exhaustion   An acute weakness precipitated by exposure to heat producing slight elevation of body temperature, profuse sweating, headache, and cool, wet skin.

heat of combustion   Heat produced when a given amount of a substance is oxidized or burned.

heat of formation   Heat required or given off when a compound is formed.

heat of fusion   Heat required to convert a substance from a solid to a liquid.

heat of solution   Heat required or given off when a compound is dissolved in a solvent.

heat of vaporization   Heat required or given off when a substance goes from a liquid to a gas.

heat rash   Irritation and reddening of the skin because of exposure to elevated temperatures.

heat sink   Anything that absorbs heat. Environmental examples include the air, a river, or outer space.

heat stress index (HSI)   A system used to indicate heat stress.

heat stroke   The rapid rise in body temperature, accompanied by hot, dry skin, convulsions, and the absence of sweating; can result in death. Sun or extreme heat can induce an imbalance in the body's heat control, i.e., the heat loss processes.

heat syncope   Fainting while standing erect in a hot environment caused by pooling of blood and dilation of blood vessels in the lower extremities.

heavy metal   A metal with an atomic weight greater than sodium (23) that forms soaps with fatty acids; in environmental regulations, a metal such as arsenic or barium that does not readily break down in the body or environment and can accumulate.

heavy water   Deuterium oxide (D_2O).

hecto- (h)   Prefix for 10^2.

Heisenberg, Werner Karl   German physicist (1901–1976) awarded the 1932 Nobel Prize in Physics for his uncertainty principle and work in quantum mechanics.

helium (He)   A noble gas element, atomic number 2, atomic weight 4.00; often used to inflate balloons that rise.

Helmholtz, Baron Hermann Ludwig Ferdinand von   German physician, physicist, mathematician, and philosopher (1821–1894) known for developing the law of conservation of energy.

hematite   The chief ore of iron.

hemato-   Prefix pertaining to the blood.

hematopoietic   Blood forming organs; pertaining to the production or development of blood cells.

hematuria   The presence of blood or blood cells in urine.

heme $C_{14}H_{32}FeN_4O_4$; ferrous iron bound to protoporphyrin; nonprotein part of hemoglobin.

hemoglobin   The red, iron-containing compound in the blood that binds oxygen and carries it to sites where oxygen is needed in the body.

hemoglobinuria   The presence of hemoglobin but not red blood cells in the urine.

hemolysis   The rupture of red blood cells with release of hemoglobin into the plasma.

hemorrhage   Escape of blood through vessel walls; to bleed profusely.

hemotoxin   A substance that causes destruction of red blood cells.

hemp   A tall plant whose fiber has commercial use; a narcotic derived from the plant.

henry (H)   A unit of measure of electrical inductance; Wb/A.

Henry, Joseph   American physicist (1797–1878) noted for his studies of electromagnetic phenomena; the unit of electrical induction was named for him.

Henry's law   States that the mass of a low-solubility gas dissolving in a fixed mass of solvent at a given temperature is nearly directly proportional to the partial pressure of that gas.

HEPA Filter   A High-efficiency particulate air filter.

heparin   A polysaccharide that prevents conversion of prothrombin to thrombin; a coagulation inhibitor.

hepatic   Pertaining to the liver.

hepatitis   Inflammation of the liver.

hepatitis B   A blood-borne virus (HBV) included in the OSHA blood-borne pathogen standard, 29 CFR 1910.1030, that causes inflammation of the liver.

hepatotoxin   A substance that damages the liver and eventually causes it to stop functioning.

herb   A plant used in medicine or cooking.

herbicide   A substance used to kill plants; generally used on weeds.

heredity   The genetic transmission of characteristics to offspring.

heroin   An addictive, abused drug derived from morphine; diacetylmorphine.

herpes simplex   An infectious disease of thin-walled vesicles, recurring in the same (generally mucus-skin interface) location.

Herschbock, Dudley Robert   American chemist (1932–) awarded the 1986 Nobel Prize in Chemistry for studies on the mechanisms of chemical reactions and detailing the sequence of events and energy release.

hertz (Hz)   A unit of measure of frequency; cycles per second.

Hertz, Gustav Ludwig   German physicist (1887–1975) awarded the 1925 Nobel Prize in Physics for confirming the quantum theory.

Hertz, Heinrich Rudolf   German physicist (1857–1894) who worked on electromagnetic waves; the first to broadcast and receive radio waves. The unit of frequency is named after him.

Herzberg, Gerhard   German-born chemist (1904–1999) who emigrated to Canada and was awarded the 1971 Nobel Prize in Chemistry for the detection of free radicals and studies of the energy levels of atoms.

Hess, Germain Henri   Swiss-born chemist (1802–1850) who did early studies on the conservation of energy in chemical reactions.

Hess's law   The law of constant heat sums, i.e., heat changes in a process are identical whether it occurs in one step or several.

heterocyclic   A chemical compound containing more than one kind of atom in a ring structure.

heterogeneous   Consisting of differing or nonuniformly dispersed substances.

heuristic   Helping to learn or discover something.

Hevesy, Georg Charles de   Hungarian chemist (1885–1966) awarded the 1943 Nobel Prize in Chemistry for his work in developing radioisotope tracer techniques; co-discoverer of hafnium.

hexachlorophene   An anti-infection agent used in soaps or creams and in treatment of skin disorders; an antiseptic effective against Gram-positive organisms.

Heyrovsky, Jaroslav   Czech chemist (1890–1967) awarded the 1959 Nobel Prize in Chemistry for work with mercury electrodes and development of the electroanalytical chemical technique of polarography.

hf   High frequency.

Hf   Hafnium.

HF   Hydrofluoric acid; human factors.

HFES   Human Factors and Ergonomics Society.

hfs   Hyperfine structure.

Hg   *Hydrargyrum* (Latin); mercury.

HHS   US Department of Health and Human Services, formerly the Department of Health, Education, and Welfare.

high-efficiency particulate air (HEPA) filter   A filter able to remove $\geq 99.97\%$ of particles 0.3 m or greater.

high explosive (HE)   A term used by the US Bureau of Alcohol, Tobacco and Firearms for materials that, when confined, can be detonated with a blasting cap. *See* low explosive.

high-explosive equivalent   The ratio of the weight of TNT (2,4,6-trinitrotoluene) to that of another explosive when both quantities produce equivalent blast effects the same distance from their detonations. The ratio is expressed as a percentage. Same as explosive equivalent.

high-frequency loss   A hearing loss at 2000 Hz or higher.

highly toxic substance   A substance lethal to humans or laboratory animals at very low concentrations, e.g., 10 to 100 ppm. *See* acute toxicity scale, toxic chemical, poison.

High-Pressure Liquid Chromatography (HPLC)   An analytical method of separating nonvolatile substances.

high-radiation area   An area accessible to personnel in which a major portion of the body could receive a radiation dose of 100 millirem (0.1 rem) in one hour. These areas are posted as high-radiation areas, and access to these areas is to be limited.

Hinshelwood, Sir Cyril Norman   English chemist (1897–1968) who shared the 1956 Nobel Prize in Chemistry for work on histamines and on the effect of drugs on bacterial cells; also investigated chemical reaction kinetics.

Hippocrates   Greek physician (~ 377 BC) known as "the father of medicine," associated with the medical profession's Hippocratic Oath.

hippuric acid   An acid formed from benzoic acid and glycine in the liver or kidney and excreted by the kidneys.

histamine   A substance that increases gastric secretions, constricts bronchial muscle, and dilates capillaries.

histolysis   Disintegration of tissue.

histopathology   The study of abnormal or diseased tissue.

histoplasmosis   A systemic fungal respiratory disease.

HIV   Human immunodeficiency virus.

Hivol   High-volume air sampler.

HLW   High-level waste, a radioactive waste with some specific activity.

HMIS   Hazardous materials identification system.

HMTA   Hazardous Materials Transportation Act.

Ho   Holmium.

Hodgkin, Dorothy Mary Crowfoot   Egyptian-born chemist who lived in England (1910–1994); awarded the 1964 Nobel Prize in Chemistry for the use of X-ray techniques in determining the structure of certain molecules, including penicillin, vitamins B1 and B2, and insulin.

Hoffman, Roald   Polish-born chemist (born Roald Safran, 1937–) who lived in the United States and shared the 1981 Nobel Prize in Chemistry for the development of mathematical rules to predict the results of chemical reactions, hence, altering the way chemical experiments were designed.

Hoffman, August Wilhelm von   German chemist (1818–1892) who produced aniline from coal products, discovered formaldehyde and other compounds, and developed a theory for classifying different types of chemicals.

holdup   The quantity of material remaining in process equipment and facilities after the in-process material, stored materials, and/or product have been removed.

Holmium (Ho)   A metallic lanthanide element, atomic number 67, atomic weight 164.93.

holography   The technique of producing three-dimensional images by wavefront reconstruction, especially through the use of lasers.

homogeneous   The same or uniform.

homogenization   The production of a uniform emulsion or suspension of immiscible substances.

homologous series   Chemical compounds with similar structures and properties.

homostasis   The maintenance of stability by an organism when exposed to external factors or conditions.

hood   The point where air is taken into the ventilation system to capture or control contaminants; A part of a respirator that completely covers the head, neck, and portions of the shoulders. *See* canopy hood, laboratory hood, snorkel.

hood entry loss   Pressure loss as air enters a hood caused by turbulence and friction.

hopcalite   A mixture of copper, cobalt, magnesium, and silver oxides used as a catalyst in gas masks to convert carbon monoxide to carbon dioxide; also used in some carbon monoxide monitors.

hormesis   The beneficial stimulation of a small dose of any substance or exposure that normally is considered toxic in higher doses.

hormone   A substance formed by an organ and conveyed to another that is thereby stimulated chemically to function.

horsepower (hp)   A unit of power equal to 745.7 watts or 33,000 footpounds/minute; the power exerted by a horse in pulling.

host   The employee responsible for a visitor(s) to a hazardous area or facility.

hot   A colloquial term for highly radioactive; also thermally high in temperature.

hot spot   The region in an area or any substance in which the level of contamination or heat is noticeably greater than in areas neighboring the region.

hot zone   The area in a hazardous waste operation where contamination occurs.

housekeeping   A term applied to maintaining a neat, clean, and orderly work place.

hp   Horsepower.

HP   Health physics.

HPGe   High-purity germanium detector; also called intrinsic.

HPLC   High-pressure liquid chromatography; high-performance liquid chromatography.

HPS   Health Physics Society.

HR   Human resources (personnel) office, department, etc.; hot-rolled.

HRS   Hazard rating system.

HS & E   Health, safety & environment.

HSI   Heat stress index.

HSM   Hospital, surgical, medical (insurance plan).

HSWA   Hazardous and Solid Waste Amendments.

HT   Heat treatment.

Huber, Robert   German chemist (1937–) who shared the 1988 Nobel Prize in Chemistry for the three-dimensional structure of proteins essential to photosynthesis.

human factors (HF)   The study of humans in the work place; also called ergonomics.

Human Factors and Ergonomics Society (HFES)   PO Box 1369, Santa Monica CA 90406.

Human immunodeficiency virus (HIV)   Human T-cell lymphotropic virus type III; a cytopathic retrovirus; the etiologic agent of acquired immunodeficiency syndrome (AIDS).

humectant   A moistening agent.

humic acid   Any of the organic acids found in decayed vegetable matter.

humidity   A measure of dampness, moisture, or water vapor in air.

humus   Organic material found in decayed vegetable matter.

hv   High voltage.

HVAC   Heating, ventilating, and air conditioning.

HW   Hot water.

HWTC   Hazardous Waste Treatment Council.

HX   Heat exchanger.

hyaline   A generic term relating to microscopic structure or cell injury.

hydrargyria   Mercury poisoning.

hydration   The reaction of molecules of water with a substance in which the hydrogen to oxygen bonds (H—O—H) are not broken. Products of hydration are called hydrates. *See* hydrolysis.

hydrazine H_2NNH_2; a colorless, fuming hygroscopic liquid; explosive hazard; toxic by ingestion and inhalation and a carcinogen.

hydride   A compound of hydrogen with a more electropositive element.

hydroa   Any bullous skin eruption.

hydrocarbons (HC)   Compounds composed solely of the elements carbon and hydrogen. Hydrocarbons are the basic component of all organic chemicals. Hundreds of thousands of molecular combinations of C and H are known to exist. Hydrocarbons are derived principally from petroleum, coal, and plants.

hydrocortisone   A bitter crystalline hormone having medical applications similar to cortisone.

hydrocracking   *See* hydrogenation.

hydrocyanic acid (HCN)   A liquid with bitter almond odor; flammable and toxic by absorption, ingestion, and inhalation.

hydrofluoric acid (HF)   A colorless, toxic, highly corrosive gas, usually in aqueous solution; will attack glass. An insidious, corrosive, colorless gas that is generally utilized in aqueous solution. Atypically for acids, the solution does not always give pain on contact with the skin and penetrates deeply.

hydrogen (H)   The lightest element, atomic number 1, atomic weight 1; (H2) a lighter-than-air flammable and explosive gas.

hydrogenation   The reaction of hydrogen with another (usually organic) compound.

hydrogen cyanide (HCN)   *See* hydrocyanic acid.

hydrogen peroxide (H_2O_2)   A colorless oxidizing liquid.

hydrogen sulfide (H_2S)   A colorless, flammable, toxic gas with a characteristic odor of rotten eggs.

hydrology   The study of the effects and distribution of water in, on, and above the earth.

hydrolysis   A chemical reaction in which the hydrogen to oxygen bonds of water are broken and react with another substance to form new substances.

hydrophilic   Describes materials having a tendency to absorb, bind, or retain water. The word is derived from the Greek and literally means "water loving." Swelling is often a property of carbohydrates and complex proteins.

hydrophobic   Water repelling; not soluble in water; literally means "water fearing."

hydroquinone   A weak topical depigmenting agent.

hydrostatics   The study of pressure and equilibrium for nonelastic fluids, e.g., water.

hydrostatic testing   A procedure to test pressurized tanks, e.g., SCUBA, SCBA, compressed gas cylinders, fire extinguishers, etc., by filling them with air under high pressure while the tanks are under water.

hydrotrope   A compound that increases water solubility of slightly soluble chemicals.

hygiene   The study of health; promulgation of rules or methods to improve or preserve health.

hygrometer   An instrument to measure humidity.

hygroscopic   The ability to readily adsorb moisture from the air. *See* deliquescent.

hyper-   Prefix meaning an increased or higher level or amount than normal, standard, or usual.

hyperbarism   Diseases of the body resulting from the pressure of ambient gases at greater than 1 atmosphere; e.g., nitrogen narcosis, bends, etc.

hyperemia   The congestion of blood in a body part.

hypergolic   Term describing the ability to ignite spontaneously on contact without a spark or external aid. Examples are rocket fuels consisting of a combination of fuel and oxidizers.

hyperplasia   Proliferation of normal cells in an organ.

hypo-   Prefix meaning a decreased or lower level or amount than normal, standard, or usual.

hypocalcemia   A calcium deficiency of the blood.

hypocholesteremia   A condition of decreased blood cholesterol.

hypoglycemia   A condition of sugar concentration in the blood being too low and causing fatigue, irritability, etc.

hypothalmus   The part of the brain that regulates temperature and other automatic functions.

hypothermia   Below-normal body temperature.

hypoxia   A low oxygen content or deficiency in inspired air.

Hz   Hertz (s^{-1}).

HZMD   Hazardous waste management division.

HZTM   Hazardous material response team.

I

I Iodine; moment of inertia; intermittent.

IACUC Institutional Animal Care and Use Committee.

IAEA International Atomic Energy Agency.

IAIABC International Association of Industrial Accident Boards and Commissions.

IAG Interagency agreement.

IAM International Association of Mechanics and Aerospace Workers.

IAP Indoor air pollution.

IAQ Indoor air quality.

IARC International Agency for Research on Cancer.

IATA International Air Transport Association.

IBC Institutional Biosafety Committee.

ibid. Abbreviation for the Latin word *ibidem* meaning in same place.

ibuprofen   A nonsteroidal anti-inflammatory agent; also antipyretic and analgesic.

I & C   Instrumentation and control.

IC   Ion chromatography; inductance-capacitance; integrated circuit; installed cost.

ICAP   Inductively coupled atomic plasma emission spectroscopy.

ICBO   International Conference of Building Officials.

ICC   Interstate Commerce Commission.

ICE   Institute of Civil Engineers.

ICP   Inductively Coupled Plasma.

ICRP   International Commission of Radiological Protection.

ICRU   International Commission of Radiological Units and Measurements.

ICT   International Critical Tables.

icterus   Jaundice; pigmentation of tissues or secretions.

ichthyology   The study of fish.

ID   Identification; inside diameter; inventory difference.

Idaho National Engineering Lab (INEL)   PO Box 1625, Idaho Falls ID 83415.

IDC   Item description code; initiating device circuits.

ideal gas   A gas that perfectly obeys the ideal gas law; such substances do not exist in actuality.

ideal gas law   A theoretical thermodynamic equation that expresses the relationship between pressure, volume, temperature, and amount of an ideal and perfect gaseous chemical at equilibrium; $PV = nRT$.

idiosyncratic   Describes the susceptibility of the body to a specific substance introduced into the body.

IDL   Instrument detection limit.

IDLH   Immediately dangerous to life and health.

IDN   Identification number.

IE   Industrial engineering.

I & E   Inspection and evaluation.

i.e.,   From the Latin *id est*, meaning "that is."

IEEE   Institute of Electrical and Electronics Engineers.

IESNA   Illuminating Engineering Society of North America.

IES   Institute of Environmental Sciences.

IGCI   Industrial Gas Cleaning Institute.

ignition   The act of catching fire.

ignition source   A source that can cause initial combustion of a substance.

ignition temperature   The minimum temperature required to initiate or cause self-sustained combustion independent of a heat source; the lower the ignition temperature, the greater the potential hazard.

IGT   Institute of Gas Technology.

IH   Industrial hygiene.

IHF   Industrial Health Foundation.

IHMM   Institute of Hazardous Materials Management.

IH&S   Industrial hygiene & safety.

IIE   Institute of Industrial Engineers.

Illium^{TM} A series of stainless steel and nickel alloys noted for corrosion resistance.

illness   A deviation from normal health; a state of sickness; if caused by working conditions; called an occupational illness.

Illuminating Engineering Society of North America (IESNA)   120 Wall St., New York NY 10005.

illumination   The density of incident light flux.

I/M   Inspection & maintenance.

IMDG   International maritime dangerous goods.

IME   Institute of Makers of Explosives.

IMECHE   Institute of Mechanical Engineers.

imine   An organic compound with a C$=$N bond.

Immediately dangerous to life and health (IDLH)   Maximum level that one can endure for 30 minutes without any impairing symptoms or irreversible health effects; used to help determine respirator selection.

imminent danger   An existing condition that could cause injury to personnel, property, or the environment at any time if the condition is allowed to continue without interruption.

immiscible   Describes a liquid that will not mix with another; two layers, cloudiness, or turbidity results.

immune body   Antibody.

immunodeficient   Lacking the ability to produce antibodies.

immunoglobulin   Any protein that can act as an antibody.

immunosupressant   Describes a substance that has a negative effect on the production of antibodies.

impaction   The forcible contact of a particle with a solid or liquid.

impinger   A device, usually containing a liquid, that is used to collect gaseous or particulate air samples.

implode   To explode inwardly; e.g., Dewar or vacuum flasks implode if ruptured.

impervious   Describes the ability of a substance to act as a barrier by restricting entrance or passage. The term is normally associated with personal protective equipment such as impervious gloves.

In   Indium.

incendiary   A substance used to start fires.

incident   The occurrence of an event.

incident commander   The person in charge at the site where a hazardous materials incident occurred.

incident rate   A measure of the frequency of specific incidents, e.g., accident occurrence; equal to the total number of events divided by the product of the number of employees and the hours they worked in a year; according to OSHA, incident rate $I = N/EH \times 200{,}000$ (the number of hours 100 full-time employees work in 1 year, assuming 40 hours/week, 50 weeks/year).

incineration   Destruction by high-temperature fire.

incompatible chemicals   Chemicals that, when brought into contact with each other, could result in dangerous reactions, formation of hazardous substances, and/or the release of energy (e.g., explosions).

incubation   The growth of microorganisms; warming; developing.

indemnify   To compensate for damage; a legal exemption against damages due to liability.

independent verification   Visually monitoring a condition (such as a valve position) independently from using control indicators, to verify the condition of a system or position of a component; ensures that the system complies with working condition as per specified requirements.

indeterminant   Imprecise and not able to be accurately measured.

indicator   A sign of an event, such as the indicator used in titrations to show when the end point is reached.

indictment   A legal statement charging a person with a crime.

indigo   A natural or synthetic blue dye originally from plants.

indirect costs   A manufacturer's directly identifiable support functions such as operational manpower and costs to support systems and/or end uses, plus other costs including management, administrative and protective services; plant maintenance and utility services; and defective products returned by next user for disposition, rework, or deviation.

indirect discharge   Introduction of industrial pollutant into a public waste system.

indium (In)   A soft, silver-white, metallic element, atomic number 49, atomic weight 114.82; used for mirrors and electronics.

Indoor air quality (IAQ)   The quality of breathing air in an inhabited building; a measure of the pollutants or contaminants in the breathing air of a building.

induced radioactivity   *See* activation.

inductance   Measured by a derived SI unit called the henry; units are Wb/A.

induction   The electrical current produced when current passes through a magnetic field; reasoning from the specific(s) to the general.

inductively coupled atomic plasma emission spectroscopy (ICAP)   A type of spectroscopy used for the analytical determination of inorganic metals in which a plasma of inert gas excites atoms of an injected sample causing them to emit light of a characteristic wavelength.

Industrial Health Foundation (IHF)   43 Penn Circle W, Pittsburgh PA 15206.

industrial hygiene (IH)   Defined by the American Industrial Hygiene Association (AIHA) as the science and art devoted to anticipation, recognition, evaluation, and control of workplace environmental factors that may result in illness, injury, or impairment or affect the well-being of workers and members of the community.

Industrial Risk Insurers (IRI)   85 Woodland St., PO Box 5010, Hartford CT 06102.

Industrial Safety Equipment Association (ISEA)   1910 N. Moore St., Arlington VA 22209.

INEL   Idaho National Engineering Laboratory.

inert   Not having active chemical properties, i.e., not capable of reacting or combining with other substances. With respect to the inert gases, a term used to denote general chemical inactivity. Also substances added to mixtures to increase bulk and weight without having active properties of their own. Certain inert substances, such as aerosol propellants, can have hazardous properties such as flammability.

inert gas   The following chemical elements: helium, neon, argon, krypton, xenon, radon; originally named for their inability to combine with other elements to form compounds; since 1962 it has been known that these elements can form a limited number of compounds.

inert gas welding   An electric welding process using helium to flush unwanted by-products of the welding into a local exhaust system.

inertial separator   A device that uses centrifugal force to separate particles.

infectious   Capable of invading a susceptible host, replicating, and causing a disease.

infectious waste   Any waste capable of invading a susceptible host and causing a disease. Included are blood; certain body fluids; infectious stocks or cultures; items used to handle infectious stocks or cultures; and animal carcasses, body parts, and blood from research animals exposed to infectious agents. Sharps (e.g., needles, etc.) are not included. *See* sharps; etiological agents.

inflammable   Capable of being easily set on fire and continuing to burn. *See* also flammable.

inflammation   A localized, protective response of body tissue elicited by injury either by infection or trauma. Usually characterized by some degree of pain, heat, redness, swelling, and/or loss of function.

influent   A fluid that flows in as opposed to an effluent that flows out.

infrared radiation   Electromagnetic energy with wavelengths between 770 and 12,000 nm.

infrared spectroscopy   An analytical tool used to measure the characteristic infrared spectra of organic molecules.

infusion   Injection of a liquid into the body through the veins.

ingestion   The process of taking substances, such as food, medicine, etc., into the stomach.

inhabited building distance   The separation between explosives locations (potential explosive sites, PES) and nonassociated locations (exposed sites, ES) that require a high degree of protection from an accidental explosion.

inhalation   The taking in of a substance (i.e., gas, vapor, fume, mist, or dust) through the respiratory system.

inhalation valve   The device on a respirator that allows air to be breathed in to the facepiece but does not permit exhaled air to exit by the same path.

inhibitor   An agent that slows or prevents an unwanted chemical reaction, such as polymerization. Also, a material used to prevent or retard rust or corrosion.

initiator   A substance used to start a reaction, often a chain reaction; a substance added to prevent or retard unwanted growth or change; an agent that might cause carcinogenicity to occur.

injury   The occurrence of physical harm.

INMM   Institute of Nuclear Material Management.

innocuous   Describes something that is harmless.

inorganic   Term used for compounds that generally do not contain carbon. Examples are sulfuric acid, sodium hydroxide, and table salt.

insol   Insoluble.

insoluble   Term used for a substance incapable of being dissolved in a solvent.

insomnia   Inability to sleep properly.

Institute of Electrical and Electronics Engineers (IEEE)   345 E. 47th St., New York NY 10017.

Institute of Environmental Sciences (IES)   940 East Northwest Highway, Mt. Pleasant IL 60056.

Institute of Gas Technology (IGT)   1700 South Mount Prospect Rd., Des Plaines IL 60018.

Institute of Hazardous Materials Management (IHMM)   11900 Parklawn Dr., Rockville MD 20852.

Institute of Industrial Engineers (IIE)   25 Technology Park, Atlanta GA 30092.

Institute of Makers of Explosives (IME)   1120 19th St., NW, Washington DC 20036.

Institute of Medicine (IOM)   2101 Constitution Ave., NW, Washington DC 20418.

Institute of Nuclear Material Management (INMM)   60 Revere Dr., Northbrook IL 60062.

Institute of Transportation Engineers (ITE)   525 School St. SW, Washington DC 20024.

Instrument Society of America (ISA)   *See* International Society for Measurement and Control.

insulin   A hormone secreted from the pancreas; essential for metabolism of glucose and maintenance of proper blood sugar level.

insure   To arrange for indemnification.

integration   To organize, blend, or add together, e.g., to add events observed over a given time period.

Interagency Testing Committee (ITC)   An independent advisory committee to the US EPA created in 1976 under TSCA; includes representatives of Council on Environmental Quality (CEQ), DOC, EPA, NCI, NIEHS, NIOSH, NSF, OSHA; liaison members include ATSDR, CPSC, DOD, DOI, FDA, NLM, NTP, and USDA.

interferon   A protein with antitumor activity.

interlock   A device that interacts with another device or mechanism to govern, control, or prevent succeeding operations.

internal audit   A review of an organization's quality assurance program.

internal control system   Administrative and accounting policies and procedures to account for and maintain control of material; included are checks and balances in the division of duties so the work of one will verify the work of another.

International Agency for Research on Cancer (IARC)   150 Cours Albert Thomas, F-69372 Lyon CEDEX 08, France.

International Air Transportation Association.(IATA)   IATA Centre, Route de l'Aeroport 33, PO Box 416, 15 - Airport, CH-1215 Geneva, Switzerland.

International Association of Industrial Accident Boards and Commissions (IAIABC)   1575 Aviation Ctr. Parkway, Daytona FL 32826.

International Association of Mechanics and Aerospace Workers (IAM)   9000 Machinists Place, Upper Marlboro MD 20722.

International Atomic Energy Agency (IAEA)   Karbster Ring 11, PO Box 590, A-1011, Vienna, Austria.

International Commission of Radiological Units and Measurements, Inc. (ICRU)   791 Woodmont Ave. Bethesda MD 20314.

International Commission on Non-Ionizing Radiological Protection (ICNRP)   Bundesamt fr Strahlenschutz, Institut für Strahlenhygiene, Ingolstadter Landstrasse 1, D-85764 Oberschleissheim, Germany.

International Commission on Radiological Protection (ICRP)   SE 171, 16 Stockholm, Sweden.

International Conference of Building Officials (ICBO)   5360 South Workman Mill Rd., Whittier CA 90601.

International Occupational Hygiene Association (IOHA)   Georgian House, Great Northern Rd., Derby DE1 1LT, United Kingdom.

International Organization for Standardization (ISO)   1, rue de Varembe, Case postale 56, CH-1211 Geneva, Switzerland.

International Society for Measurement and Control   Also known as the Instrument Society of America (ISA), 67 Alexander Drive, Research Triangle Park NC 27709.

International System of Units (Le Système International d'Unités)   The internationally agreed upon system of units most often used in science; basically the metric system with time, temperature, and electrical terms, etc. added.

International Transportation Safety Board (ITSB)   800 Independence Ave., SW, Washington DC 20594.

International Union of Pure and Applied Chemistry (IUPAC)   IUPAC Secretariat, PO Box 13757, Research Triangle Park NC 27709-3757.

interpolation   Any process for estimating intermediate values between two or more known points.

Interstate Commerce Commission (ICC)   Congress abolished the ICC in 1995 and created a Surface Transportation Board within DOT to perform the small number of regulatory tasks that had remained in the ICC.

interstitial   Located between.

interstitial space   The space between floors or walls in a building or structure.

interstitial fibrosis   Scarring of the lungs.

intoxication   A term meaning drunkenness or poisoning.

Intravenous (IV)   A term describing a substance put into or inside a vein.

intrinsically safe   Term used to describe a device, instrument, process, or building that is inherently safe and cannot malfunction, or if malfunction occurs the process will self-correct or result in only minor, inconsequential events.

inverse square law   A mathematical expression that one value is inversely proportional to the square of another. For example, an expression used to express the relationship between energy and distance traveled.

inversion   Reversing or turning upside down, often used to refer to atmospheric conditions in which a layer of warm air is prevented from rising thus preventing dispersion of pollutants.

in vitro   Latin term meaning literally "in glass"; the term is used to refer to experiments done in a nonliving container

in vivo   Latin term meaning literally "in life"; the term is used to refer to experiments done in or on a living system.

I/O   Input/Output.

IOC   Initial operational capacity.

iodine (I)   A shiny, corrosive, highly toxic, sublimable solid group VIIA halogen element, atomic number 53, atomic weight 126.90; used in medicine, dyes, and photography.

iodoform CHI_3; yellow crystalline compound with a disagreeable odor; has topical use as a mild bactericide.

IOHA   International Occupational Hygiene Association.

IOM   Institute of Medicine.

ion   An atom or group of atoms (radical) with a positive or negative charge as a result of having lost or gained electrons. Also an electron not associated with a nucleus. *See* ionization.

ion exchange   A reversible chemical reaction between a solid (ion exchanger) and a solution whereby ions may be transferred from one substance to the other.

ion exchange chromatography   A type of liquid chromatography used for chemical analysis and separation of chemicals based on the exchange of ions between the liquid phase and the solid phase.

ionization   The separation or dissociation of molecules into ions of opposite electrical charge that sometimes occurs spontaneously in many salts when dissolved in water or melted. Also, the process of adding electrons to or removing them from atoms or molecules, thereby creating ions. High temperatures, electrical discharges, and nuclear radiation can cause ionization.

ionization potential (IP)   The energy required, generally measured in eV, to remove an electron from a given moiety.

ionizing radiation   Radiation capable of producing charged particles (alpha, beta), nonparticulate radiation (X rays), and neutrons; also any radiation with sufficient energy to displace electrons from atoms or molecules, thereby producing ions. Examples include alpha, beta, gamma, X rays, neutrons, and ultraviolet light. High doses of ionizing radiation may produce severe skin or tissue damage.

ion pair   Two oppositely charged particles formed as a result of incident radiation on an outer orbital electron.

IP   Ionization potential (measured in eV); inhalable particles; interproject.

ipecac   A common emetic.

IPS   Interruptible power supply; inside pipe size; in-process

Ir   Iridium.

IR   Infrared.

IRI   Industrial Risk Insurers.

IRDS   Primary irritation dose.

iridium (Ir)   A silver-white, brittle, corrosion-resistant group VIII metallic element, atomic number 77, atomic weight 192.2; used for cancer therapy.

iridocyclitis   The inflammation of the iris of the eyes and ciliary areas.

iron (Fe)   A common, silvery, malleable, magnetizable group VIII metallic element, atomic number 26, atomic weight 55.85; widely used in structural materials in the form of alloys.

irradiate   To expose or treat with radiation.

irritant   A substance, not a corrosive, that on immediate, prolonged, or repeated contact with living tissue induces local irritation. A reversible inflammatory effect caused by chemical action at the site of contact, usually the skin, eyes, nose, or respiratory system; also a chemical producing an irritating effect at the site of skin contact and therefore defined as a primary irritant. Chemical irritants affect everyone but do not produce the same degree of irritation. Irritants are classified by increasing tendency to irritate, as follows: nonirritating, practically nonirritating, minimally irritating, slightly irritating, mildly, moderately, severe, and extremely irritating. Extremely irritating materials are considered corrosive.

irritation   A condition of soreness, itching, roughness, or inflammation of a body part.

ISA   Instrument Society of America.

ischemia   Insufficient blood in part of the body often due to constriction or blockage.

ISEA   Industrial Safety Equipment Association.

ISO   International Organization for Standardization.

isoamyl acetate   A colorless, oily alcohol with a banana-like odor used to qualitatively fit test respirators.

isobar   Elements with the same mass number and different atomic numbers.

isocyanate   Any compound that contains the radical O=C—N—

isokinetic   At the same speed or flow rate.

isokinetic sampling   An air sampling technique in which the incoming air being sampled in the instrument has the same flow as the effluent being sampled.

isomer   Any of two or more chemical compounds having the same number and kind of constituent atoms but having different molecular structures.

isopropyl alcohol (IPA) C_3H_7OH; a colorless, flammable liquid, often called rubbing alcohol; toxic by ingestion.

isothermal   Under conditions of constant temperature.

isotonic   Term applied to solutions with the same osmotic pressure.

isotope   An atom of an element with an integer atomic weight; isotopes of a given element have the same number of protons but different numbers of neutrons in their nuclei. For example, uranium-235 (92 protons and 143 neutrons) and uranium-238 (92 protons and 146 neutrons). Isotopes have the same chemical properties but often different physical properties (for example, carbon-12 and carbon-13 are stable, carbon-14 is radioactive).

isotope effect   The effect that different isotopes of the same element exhibit in a chemical reaction, caused by the difference in mass.

isotope separation   The process of separating isotopes from one another or changing their relative abundance, usually by gaseous diffusion or electromagnetic separation. Isotope separation is a step in the isotopic enrichment process.

isotropic   A term for the property of transmitting light equally in all directions.

isozyme   Isoenzyme.

ISTM   International Society for Testing Materials.

Itai-itai   A disease first observed in Japan, meaning "ouch-ouch," caused by exposure to cadmium.

ITC   Interagency Testing Committee.

ITE   Institute of Transportation Engineers.

item   A single piece or container of material having a unique identification and previously determined material mass whose integrity can be visually verified. Also an all-inclusive term used in place of any of the following: appurtenance, assembly, component, equipment, material, module, part, structure, subassembly, subsystem, system, or unit.

-itis   Latin suffix meaning inflammation; usually used in reference to anatomy.

ITSB   International Transportation Safety Board.

IUPAC   International Union of Pure and Applied Chemistry.

IV   Intravenous.

IW   Industrial waste.

J

J   Joule.

J/mol   Joule/mole; molar energy.

JA   Job analysis.

jaundice   A sign of a disease that causes yellow discoloration of skin, whites of eyes (sclerae), mucous membranes, and body fluids with bile pigment (bilirubin); caused by pathological conditions that interrupt the liver's normal production and discharge of bile.

Jenner, Edward   English physician (1749–1823) who observed that dairymaids had immunity to smallpox and discovered how to vaccinate against the disease.

jet   A high-velocity stream of fluid emitted from a narrow opening; a type of airplane.

jet fuel   One of several types of fuel for jet airplanes similar to the petroleum distillate kerosene.

JHA   Job hazard analysis.

Job analysis (JA)   A technique used in system safety engineering to obtain a detailed list of the duties and tasks for a specific job. The first step in obtaining the data required for task analysis.

Job hazard analysis (JHA)   A technique used in system safety engineering for identifying the hazards associated with a particular task or job that may not be obvious from a routine examination of the operation. It is commonly a three-part process involving identification, analysis, and recommendations for correcting or eliminating hazards.

job rotation   Shifting people periodically during their work-shift to avoid prolonged exposure to potential hazards or discomforts.

job safety analysis (JSA)   *See* job safety and health analysis (JSHA).

Job safety and health analysis (JSHA)   A four-part system of system safety engineering used to identify potential sources of occupational injury and exposure by observing the tasks being performed. The analysis includes identification of the job; separating the job into steps; identifying key components of each step; and conducting an efficiency check. Also called job safety analysis.

jock itch   An infection caused by ringworm fungus; also called ringworm and athlete's foot.

joint   A crack or opening; part of the body between two moving parts.

Joliot-Curie, Irène   French chemist (1897–1956), daughter of Pierre and Marie Curie, who, together with her husband, Jean Frédéric Joliot-Curie, was awarded the 1935 Nobel Prize in chemistry for their work on the production of artificial radioactive elements.

Joliot-Curie, Jean Frédéric   French chemist (1900–1958), World War II resistance fighter, and communist who worked with Marie Curie and married her daughter Irene, with whom he shared the 1935 Nobel Prize in Chemistry for their work on the production of artificial radioactive elements.

joule (J)   The international system unit of energy equal to the amount of work done when one ampere is passed through one ohm for one second; also the heat produced when one watt flows for one second.

Joule, James Prescott   English physicist (1881–1889) who worked with Dalton and Kelvin; demonstrated that heat is a form of energy, established the theory of the conservation of energy, and studied temperature changes in gases and refrigeration.

Joule-Thomas coefficient   The change in temperature per atmosphere change of pressure on a fluid at constant enthalpy.

JSA   Job safety analysis.

JSHA   Job safety and health analysis.

jute   A type of fiber used in burlap with low strength and durability.

juvenile hormone   One of many hormones that retard the growth of insect larvae; obtained from silk moths and produced naturally; a possible component of insecticides.

K

k   Kilo- (10^3).

K   *Kalium* (Latin); potassium; Kelvin; kayser; uppercase Greek letter kappa.

kaolin   A form of aluminum silicate found in certain clays used for refractories, cosmetics, and ceramics; also known as China clay and used in the making of porcelain.

kaolinosis   A form of pneumoconiosis caused by inhalation of kaolin dust, usually a result of grinding, milling, or other operations.

kapok   A light, cotton-type fiber used in life jackets, insulation, and upholstery.

Kaposi, Moritz   Austrian dermatologist (1837–1902) who named a specific type of malignant tumor commonly associated with AIDS.

Kaposi's sarcoma   A malignant, metastasizing tumor, usually involving the skin. It produces reddish-blue or brown disfigurations, often appearing first at the extremities. Named after Moritz Kaposi.

Karl Fischer reagent   A solution of iodine, sulfur dioxide, and pyridine in methyl alcohol used to determine the amount of water present in a substance.

Karrer, Paul   Swiss chemist (1889–1971) who shared the 1937 Nobel Prize in Chemistry for his investigations of carotenoids, flavins, and vitamins A and B_2.

kcal/mole   Kilocalories per mole.

K-capture   A radioactive decay process in which an electron from the K shell of an atom is captured by the nucleus and combines with a proton to form a neutron with the emission of X-radiation.

KE   Kinetic energy.

Kekule von Stradonitz, Frederich August   German chemist (1829–1911) who began his studies in architecture but was influenced by Liebig to study chemistry; best known for his study and depiction of the structure of benzene and his four-volume work on organic chemistry.

kelthane   A chlorinated hydrocarbon pesticide; also known as decofol.

Kelvin, William Thompson   British mathematician and physicist (1824–1907) born in Ireland who studied heat, temperature, magnetism, and electricity. Proposed the absolute Kelvin scale of temperature measurement.

Kelvin temperature scale (K)   A system for measuring absolute temperature obtained by adding 273 to degrees centigrade if the value is above 0 degrees centigrade or subtracting 273 if the value is below 0 degrees centigrade.

Kendrew, Sir John Cowdery   English chemist (1917–) who shared the 1962 Nobel Prize in Chemistry for studies of the structure of globular proteins.

kepone $C_{10}Cl_{10}O$; a carcinogenic, chlorinated hydrocarbon insecticide and fungicide toxic by skin absorption and inhalation used for the control of fire ants. Implicated in a large spill on the James River in Virginia that caused harm to many workers, residents, and the environment; also known as chlordecone.

keratin   A class of natural fibrous proteins in animals and humans; components of skin, hair, feathers, wool, nails, and hooves.

keratitis   Inflammation of the cornea of the eye.

keratoconjunctivitis   Inflammation of the conjunctiva and of the cornea; may result from overexposure to UV laser radiation.

keratolytic fungus   Lesions of the nails or hair as a result of exposure to the ringworm fungus.

kerosene   A white, oily petroleum distillate with a strong odor; toxic by inhalation; used as a solvent, fuel, jet fuel, and diesel fuel; also known as kerosine.

kerosine   *See* kerosene.

ketone   A class of organic compounds commonly used as solvents and reagents with the general structure R_2—C=O; acetone is a member of this class.

ketosis   A metabolic disorder in which excessive production or accumulation of ketones in the body occurs.

Kev, keV   kilo electron volt; a unit of energy; one Kev equals 1000 electron volts.

Kevlar™   An extremely resistant, energy-absorbing synthetic fiber with a high tensile strength; used for radial tires and bullet-proof vests.

kg   Kilogram, i.e., 1000 grams (g).

Kick's law   A rule that states the amount of energy required to crush a specific quantity of material to a certain fraction of its original size is always the same regardless of the original size.

kieselguhr   A soft, material composed of the skeletal remains of small prehistoric animals; used for filtration, clarification, decolorizing, chromatography, and many other applications.

kiln   A high-temperature furnace used for ceramics and incineration.

kiln workers   Workers at risk from exposure to infrared radiation and heat exposure.

kilo- (k)   A prefix that multiplies a basic unit by 1000; e.g., 1 kilometer = 1000 meters.

kilocalorie (kcal)   The amount of heat necessary to raise the temperature of 1000 grams of water 1°C.

kilogram (kg)   1000 grams; the mass of a liter of water at 4°C.

kinematics   The study of bodily motions in space.

kinesiology   The study of movement.

kinetic energy (KE)   Energy of motion; $1/2\ mv^2$.

kinetics   The study of chemical reactions that deals with the rate of change going from the initial to the final state. For example, the time necessary for two substances to react and form a third or the time required for the human body to absorb and/or metabolize a substance.

Kirchhoff, Gustav Robert   German physicist (1824–1887) who was a student of Gauss and made important contributions to the theory of circuits using topology. Best known for his publication of Kirchhoff's law in 1854.

Kistiakowsky, George Bogdan   American chemist (1900–1982) born in

Russia who was a member of the Manhattan Project at Los Alamos National Laboratory and was a world-renowned authority on high explosives.

Kjeldahl, Johan Gustav Christoffer Thorsager   Danish chemist (1849–1900) known for his expertise in analytical chemistry, most notably for the method of determining the amount of nitrogen in a substance, which can be converted to the amount of protein present. The special flask used for nitrogen-Kjeldahl determinations is named after him.

Klug, Sir Aaron   Lithuanian-born biochemist (1926–) who emigrated to South Africa then to the United Kingdom; awarded the 1982 Nobel Prize in Chemistry for his studies of X-ray diffraction, modeling, and the structure of viruses in combination with proteins and for the development of crystallographic electron microscopy.

klystron   An electron tube making use of the controlled speed of a stream of electrons that produce microwave oscillations; used in electron spin resonance spectroscopy (ESR) to study free radicals.

knock   A power loss in a gasoline-powered engine due to improper ignition of part of the gasoline at the cylinder head; related to octane number.

Knoop hardness   A hardness scale for material ranging from 300–600 for glass to 6000-6500 for diamond, calculated by measuring the indentation produced by a diamond tip pressed onto the surface of the sample.

Koch, Heinrich Hermann Robert   German bacteriologist and physician (1843–1910) who discovered tuberculosis bacillus and cholera bacillus; awarded the 1905 Nobel Prize in Medicine or Physiology.

Kohlrausch, Friedrich Wilhelm Georg   German physicist (1840–1910) who studied electrolytic conduction of ions in solution.

Kohlrausch's law   States that ions in solution migrate independently and their conduction is equal to the sum of the conductance of the anodes and cathodes.

Kr   Krypton.

Krebs, Sir Hans Adolf   German physiologist (1900–1981) who emigrated to England and was awarded the 1953 Nobel Prize in Medicine or Physiology for his work on metabolism.

Krebs cycle   A series of aerobic enzymatic reactions in aerobic organisms involving the oxidative metabolism of acetyl groups, especially during respiration, that provide the main source of cellular energy.

krypton (Kr)   A colorless, odorless group VIII noble gas element, atomic number 36 and atomic weight 83.80; used in bulbs, lasers, and lights.

Kuhn, Richard   Austrian-born biochemist (1900–1967) who worked in

Germany and Switzerland and was awarded the 1938 Nobel Prize in Chemistry for his work on the structure and synthesis of vitamins A and B_6 and carotinoids. Nazi Germany did not allow him to accept the award, which was presented after World War II.

kwashiorkor   A nutritional disease caused by lack of protein.

L

λ   Lower case Greek letter lambda; wavelength.

l   Levo, levorotatory.

L   Liter; lambert.

l-   Prefix indicating a compound that rotates light to the left, i.e., levorotatory.

La   Lanthanum.

label   A word or phrase used for identification; the use of a radioactive atom or isotope to act as a tracer to follow the mechanisms, reactions, and degradations of chemical or biological processes.

labile   Susceptible of undergoing immediate, rapid, or continual chemical, physical, or biological change, transformation, or breakdown.

laboratory   A room or building in which scientific study, work, or research is performed.

laboratory chemical hood   *See* laboratory hood.

laboratory hood   According to ANSIZ 9.5, Laboratory standard: a box like structure with one open side intended for placement on a table or bench. The bench and the hood may be one integral structure. The open side is provided with a sash or sash that moves vertically and/or horizontally to close the opening. Provisions are made for exhausting air from the top or back of the hood, and adjacent or fixed internal baffles may be provided to obtain proper airflow distribution across the open face. Provisions may be made for utilities and lighting. Also known as a laboratory chemical hood, chemical hood, or fume hood.

laboratory scale   Describes scientific work that can be performed by one person; generally, small-scale operations or procedures as opposed to production work.

lachrymator   A substance, usually gaseous or volatile liquid, that irritates the eyes and mucous membranes and produces an increase in the flow of tears. Also known as a lacrimator.

lacquer   A coating dried by the evaporation of the solvent.

lacrimator   *See* lachrymator.

lactic acid $CH_3CHOH_2O_2H$; a colorless, odorless liquid intermediate in the fermentation (oxidation, metabolism) of sugar obtained by the reaction of lactic acid bacillus on milk or milk sugar. After strenuous exercise, lactic acid may accumulate in muscles and cause cramps.

lactose   Milk sugar, found in cow's milk and whey. Some people develop allergies to lactic acid-fixing bacteria present in dairy products.

LAER   Lowest achievable emission rate.

lagoon   A pond in which sewage or other organic wastes are decomposed by the action of sunlight, oxygen, algae, and biological microorganisms to restore water purity. Also called an oxidation pond. They are often used after the process of activated sludge treatment.

Lambert (L)   A unit of measure in the centimeter-gram-second system to denote brightness equal to the brightness of a perfectly diffusing surface that radiates or reflects one lumen per cm^2; named after the eighteenth century German physicist Johann H. Lambert.

Lambert, Johann Heinrich   German physicist and mathematician (1728–1777) who studied light and showed that π was an irrational number.

laminar   Nonturbulent, streamlined, or along parallel lines; laminar air flow means that the entire body of air contained in a designated area moves uniformly in the same direction along parallel flow lines.

laminate   To layer with adhesive between layers.

lampblack   A black or gray pigment made by inefficiently burning heavy oils in a closed system and collecting the soot.

landfill   A location used for the disposal of waste material(s); such sites may be for hazardous (i.e., permitted) or nonhazardous waste.

Langmuir, Irving   American chemist (1881–1957) awarded the 1932 Nobel Prize in Chemistry for his studies of surface chemistry. Invented the mercury pump enabling the attainment of very low pressures needed for vacuum tubes; made theoretical contributions to the study of adsorption and thermonuclear fusion; and coined the term "plasma" for ionized gas. Worked on problems of ice formation on aircraft wings that led to the discovery of producing rain by seeding clouds with dry ice and silver iodide.

LANL   Los Alamos National Laboratory; formerly LASL

lanthanum (La)   A malleable, ductile, lanthanide group IIIB metallic element, atomic number 57, atomic weight 138.91.

larvicide   An insecticide specific to insect larvae.

laryngitis   Inflammation of the larynx; often accompanied by dryness, hoarseness, cough, and soreness.

larynx   Part of the respiratory tract between the pharynx and the trachea; the upper part of the trachea or windpipe; the organ of voice production housing the vocal cords.

laser   Acronym for "light amplification by stimulated emission of radiation"; a device that emits highly amplified and coherent light at a discrete frequency.

Laser Institute of America (LSA)   12424 Research Parkway, Orlando FL 32826.

Laser safety officer (LSO)   The individual specified in the ANSI laser standard who oversees the laser program.

LASL   Los Alamos Scientific Laboratory, an obsolete name for the Los Alamos National Laboratory. *See* LANL.

latent heat   The energy, measured in calories per gram, absorbed or given off when a substance undergoes a change in physical state, e.g., freezing, melting, boiling.

latent heat of fusion   The amount of heat necessary to convert a solid to a liquid at a constant temperature.

latent heat of vaporization   The amount of heat necessary to convert a liquid to a gas at a constant temperature.

latent period   The time between exposure and the first signs of the damage or disease.

latex   A naturally occurring polymer containing about 60% water and 35% hydrocarbons that can be made synthetically; commonly used in the manufacture of thin rubber products like surgeon's gloves.

lattice   An open framework; the structural arrangement of atoms or ions in a solid or the arrangement of objects in space.

lavage   Washing out of a cavity or biological organ, as in gastric lavage.

Lavoisier, Antoine Laurent   French chemist (1743–1794) called the founder of modern chemistry; wrote *Traité élémentaire de chimie*; showed that air is a mixture of gases disproving the phlogiston theory; helped devise a modern method of naming chemical compounds and the metric system; guillotined in Paris during the French Revolution.

Lawrence, Ernest Orlando   American physicist (1901–1958) awarded the 1939 Nobel Prize in Physics for inventing the cyclotron. The element lawrencium is named after him.

Lawrence Livermore National Laboratory (LLNL)   7000 East Ave., Livermore CA 94550-9234.

lawrencium (Lr)   A radioactive transuranic element, atomic number 103 and atomic weight 260.

LC   Liquid chromatography; lethal concentration.

LC_{50}   The concentration of a toxic material in the gas phase that induces death in half of the exposed population within a specified time.

lcd   Liquid crystal display; lowest common denominator; least common divisor.

LCL   Lower confidence limit; lower control limit.

LC_{Lo}   The lowest concentration in inhalation studies of a substance at which an adverse health effect is noted.

LC_x   The concentration of a toxic material in the gas phase that induces death in $x\%$ of the animal population within a specified time.

LD_{50}   The dose that induces death in 50% of the exposed population within a specified time, usually 14 days.

$LD_{50/30}$ The acute dose expected to induce death within 30 days in 50% of those exposed without medical intervention.

LD_{lo}   The lowest dose necessary to kill test animals.

LD_x   The dose that induces death in $x\%$ of the animal population tested in a specified time, usually 14 days.

leach   To remove a solid but soluble substance from a solid matrix by percolating a liquid through the matrix.

lead (Pb)   A toxic, heavy, ductile, soft group IVA metallic element, atomic number 82 and atomic weight 207.19; responsible for many poisonings, particularly among children; formerly widely used in gasoline, paint, and plumbing.

lead dross   Waste scrap from sulfuric acid tanks containing lead and sulfuric acid.

Lead Industries Association (LIA)   292 Madison Ave., New York NY 10017.

lead poisoning   Both organic and inorganic lead compounds are poisonous if ingested or inhaled by humans.

leak test   A test given to determine whether a device or connection is leaking, e.g., soap solution on a gas tank fitting.

learning objective   A well-defined, concise statement that describes a specific behavior, usually containing an action, a condition, and/or a standard.

Le Chatelier, Henry   French chemist (1850–1936) who discovered the law of reactions and the effects of pressure and temperature on equilibrium called Le Chatelier's principle. Also studied metallurgy, ceramics, and combustion and devised a railway water-brake and an optical pyrometer.

Le Chatelier's principle   States that if a system at equilibrium is stressed in some fashion, the equilibrium will shift to offset the stress.

LED   Light-emitting diode.

legal chemistry   Also called forensic chemistry; The application of chemistry to civil or criminal law in which the control, use, or action of chemicals, chemical products, byproducts, or chemical processes is studied and investigated.

Legionnaire's disease   A disease caused by a type of bacteria (*Legionella*) found in the cooling towers of ventilation systems; known to cause a specific type of pneumonia; first identified at an American Legion convention in Philadelphia in 1976.

LEL   Lower explosive limit; also known as the lower flammability limit (LFL).

leptospirosis   A bacterial infection caused by *Leptospira interrogans* through a break in the skin. At risk are farmers, sewer workers, miners, livestock handlers, veterinarians, and laboratory workers in contact with animals. Also known as Weil's disease, swineherd's disease, and canicola fever.

lesion   A wound or injury; discontinuity or loss of tissue or body part.

LET   Linear energy transfer.

lethal   Capable of causing death.

leukemia   A progressive, malignant disease of the blood or blood-producing organs characterized by a change in the production or development of white blood cells in the blood and bone marrow.

leuko-   A prefix meaning white, as in white blood cells.

leukocyte   A white blood cell or, in general, any colorless cell mass.

leukocytosis   An abnormally large number (15,000–25,000 per mm^3) of leukocytes, often observed in acute infections.

leukopenia   A condition in which the total number of leukocytes in the circulating blood is less than normal (4000–5000 per mm^3).

level of concern (LOC)   The air concentration of a hazardous substance that may cause an immediate health threat.

level of confidence (LOC)   The probability that a value is within the specific range being estimated.

levorotatory (l)   Describes the property of a chemical substance in solution to rotate the plane of polarized light passed through its solution to the left or counterclockwise; such chemical isomers are denoted by the prefix l- or a minus sign.

Lewis acid   A system that defines acids as substances that are electron acceptors.

Lewis base   A system that defines bases as substances that are electron donors.

Lewis, Gilbert Newton   American physicist and chemist (1875–1946) who developed the valence theory of chemical reactions, discovered "heavy" water, invented the cyclotron, and established an acid-base theory.

LFL   Lower flammability limit; also known as the lower explosion limit (LEL).

Li   Lithium.

LIA   Lead Industries Association.

Libby, Willard Frank   American chemist (1908–1980) awarded the 1960 Nobel Prize in Chemistry for his role in developing ^{14}C dating; worked on the atomic bomb and served on the Atomic Energy Commission (AEC).

Library of Congress (LOC)   101 Independence Ave., SW, Washington DC 20540.

lichenification   Leathery thickening of the skin caused by scratching atopic or chronic contact dermatitis.

lidocaine   A common anesthetic.

Liebig, Baron Justus Freiherr von   A German chemist (1803–1873) who worked with Gay-Lussac; studied organic and agricultural chemistry and developed new analytical techniques, such as distillation. Developed the condenser that carries his name.

lifeline   A device, usually a cable, rope, or line, used to ensure a workers safety while working in a hazardous location, e.g., high places.

ligament   A band of tissue joining bones, cartilage, or supporting organs or muscles.

ligand   An atom, molecule, or ion attached to the central atom of a coordination compound, chelate, or complex, such as EDTA.

light microscopy   Optical microscopy.

ligroin   A toxic, highly flammable, volatile fraction of petroleum with a boiling point range of 60–110°C.

lime   Calcium oxide; sometimes used in cement; a high-volume chemical.

limestone   Calcium carbonate, $CaCO_3$, also called marble or dolomite; used to remove sulfur dioxide from stack exhaust.

limestone scrubber   A device used to remove sulfur-containing gases from air emissions.

limiting conditions for operation (LCO)   Administratively established constraints on safety-related facility equipment and operation characteristics that are adhered to during operation of a facility. Used to specify the minimum performance level required for safe operation of the facility.

limiting safety system settings (LSSS)   Limiting values for settings in safety channels by which point protective action must be initiated or chosen so that automatic protective action will terminate an abnormal situation before a safety limit is reached, except for certain uncontrolled accidental conditions.

limit of detection (LOD)   The lowest concentration that an instrument or analytical technique can determine.

limits of error (LE)   Boundaries within which the value of the attribute being determined lies within a specified probability. Boundaries are defined to be plus or minus twice the standard deviation of the attribute unless otherwise stipulated.

Lindane™ $C_6H_6Cl_6$; a toxic, chlorinated hydrocarbon insecticide whose use is restricted.

linear absorption coefficient   The amount of radiation absorbed by a given thickness of substance.

linear accelerator   A device used to speed up atomic and nuclear particles so that their energy, reactions, and effects can be studied.

linear energy transfer (LET)   The amount of energy lost linearly when locally absorbed by an ionizing particle as it passes through a specific medium.

line supervision   The direct supervisor of individuals who handle hazardous material or of individuals who work in any support function, i.e., maintenance, radiation protection, etc.

liniment   A medicinal fluid applied to the skin.

linseed oil   An oil used in paints and pharmaceuticals, also called flaxseed oil.

lipase   An enzyme that changes fat into fatty acids and glycerol.

lipid   A general term used to include all fats and fat-derived materials; a major component of all living cells.

lipid solvent   A water-insoluble solvent that removes the surface film on skin and disturbs the water-holding properties of keratin cells, allowing injury to occur to epidermal cells.

lipo-   A prefix denoting a relationship to fat or lipid.

liquefaction   Transformation into a liquid.

liquefied natural gas (LNG)   The compressed or liquefied form of the substance known as natural gas that is a mixture of low-molecular-weight hydrocarbons, primarily composed of 85% methane and 10% ethane.

liquefied petroleum gas (LPG)   The compressed or liquefied form of a gas obtained as a by-product in petroleum refining or natural gas production that is a mixture of several low-molecular-weight hydrocarbons.

liquid   An amorphous state of matter in which the substance has a definite volume but assumes the shape of the container; a fluid but not a gas.

liquid air   Air cooled below $-189°C$, potentially dangerous because of the low temperature and also because liquid oxygen may fractionate and separate out as it warms.

liquid chromatography (LC)   An analytical method based on the selective separation of mixture components in solution by selective adsorption through various media. *See* high-performance or high-pressure liquid chromatography.

liquid crystal   An organic compound that is an intermediate between a solid and a liquid and changes color on heating or cooling.

liquor   Distilled spirits; a concentrated solution; digested waste.

Lister, Baron Joseph   English surgeon (1827–1912) who studied inflammation of wounds and used phenol to prevent infection. Considered the founder of antiseptic surgery.

listeriosis   A disease of animals and humans, particularly if immunocompromised or pregnant, caused by the bacterium *Listeria monocytogenes*. Symptoms may include septicemia and necrosis of the liver.

liter (L)   The metric unit equal to the volume of one kilogram of water at 4°C and standard atmospheric pressure, 760 mmHg.

lithium (Li)   A light, flammable group IA alkali metal element, atomic number 3 and atomic weight 6.94.

litmus   A blue powder, soluble in water, that changes color depending on pH (red at pH ≤ 4.5, blue at pH ≥ 8.3).

litmus paper   Paper containing the chemical litmus; used to indicate, test, and determine pH.

Little, Arthur Dehon   American chemical engineer (1863–1935) who was an authority on paper technology and chemistry; founded Arthur D. Little, Inc. in 1909.

liver   A large red organ in the upper right side of the human abdomen. Its function includes the storage and filtration of blood, conversion of sugar into glycogen, excretion of bilirubin and other substances formed in the body, secretion of bile, and many metabolic functions.

LLD   Lower-level discriminator.

LLNL   Lawrence Livermore National Laboratory.

LLW (LLRW)   Low-level (radioactive) waste.

lm   Lumen.

LNG   Liquefied natural gas.

LOC   Level of concern; level of confidence.

local exhaust   The use of ventilation situated at a specific point, spot, or location to capture and remove air contaminants generated or produced at that area before they spread to the surrounding environment.

localized   Specific to one location or position rather than spread throughout; opposite of systemic.

loc. cit.   Abbreviation for the Latin phrase *loco citato*, "in place cited".

lockjaw   A symptom of tetanus in which the jaw is locked closed because of a spasm of the mastication muscles; also known as trimus.

lockout device   A device utilizing a lock and key, or other mechanical means such as a chain, hasp, or bar secured by a lock and key, to secure an energy-isolating device in the safe position, i.e., to de-energize.

lockout/tagout coordinator (LTC)   The person authorized to approve, issue, and administratively control all lockout/tagouts in his/her area of responsibility.

lockout/tagout system (L/T)   The procedures used for the authorized placement, removal, and administrative control of tags and locks to ensure the protection of personnel, the environment, and equipment.

LOD   Limit of detection.

LOE   Level of effort.

LOEL   Lowest observable effect level.

Löffler, Friedrich August Johannes   German bacteriologist and surgeon (1852–1915) who was the first to culture diphtheria bacillus, discovered the cause of glanders and swine erysipelas, isolated an organism causing food poisoning, and prepared a vaccine against foot-and-mouth disease.

Löffler's syndrome   Pulmonary eosinophilia named after F. A. J. Löffler, a German bacteriologist and surgeon (1852–1915); caused by the sensitizing properties of nickel.

log   Logarithm; a record used by operating and support personnel to describe or record information and events necessary for evaluating conditions.

logarithm   The exponent denoting the power to which a number, or base, is raised to produce another number. Usually common logarithms are raised to the base 10; natural logs are raised to the base e (2.1714).

log normal distribution   The logarithmic expression of a normal distribution pattern; used to express, e.g., particle size distribution; characterized by a gaussian or bell-shaped curve.

log sheets (round sheets)   A record of parameters to be recorded for equipment, operations, or areas located within the responsibility of a particular shift station or administrative position that includes maximum and minimum acceptable operating parameter values and comments. Log sheets often include process run sheets, utilities round sheets, and lockout/tagout logs.

lone star tick   A type of tick believed to be one of several ectoparasites that may be responsible for the transmission of disease to humans by their bite. A hazard to scientists engaged in field work.

Long, Crawford Williamson   American surgeon (1815–1878) who claimed to be the first to use diethyl ether as an anesthetic.

longitudinal study   An epidemiological study in which the data are collected from the same group over a time period.

Los Alamos National Laboratory (LANL)   PO Box 1663, Los Alamos NM 87545-1362.

loss   In ventilation, the conversion of static pressure into heat; in insurance, that which is missing, damaged, or no longer capable of normal use.

loss control   An insurance term used to denote procedures used to minimize insurance company losses.

low explosive   An explosive that deflagrates but does not detonate.

lower explosion limit (LEL)   *See* lower flammability limit.

lower flammability limit (LFL)   The lowest concentration of an ignited substance or mixture, at ordinary temperature and pressure, that will burn indefinitely. Expressed as a percent by volume of the substance in air. Also known as the lower explosive limit (LEL).

lowest toxic dose (LTD)   The lowest dose of a substance capable of causing observable toxic effects in test animals.

LOX   Liquid oxygen.

LPA   Local planning authority.

LPG   Liquefied petroleum gas.

Lr   Lawrencium.

LSA   Laser Institute of America.

LSD   Lysergic acid diethylamide.

LSO   Laser safety officer.

LTD   Long-term disability; lowest toxic dose.

Lu   Lutetium.

lumen (lm)   The flux or flow of light on one square foot of a sphere, with a radius of one foot, emanating from a one-candle light source in the center of the sphere, which radiates in all directions.

luminescence   A general term used to denote the emission of visible or invisible nonionizing radiation unaccompanied by high temperature as a result of the absorption of excited energy. Examples include bioluminescence, chemiluminescence, fluorescence, and phosphorescence.

luminous dial painters   Experienced an increase in carcinomas because they smoothed the fine brushes used to paint radium on watch dials on their tongues.

lung counter   An instrument used to identify and measure radioactivity in human lungs.

lung squeeze   Compression of the lungs forcing blood and tissue fluids into the lungs and respiratory system; often caused by hyperbaric environments but also seen in swimmers who dive while holding their breath.

lupus erythematosus   An inflammatory dermatitis disease that may be chronic, subacute, or systemic; in some cases thought to be caused by exposure to ultraviolet radiation and certain chemicals.

LUST   Leaking underground storage tank; an obsolete term. *See* UST, underground storage tank

luster   Soft reflected light; the surface appearance judged by brilliance and ability to reflect light compared to other materials, e.g., metals.

lutetium (Lu)   A soft, malleable, group IIIB metallic lanthanide element, atomic number 71, atomic weight 174.97; used in the nuclear industry.

lux (lx)   The international system unit of illumination equal to the direct illumination on a surface one meter from a uniform point source of one candle intensity or equal to one lumen per square meter.

lx   Lux.

lye   Potassium or sodium hydroxide; made by leaching wood ash.

lymph   Fluid containing white blood cells collected from the tissues throughout the body, flowing in the lymphatic vessels, and eventually added to the venous blood circulation.

lymphadenitis   Inflammation of a lymph node(s).

lymphadenopathy   Any disease affecting a lymph node(s).

lymphocytosis   A form of leukocytosis in which an increase in the number of lymphocytes occurs.

lymphogranuloma venereum   A venereal infection frequently caused by *Chlamydia trachomatis*, characterized by a transient genital ulcer in males and perirectal lymph nodes in females; also known as venereal lymphogranuloma.

lymphoma   A general term for many of the various types of abnormal diseases of lymph tissue.

lysergic acid diethylamide (LSD) $C_{20}H_{25}N_3O$; a commonly used hallucinogenic substance of abuse.

lysis   The bursting of a cell by the destruction of its cell membrane.

Lysol^{TM} A commercial disinfectant containing phenol.

lysozyme   An enzyme occurring in tears with mild antiseptic properties.

M

μ   Lower case Greek letter mu; micro- (prefix $= 10^{-6}$); one-millionth; dipole moment.

m   Mass; meter; milli- (prefix $= 10^{-3}$); meta-; minem (0.06 mL); molal.

m^{-1} Reciprocal meter(s); wave number.

M   Molar; mega- (prefix $= 10^{6}$); moment; uppercase Greek letter mu; Mach number.

MA   Maritime Administration; aerodynamic moment; mental age.

M&S   Materials and supply.

MAAM   Mobile ambient air monitoring.

MAC   Maximum allowable concentration

Mace™   A gas dispersed as an aerosol used to control riots; also known as chemical mace; chloroacetophenone.

macerate   To break up or soften a substance by prolonged soaking in water.

machine guard   A device to protect workers from being caught or entangled in a pinch point of a piece of machinery, e.g., belt guard on a vacuum pump.

macrobiosis   Longevity.

macrocyclic   In organic chemistry, a large (>15 C atoms) ring structure.

macromolecule   A large, usually organic, molecule comprised of hundreds or thousands of atoms; a colloid, e.g., protein, nucleic acid; a polymer with more than 100 monomers.

macrophage   Immune cells used to remove foreign bodies.

macroscopic   Visible without the aid of a microscope; large.

MACT   Maximum achievable control technology.

macuke, macule   The spot on the retina of the eye devoid of blood vessels responsible for color detection. A discolored spot or area, not elevated or depressed; any color or size.

magazine   A building or structure used for storage of explosives or military ordnance.

magic acid   Equivalent concentrations of SbF_5 and FSO_2OH.

magic numbers   8, 20, 28, 50, 82, 126; those nuclei with these numbers of protons, neutrons, or both exhibit greater stability.

magnesia   Magnesium oxide.

magnesium (Mg)   A hard, silvery, light, flammable group IIA metallic element, atomic number 12, atomic weight 24.30.

magnetic field   The volume surrounding a magnet or electricity carrying body that has a detectable magnetic force.

major component (MC)   A specially designed item of piece parts, hardware, material, etc., designed to perform a specific operation.

MAK   Maximum allowable concentration (German PELs).

makeup air   Outdoor air, which is usually tempered and clean, supplied to replace air removed by exhaust ventilation such as hoods or industrial processes.

make-or-buy decision   A recommendation reviewed and approved by management and operating contractor management.

mal-   A prefix meaning bad, ill, or poor.

malaise   An indefinite feeling of bodily discomfort.

malathion   A widely used insecticide.

malignant   Often pertaining to a tumor that tends to become progressively worse, invade adjacent tissues, metastasize, and may result in death. *See* cancer.

maltose $C_{12}H_{22}O_{11} \cdot H_2O$; nutrient or sweetener; a reducing disaccharide.

Management Information System (MIS)   A DOE-wide integrated, computer-based data processing system designed to plan for the economical and effective management of data resources.

manganese (Mn)   A brittle, silvery, flammable in powdered form group VIIB metallic element; atomic number 25, atomic weight 64.94; used in alloys and metal production.

manifest   A list of items being transported or carried on board; especially important when transporting waste.

manifold   Multifold; an aperture with many openings; a device for making many connections.

mannitol $C_6H_8(OH)_6$; a naturally occurring, straight-chain hexahydric alcohol sugar used in dietetic foods.

manometer   An instrument for measuring pressure.

manufacturing technology development (MTD)   Associated with process technology development to identify new processes, equipment needs and potential problems pertaining to capability requirements. Used to establish production milestones, program surveillance operations, and team representatives to analyze design information; to contribute to formulation of product trees, assessments, and operational sequences; to review for accuracy and completeness; and to prepare process labor and material estimates by fiscal year.

Marcus, Rudolph Arthur   American chemist (1923–) awarded the 1992 Nobel Prize in Chemistry for his electron transfer theory.

marijuana   A product of *Cannabis sativa* containing tetrahydrocannabinol, an hallucinogen; the dried flowers and leaves of *C. sativa* are sometimes smoked for their intoxicating effect.

Maritime Administration (MA)   USDOT, 400 7th St., SW, Washington DC 20590.

marker   A chemical added as a reference point; a substance used as an indicator or sign of an event or occurrence.

marsh gas   *See* methane.

Martin, Archer John Porter   English chemist (1910–1994) who shared the 1952 Nobel Prize in Chemistry for work with partition chromatography.

maser   Microwave amplification by simulated emission of radiation.

mash   A mixture of grains; barley, or other grain with water to prepare wort for brewing.

mass (m)   The quantity of matter, usually measured in grams or pounds.

mass-energy equation   An equation, $E = mc^2$, developed by Albert Einstein to indicate that the energy of a body varies with the product of its mass and the square of the speed of light in a vacuum; the equation predicts the possibility of releasing enormous amounts of energy by the conversion of mass to energy.

mass number   The sum of the number of protons and neutrons in a nucleus.

mass spectrometry (MS)   An analytical device based on the separation and acceleration through a magnetic field of charged particles to obtain a "fingerprint" of the substance analyzed.

mastic   A puttylike sealant.

material   A nonspecific term often used in engineering to denote a product of which something is composed or made. For example, any raw, in-process, or manufactured commodity, equipment, component, accessory, part, assembly, or product of any kind.

material balance   The comparison of the input and output of quantities of materials used for a process. Generally, a comparison between the beginning and ending inventory plus quantities used or discarded.

material safety data sheet (MSDS)   A document that gives information about hazardous chemicals; legally, must be made available for every hazardous substance.

materials handling   A general term used in industry to denote procedures used for transportation, storage, handling, and distribution of products and materials.

material status report (MSR)   An inventory status report, often pertaining to nuclear material, of materials received, produced, possessed, transferred, consumed, disposed of, or lost during a specified period.

matter   Anything that has mass and occupies space; three forms or states exist: gas, liquid, and solid. *See* mass.

maximum permissible concentration (MPC)   The amount of a substance in air and water that, when inhaled or ingested, based on current knowledge, produces no measurable adverse effect.

maximum permissible dose (MPD)   The dose or concentration of a substance that, based on current knowledge and research, produces no measurable adverse effects during a person's lifetime. *See* burden.

Maxwell, James Clark   Scottish physicist (1831–1879) who is credited as the father of electromagnetic theory.

may   A term used in legal contracts to denote permission; neither a requirement nor a recommendation.

MC&A   Material control and accountability.

MCA   Maximum credible accident; multichannel analyzer.

McCready, Benjamin W.   American physician who wrote the first book on occupational medicine in 1837.

MCL   Maximum contaminant level.

McMillan, Edwin Mattison   American physicist (1907–) who shared the 1951 Nobel Prize in Chemistry for his work in nuclear physics and the discovery of neptunium and plutonium.

MCS   Multiple chemical sensitivity.

Md   Mendelevium.

MD   Medical doctor; medical.

MDA   Minimum detectable amount.

MDC   Minimum detectable concentration.

Me   Shorthand for methyl.

mean   The average; sum of results divided by the number of results.

mean free path   The average distance particles travel before collision.

means of egress   The way to exit, comprised of three components: the access to, the exit, and the means of exit, from an area.

measurement control   The procedures and activities used to ensure that a measurement process generates values of sufficient quality for their use.

measurement control sample   A sample of known accuracy, prepared to mimic the analyte and matrix, as closely as possible, of unknown samples. A control sample used for measurement must be traceable to the national standard base or be well-characterized by approved methods. The performance of a measurement system is evaluated by comparing the measured value of the control sample to its known value.

measuring and test equipment (M&TE)   Devices or systems used to calibrate, measure, gauge, test, inspect, control, or acquire data to verify conformance to specifications and/or requirements.

mechanical advantage   The ratio of the output to input in a machine.

meclizine hydrochloride   An antihistamine.

MED   Minimum erythemal dose.

MEDLARS®   Medical Literature Analysis and Retrieval System.

median   A term, often used in statistics, to denote the middle, mid-, or half-way point or location in a series of values.

Medical Literature Analysis and Retrieval System (MEDLARS®)   The computerized databases offered by the National Library of Medicine.

medical monitoring   The use, measurement, observation, and detection by physical examination and use of biological or chemical indicators or parameters by medical professionals to determine whether there has been exposure to a disease or otherwise undesired agent or act.

medical screening   The performance of medical tests to attempt to identify symptoms of exposure to undesired events or agents.

medical surveillance   Participation in a medical program or system designed to screen employees for exposure to undesired effects, events, or agents.

medical testing   Biological and chemical medical tests and determinations conducted to determine whether the subject's biological and chemical parameters are outside the norm.

medical waste   Items defined by the US EPA as associated with bloodborne pathogens.

medium   A substance used to grow microorganisms, often in petri dishes.

MEDLINE   MEDLARS on-line. i.e., on the Internet.

medulla   Marrow; the inner or central portion of an organ.

mega- (M)   A prefix $= 10^6$; multiplies a basic unit by 1,000,000; large.

meiosis   The division of chromosome pairs as the germ cell matures.

MEK   Methyl ethyl ketone.

mel-, melo-   A prefix meaning cheek or extremity.

melanin   A dark pigment, found in skin, hair, etc.

melanoma   A type of skin tumor or mole, often malignant, composed of melanin-pigmented cells.

melting point   The temperature at which the crystals of a pure substance are in equilibrium with its liquid phase; also known as the freezing point.

membrane   The thin layer of tissue covering a surface, lining an organ, or dividing an organ.

mendelevium (Md)   A transuranic element, atomic number 101, atomic weight 258.

Mendeleyev, Dmitri Ivanovich   Siberian chemist (1834–1907) who in 1869 published the principle of periodicity among the elements.

meniscus   The curve of a liquid at the container interface, caused by surface tension.

menthol $CH_3C_6H_9(C_3H_7)OH$; A white, cool-tasting compound used as a fragrance and flavoring in food and tobacco.

mercaptan   *See* -thiol.

mercury (Hg)   A group IIB metallic element, atomic number 80, atomic weight 200.59; a silver-colored, dense liquid at room temperature; somewhat volatile; the vapors are extremely toxic by inhalation or skin absorption. Spills are hazardous and require special clean-up procedures.

mercury fulminate   A highly toxic, gray solid compound that explodes readily when dry; it is commonly used in the manufacture of detonators and blasting caps.

Merrifield, Robert Bruce   American chemist (1921–) awarded the 1984 Nobel Prize in Chemistry for his work with peptide and protein syntheses.

mescaline   A highly toxic, hallucinogenic, solid alkaloid derived from a cactus plant.

mesh   The number of openings per unit area of a screen or sieve.

meso-   A prefix meaning middle or intermediate; an optically inactive isomer composed of equal parts of dextro- and levorotational isomers, thereby canceling the optical activity.

meson   An unstable subatomic particle that can be produced artificially and weighs more than an electron but less than a proton.

mesothelioma   A squamous cell tumor, usually malignant, of the epithelium tissue of the membranes that constitute the lining of the chest, lungs, and abdomen.

meta-   A prefix, originally meaning "beyond"; used to denote skipping a position from a functional group in aromatic compounds.

metabolism   The physical and chemical changes that occur in a living organism to produce and maintain the energy used by the organism.

metabolite   Any substance produced or used by the metabolic process.

metal   A hard, often shiny material that conducts heat and electricity; an element whose compounds form positive ions in solution or whose oxides form hydroxides with water.

Metal Casting Society (MCS)   455 State St., Des Plaines IL 60016.

metal fume fever   Flulike symptoms caused by the inhalation of fumes from heated metals or their oxides.

metastable   Describes an unstable or transient but significant chemical or physical state.

metastasis   The spread of disease from one organ to another; often referring to the spread of malignancy from the site of primary cancer to secondary sites.

metathesis   Double decomposition.

meter   The basic unit of length in the SI unit system; 1,650,763.73 wavelengths of the orange-red line of the krypton isotope; 39.37 inches; a device to measure a particular parameter.

methane CH_4; an odorless, tasteless, colorless, naturally occurring gas that is lighter than air. It is a severe fire and explosion hazard. *See* marsh gas.

methanol   *See* methyl alcohol.

methemoglobinemia   The presence of hemoglobin in the oxidized state in the blood.

methyl alcohol CH_3OH; a highly flammable, colorless liquid that causes blindness and is very toxic by ingestion.

methylene chloride CH_2Cl_2; a colorless, volatile toxic liquid with etherlike odor; used in paint removers, solvents, etc.; a narcotic.

methyl ethyl ketone (MEK)   A flammable, colorless liquid, toxic by inhalation, with an odor like acetone; used as a solvent.

methyl violet   A green powder used as an indicator and topical bactericide.

MeV   Million electron volts.

mf   Medium frequency.

MFD   Minimum fatal dose.

mfp   Mean free path.

Mg   Magnesium.

mg   Milligram.

mg/kg   Milligrams of substance administered per kilogram of body weight; a measure of dose.

mg/m³   Milligrams per cubic meter; the unit used to measure air contamination.

mH   Millihenry, the meter-kilogram-second unit used to measure induction; named after the nineteenth century American physicist Joseph Henry.

mho   A unit of conductivity; reciprocal ohm.

MHW   Mixed hazardous waste.

mica   A group of several silicates of varying composition that have similar physical properties and crystalline structure. They are irritants by inhalation and may cause lung damage.

micelle   An electrically charged colloidal aggregate of large (usually organic) molecules.

Michel, Hartmut   German chemist (1948–) who shared the 1988 Nobel Prize in Chemistry for work on the structure of certain proteins essential for photosynthesis.

micro-   A prefix that divides a basic unit into one million (10^{-6}) parts; very small.

microbar   A unit for measurement of pressure used in acoustics = one dyne/cm².

microbe   A microscopic living organism, e.g., bacteria, fungi, etc.

microcurie (μCi)   One-millionth of a curie.

microencapsulation   Enclosure of a material in hollow spheres or capsules in the micrometer size range.

microfilm   A film containing a photographic record of information on a reduced scale; used to compact data and information.

micrometer   One-millionth (10^{-6}) of a meter; 10,000 Angstrom; also called micron.

Micron (μ)   *See* micrometer.

micronucleus   The smaller nucleus in protozoans containing the inheritable germ substance.

micronutrient   Essential substance required in small amounts for the organism; a trace element.

microorganism   An organism that can only be viewed using a microscope, e.g., bacteria, fungi.

microscope   An optical instrument in which lenses are used to enlarge images of minute objects. Used in chemistry to study physical structure and to identify materials. Commonly used in legal, forensic, or police chemistry.

microwave   Electromagnetic radiation in the 1 mm to 1 m range; between infrared and short-wave radio wavelengths.

Midgley, Thomas Jr.   American chemist (1889–1944) noted for his work with synthetic rubber and antiknock gasoline.

MIG   Metal inert gas.

mil   1/1000 inch; used to measure wire, cables, coatings, thickness.

milk of magnesia   A white suspension of magnesium hydroxide in water used medically as a laxative.

mill   A specific type of mechanical process, e.g., ball mill; a cost factor equal to 1/1000 of a dollar.

mill tailings   The residue from mineral processing.

milli- (m)   A prefix $= 10^{-3}$; divides a basic unit by 1000.

millimeter of mercury (mmHg)   A unit used to denote pressure (or vacuum) by measuring the height in millimeters of a column of mercury in a tube.

Mine Safety and Health Administration (MSHA)   US DOL, 4015 Wilson Blvd., Arlington VA 22203. The part of the Department of Labor responsible for safety in mines and related industries.

Mine Safety and Health Review Commission (MSHRC)   1730 K St., NW, Washington DC 20006.

mineral spirits   A petroleum distillate often used as a solvent for paint and varnish; a grade of naphtha.

minimum detectable level   The smallest concentration of a given substance that can be reproducibly determined.

minimum erythemal dose (MED)   The smallest dose of ultraviolet radiation to cause reddening of the skin.

miosis   Unusual contraction or smallness of the pupil of the eye.

MIPS   Millions of instructions per second.

miscella   A solution or mixture containing oil or grease.

miscibility   The ability of a gas or a liquid to completely and uniformly dissolve in another gas or liquid.

mist   Liquid droplets suspended in air that are condensates from gases or liquids dispersed by splashing or atomization.

Mitchell, Peter Dennis   English biochemist (1920–) awarded the 1978 Nobel Prize in Chemistry for work in cellular energy transfer.

miticide   A class of pesticide specifically for killing mites and small animals of the spider class.

mitigation   Measures taken to reduce the effect or impact.

mitochondria   Particles of cytoplasm; slender microscopic filaments or rods.

mitomycin   A group of antitumor antibiotics (i.e., A, B, C) derived from *Streptomyces caespitosus.*

mitosis   The indirect division occurring in the nucleus of a cell that usually produces two identical new nuclei.

mixture   A blend of two or more substances that are not chemically combined and may be separated mechanically. *See* solution.

ML (ml, ML)   Milliliter.

MLD   Median lethal dose.

mmHg   Millimeter of mercury (torr).

Mn   Manganese.

Mo   Molybdenum.

MOA   Memorandum of agreement.

mobile   Capable of movement, i.e., freely flowing.

mode   In statistics, the value that occurs most frequently.

modeling   Use of a physical or mathematical representation to depict an event or structure.

moderator   A substance that modifies or controls a reaction.

Mohr, Karl Friedrich   German pharmacist (1806–1879) known for his work in volumetric analysis.

Mohr's test   A chemical test to determine whether hydrochloric acid is present in stomach contents; named after the American pharmaceutical chemist Francis Mohr.

Mohs, Friedrich   German mineralogist (1773–1839) who developed the scale for measuring hardness.

Mohs scale   A scale used to measure hardness of minerals (talc has a value of 1, diamond has a value of 10); named after the German mineralogist Friedrich Mohs.

moiety   An unspecified portion of a sample; also one of the portions into which something is divided.

Moisson, Henri   French chemist (1852–1907) awarded the 1906 Nobel Prize in Chemistry for first isolating fluorine.

mol. wt   Molecular weight.

molal (m)   A concentration in which moles of solute are given per kilogram of solvent.

molar (M)   The number of moles of solute per liter of solution.

mold   A fungus that produces a surface growth on damp, decaying organic material or on a living organism.

mole   A quantity of a pure substance whose weight in grams is equal to the formula weight of the compound.

molecular biology   The subspeciality of biology that concerns physico-chemical interactions at the molecular level.

molecular sieve   A microporous structure of the zeolite class of minerals with pore size ranging from 5 to 10 angstroms. Known for their ability to undergo dehydration. Dehydrated materials have a strong affinity to recapture water; used as drying agents for gases and liquids.

molecular weight (MW)   The weight of a compound equal to the sum of the individual atoms that compose the chemical formula of the compound. *See* formula weight.

molecule   The smallest unit of a compound that retains all the properties of the compound.

Molina, Mario   Mexican-born chemist (1943–) who shared the 1995 Nobel Prize in Chemistry for work in atmospheric chemistry.

molybdenum (Mo)   A hard, gray group VIB metallic element, atomic number 42, atomic weight 95.94; flammable if powdered.

moment   Force times the distance of application.

momentum   Mass times velocity.

monazite   A mineral source for rare earth elements, e.g., thorium and cerium.

Monel™   A registered trade mark used to denote corrosion resistant alloys of nickel and copper.

monitor   To check, track, regulate, review, observe, oversee, measure, evaluate, or control all or part of the various aspects of an operation, process, or piece of equipment. In radiation safety; the periodic or continuous determination of the amount of ionizing radiation or radioactive contamination present in an occupied region. Used for safety purposes of health protection or contamination control. *See* radiological survey.

monitoring wells   Holes bored into the ground; used for taking samples for analysis of runoff and underground distribution from waste sites.

mono-   A prefix meaning one, single, or lone.

monobasic   Acids with one displaceable hydrogen atom per molecule.

monochromatic   Composed of a single color or wavelength.

monomer   A compound with simple structures or low molecular weight that is capable of polymerization.

monosodium glutamate (MSG) $C_5H_8NNaO_4 \cdot H_2O$; a white crystalline compound with meaty taste; used in food flavoring, especially Asian food; in large doses can cause headaches, chest pain, etc.

Monte Carlo analysis   The repeated random sampling from a group of values to give an estimate of the distribution.

Moore, Stanford   American biochemist (1913–1982) who shared the 1972 Nobel Prize in Chemistry for enzyme work.

morbidity rate   The rate at which people in a defined population become ill from a particular disease during a specified time period.

mordant   A substance used to bind dyes to textiles.

morphine   A white, crystalline alkaloid, derived from opium, used as an analgesic.

morphology   A branch of biology concerned with the form and structure of animals, plants, and other organisms.

MORT   Management oversight risk tree.

mortality rate   The rate at which people in a defined population die from a particular disease during a specified time period.

Moseley, Henry   English chemist (1887–1915); killed in World War I; noted for his application of X-ray spectra to more accurately position elements in the periodic table.

Moseley's law   States that the square root of the frequency of an element's X-ray spectral line is directly proportional to its atomic number.

mottling   Colored spots.

MOU   Memorandum of understanding.

mp, MP   Melting point.

MPA   Maximum probable accident; manufacturing project approval.

MPC   Maximum permissible concentration; maintenance publication coordinator.

MPD   Maximum permissible dose.

MPE   Maximum permissible exposure.

MPH   Master of Public Health.

mppcf   Million particles per cubic foot.

MRI   Magnetic resonance instrument; magnetic resonance imaging.

MS   Mass spectrometry; mass spectrometer.

MSA   Mine safety appliance; Management Science America (computer system).

MS&C   Material scheduling and control.

MSDS   Material Safety Data Sheet (OSHA Form 174).

MSE   Molten salt extraction.

MSG   Monosodium glutamate.

MSHA   US Mine Safety and Health Administration.

MSHRC   Mine Safety and Health Review Commission.

MSK   Musculoskeletal effects.

MSST   Maximum safe storage temperature.

MTBE   Methyl-*tert*-butyl ether.

MTBF   Mean time between failures (commonly used in statistics).

MTCE   Maintenance.

MTD   Maximum tolerated dose; manufacturing technology development.

mucous membrane   The lining around a hollow area of the body such as the nose, mouth, stomach.

MUF   Material unaccounted for.

muffle furnace   An oven, kiln, or furnace that heats materials without direct contact with the heat source.

Müller, Paul Hermann   Swiss chemist (1899–1965) awarded the 1948 Nobel Prize in Medicine or Physiology for showing the usefulness of DDT as an insecticide.

Mulliken, Robert Sanderson   American scientist (1896–1986) awarded the 1966 Nobel Prize in Chemistry for work with isotope separations.

multiple chemical sensitivity (MCS)   The condition of sensitivity to a number of chemicals or substances, perceived to occur as a result of a significant exposure to something affecting the immune system.

muriatic acid   Hydrochloric acid.

mustard gas   Dichlorodiethyl sulfide, used as poison gas in World War I.

mutagen   A substance capable of reacting with genes and chromosomes to produce a change (mutation) in the genetic material of a living cell.

MW   Molecular weight.

MWe   Electrical megawatt

MWhr   Megawatt-hours.

Mx   Maxwell; centimeter-gram-second unit of electromagnetic flux; named after James Clerk Maxwell, a nineteenth century Scottish physicist (1831–1879).

my-, myo-   A prefix meaning pertaining to the muscle.

myasthenia   Muscular weakness.

mycotoxin   A highly toxic substance produced by molds and fungi, e.g., aflatoxin, fumonisin.

myelin   A sheath-type substance enclosing major groups of nerves.

Mylar™   DuPont polyester film.

myoglobin   A protein molecule, similar to hemoglobin, containing iron and porphyrin.

myosin   An essential part of muscle tissue; protein with molecular weight $\sim 500,000$.

myrrh   A gum resin whose pleasant aroma has led to use in perfumes.

N

ν Lower case Greek letter nu; frequency.

n   Neutron; nano- (10^{-9}).

N   Nitrogen; Avogadro's number (the number of molecules of a gas contained in 22.4 liters, i.e., 6.0×10^{23}); normal solution; uppercase Greek letter nu ; newton; north.

Na   *Natrium* (Latin); sodium.

NA   Not applicable; not available; nonattainment.

NAAQS   National Ambient Air Quality Standards.

NAC   National Audiovisual Center.

NAE   National Academy of Engineering.

NAFI   National Association of Fire Investigators.

NaK   Any sodium potassium alloy; liquid alloy over a useful range of temperature.

Nalgene™   Plastic material often used in laboratories.

nano- (n)   A prefix = 10^{-9}; one-billionth of the noun, e.g., nanogram, nanosecond, etc.

NAP   National Academy Press.

napalm   A flammable, incendiary agent similar to soap mixed with gasoline; sometimes called jellied gasoline.

naphtha   A general term for certain flammable liquid petroleum products; petroleum benzin.

narcosis   A stupor or unconsciousness produced by various chemical substances.

narcotic   A chemical substance that affects the central nervous system.

NAS   National Academy of Sciences.

NASA   US National Aeronautics and Space Administration.

nascent   In the process of being formed in a chemical or biological reaction.

National Academy of Engineering (NAE)   2101 Constitution Ave., Washington DC 20418.

National Academy Press (NAP)   2101 Constitution Ave., NW, Washington DC 20418.

National Academy of Sciences (NAS)   2101 Constitution Ave., NW, Washington DC 20418.

National Aeronautics and Space Administration (NASA)   600 Independence Ave., SW, Washington DC 20546.

National Association of Fire Investigators (NAFI)   PO Box 957257, Hoffman Estates IL 60195.

National Audiovisual Center (NAC)   8700 Edgeworth Dr., Capitol Heights MD 20743.

National Board of Boiler & Pressure Vessel Inspectors (NBBPVI)   1055 Crupper Ave., Columbus OH 43229.

National Bureau of Standards (NBS)   The former name of what is now called the National Institute of Standards and Technology.

National Cancer Institute (NCI)   9000 Rockville Pike, Bethesda MD 20892; One of the Institutes of the National Institutes of Health; the primary center for US government cancer research.

National Center for Devices and Radiological Health (NCDRH)   5600 Fishers Ln., Rockville MD 20857.

National Center for Toxicological Research (NTCR)   Highway 365, Jefferson AR 72079; The Food and Drug Administration's (FDA) center for conducting toxicology research.

National Clearinghouse for Alcohol and Drug Information   PO Box 2345, Rockville MD 20847.

National Contingency Plan (NCP)   The EPA plan required under CERCLA to respond to unplanned releases of hazardous substances and pollution contaminants into the environment.

National Council on Compensation Insurance (NCCI)   200 East 42nd St., New York NY 10017.

National Council on Radiation Protection and Measurements (NCRP)   7910 Woodmont Ave., Bethesda MD 20814.

National Electrical Code (NEC)   The standards and guidelines for electrical safety.

National Electrical Manufacturers Association (NEMA)   210 L St., Washington DC 20037.

National Energy Software Center (NESC)   US EPA, Enterprise Technology Services Division, MD 34, Research Triangle Park NC 27711.

National Environmental Health Association (NEHA)   720 South Colorado Blvd., Denver CO 80222.

National Fire Protection Association (NFPA)   1 Batterymarch Park, PO Box 9101, Quincy MA 02269-9101.

National Fire Sprinkler Association (NFSA)   PO Box 1000, Robin Hill Corporation Pk., Patterson NY 12563.

National Formulary (NF)   A pharmaceutical reference work of formulations.

National Highway Traffic Safety Administration (NHTSA)   DOT, 400 7th St., SW, Washington, DC 20590.

National Institute of Environmental Health Sciences (NIEHS)   PO Box 12233, Research Triangle Park NC 27709.

National Institute for Farm Safety (NIFS)   2601 Rose Ct., Columbia MO 65202.

National Institute of Occupational Safety and Health (NIOSH)   1600 Clifton Ave., NE, Atlanta GA3033; An institute in the US Public Health Service of the Department of Health and Human Services of the Communicable Disease Centers that deals with research and recommendations relating to safety and health issues; also responsible for training occupational health and safety professionals and certifying respirators; also a component of the NTP.

National Institute of Standards and Technology (NIST)   Formerly the National Bureau of Standards (NBS)   Department of Commerce, Quince Orchard Rd., Gaithersburg MD 20899.

National Institutes of Health (NIH)   9000 Rockville Pike, Bethesda MD 80222; The Institutes of the US Public Health Service of the Department of Health and Human Services that conduct and coordinate research on public health-related issues. NIH is comprised of several separate Institutes, each dedicated to the study of specialized areas of medical health.

National Labor Management Association (NLMA)   PO Box 819, Jamestown NY 14702-0819.

National Library of Medicine (NLM)   8600 Rockville Pike, Bethesda MD 20209; the information branch of the National Institutes of Health (NIH).

National Mine, Safety and Health Academy   PO Box 1166, Beckley WV 25801.

National Oceanic and Atmospheric Administration (NOAA)   600 Independence Ave., SW, Washington DC 20546.

National Paint & Coatings Association (NPCA)   1500 Rhode Island Ave., NW, Washington, DC 20005.

National Pesticide Information Retrieval System   Purdue University, Entomology Hall, West Lafayette IN 47907.

National Petroleum Engineers Association (NPEA)   1899 L St., NW, Washington DC 20036.

National Petroleum Refiners Association (NPRA)   1899 L St., NW, Washington DC 20036.

National Propane Gas Association (NPGA)   1600 Eisenhower Ln., Lisle IL 60532.

National Registry of Certified Chemists (NRCC)   815 Fifteenth St., NW, #508, Washington DC 20005.

National Renewable Energy Laboratory (NREL)   1617 Cole Blvd., Golden CO 80401-3393.

National Research Council (NRC)   2101 Constitution Ave., Washington DC 20418; research organization that consists of the National Academy of Sciences (NAS), the National Academy of Engineering (NAE), and the Institute of Medicine (IOM).

National Response Center (NRC)   The US Coast Guard center for reporting pollution incidents; 2100 Second Ave., SW, Washington DC 20593.

National Safety Council (NSC)   1121 Spring Lake Dr., Itasca IL 60143-3201.

National Safety Management Association (NSMA)   3871 Piedmont Ave., Oakland CA 94611.

National Sanitation Foundation (NSF)   PO Box 130140, Ann Arbor MI 48113.

National Science Foundation (NSF)   1800 G St., NW, Washington DC 20550.

National Slag Association (NSA)   110 W. Lancaster Ave., Wayne PA 19087-4043.

National Society for the Prevention of Blindness (NSPB)   500 East Remington Rd., Schaumberg IL 60173.

National Society of Professional Engineers (NSPE)   1420 King Ave., Alexandria VA 22314.

National Solid Waste Management Association (NSWMA)   1730 Rhode Island Ave., NW, Washington DC 20036.

National Standards Association (NSA)   5161 River Rd., Bethesda MD 20816.

National Technical Information Service (NTIS)   5285 Port Royal Rd., Springfield VA 22161.

National Toxicology Program (NTP)   PO Box 12233, Research Triangle Park NC 27709; a cooperative effort within the US Public Health Service of the Department of Health and Human Services to coordinate toxicology research and testing activities within the Department. Composed of the National Center for Toxicology Research, the National Institute of Environmental Health Sciences, and the National Institute of Occupational Safety and Health.

National Transportation Safety Board (NTSB)   DOT, 490 L'Enfant Plaza, SW, Washington DC 20594.

natrium   Latin name for sodium.

Natta, Giulio   Italian chemist (1903–1979) who shared the 1963 Nobel Prize in Chemistry for work on catalytic polymerization.

natural gas   A combustible mixture of naturally occurring gases and hydrocarbons including methane, hydrogen sulfide, and carbon dioxide.

natural radiation   *See* background radiation.

natural uranium   Uranium as found in nature, which contains 0.7% uranium-235, 99.3% uranium-238, and a trace of uranium-234.

nausea   An unpleasant sensation in the gastrointestinal system that may result in vomiting.

n.b.   Latin for *nota bene*, "note well."

Nb   Niobium.

NBBPVI   National Board of Boiler & Pressure Vessel Inspectors.

NBC   National Building Code.

NBFU   National Board of Fire Underwriters.

NBS   National Bureau of Standards; replaced by NIST. *See* NIST.

NCADI   National Clearinghouse for Alcohol and Drug Information.

NCCI   National Council on Compensation Insurance.

NCDRH   National Center for Devices and Radiological Health.

NCI   National Cancer Institute.

NCP   National Contingency Plan.

NCR   Nonconformance report.

NCRP   National Council on Radiation Protection and Measurements.

NCTR   National Center for Toxicological Research.

Nd   Neodymium.

NDA   Nondestructive assay.

NDT   Nondestructive testing.

Ne   Neon.

nebulize   To form an aerosol.

NEC   National Electrical Code.

necro-   Prefix meaning dead or death.

necropsy   The examination of a body after death.

necrosis   Destruction and death of tissue.

negative pressure check   A qualitative respirator fit check whereby the wearer covers the filter openings, inhales, listens and feels for inward air leakage around the facepiece seal, and holds the test for 5–10 seconds.

negligence   A legal term for not being prudent, using proper standard care and/or good judgment.

NEHA   National Environmental Health Association.

NEMA   National Electrical Manufacturers Association.

nematic   A linear molecular structure: substances used in some liquid crystals.

neo-   A prefix meaning new.

neodymium (Nd)   A lanthanide group IIIB metallic element, atomic number 60, atomic weight 144.24; used in lasers and electronics.

neon (Ne)   A colorless, odorless, tasteless, nonflammable noble gas element; atomic number 10, atomic weight 20.18; used in electric signs and lasers.

neoplasm   Abnormal cell growth or tumor.

neoprene   Polychloroprene; used in rubber products.

Neosporin™   An antibiotic.

NEPA   National Environmental Policy Act.

nephr-, nephro-   A prefix meaning pertaining to the kidney.

nephritis   Inflammation of the kidneys.

nephrotoxin   An agent that induces kidney damage.

neptunium (Np)   A radioactive transuranic metallic element, atomic number 93, atomic weight 237.05; used in neutron detection instruments.

Nernst, Walther Hermann   German chemist (1864–1941) awarded the 1920 Nobel Prize in Chemistry for developing the third law of thermodynamics.

nerve gas   One of several colorless, odorless, tasteless, highly toxic, rapidly absorbed agents developed for chemical warfare; cholinesterase inhibitors that therefore interfere with the nervous system.

NESC   National Electrical Safety Code; National Energy Software Center.

NESHAP   National Emission Standard for Hazardous Air Pollutants.

Nessler tube   A type of tube used to do color analysis and comparison.

neuritis   Inflammation of a nerve; can cause pain, reflex loss, and atrophy.

neuro-   A prefix meaning pertaining to the nerves.

neurotoxin   An agent that induces damage to the nervous system.

neutraceutical   An additive to food made for health benefit(s).

neutralization   A chemical reaction between an acid and a base in which the characteristic properties of each disappear. Neutralization may occur with organic as well as inorganic chemicals. Also means the equivalence point for an acid-base reaction.

neutrino   An energized atomic particle without mass or charge that results from the collision of two atomic particles.

Neutron (n)   An atomic particle found in the nucleus that has a slightly greater mass than a proton but has no electrical charge.

neutron activation analysis (NAA)   An analytical technique for detecting very low concentrations of elements, usually metals, by subjecting the sample to a high neutron flux and subsequently measuring the induced radioactivity.

neutron capture   The process in which an atomic nucleus absorbs a neutron.

neutron chain reaction   A process in which some neutrons released in one fission event cause other fissions to occur.

newton (N)   A unit of force equal to the acceleration of a one-kilogram mass accelerated at one meter per second per second; named after the English philosopher and mathematician Sir Isaac Newton.

Newton, Sir Isaac   English physicist and mathematician (1642–1727) who postulated the law of gravity, studied light, and invented differential calculus.

NF   National Formulary.

NFC   National Fire Code.

NFPA   National Fire Protection Association; National Fluid Power Association.

NFSA   National Fire Sprinkler Association.

NHTSA   National Highway Traffic Safety Administration.

Ni   Nickel.

niacin   An antipellagra vitamin; lowers cholesterol level.

nickel (Ni)   A silvery group VIII transition metal element, atomic number

28, atomic weight 58.70; used in jewelry and alloys; flammable and toxic if powdered.

nickel carbonyl $Ni(CO)_4$; a highly toxic, carcinogenic, irritating, flammable, explosive gas.

nicotine $C_5H_4NC_4H_7NCH_3$; a toxic, unstable alkaloid derived from tobacco.

NIEHS   National Institute of Environmental Health Sciences.

NIFS   National Institute for Farm Safety.

NIH   National Institutes of Health.

NIMBY   Not in my backyard.

ninhydrin $C_9H_4O_3 \cdot H_2O$; a white crystalline compound used to determine amino acids and amines.

niobium (Nb)   A gray or silvery group VB metallic element, atomic number 41, atomic weight 92.91; used in rocket fuels and superconductors; also called columbium by mineralogists.

NIOSH   National Institute for Occupational Safety and Health.

NIST   National Institute of Standards and Technology.

nitrile   Any compound containing the $—C\!\!=\!\!N$ group.

nitrogen (N)   A gaseous group VA element, atomic number 7, atomic weight 14.01; N_2 is the most abundant gas in air.

nitrogen mustard   A class of fishy smelling, irritating, chemical lachrymators, similar to mustard gas and used in medicine.

nitrogen narcosis   Euphoria, decreased coordination and motor ability resulting from high-air-pressure breathing such as for divers. Nitrogen concentrates in the body fluids and intoxicates the individual.

nitrogen-phosphorus detector   A specific type of detector used in gas chromatography for the vaporizable detection of phosphorus- and nitrogen-containing chemicals.

nitrogenase   An enzyme used for nitrogen fixation.

nitroglycerin   A pale-yellow, heat- and shock-sensitive, viscous liquid; a major substance used as an explosive and in explosive materials and as a vasodilator used to treat angina pectoris.

nitromethane CH_3NO_2; a colorless, toxic, highly flammable liquid used as a racing fuel and gasoline additive.

nitron $C_{20}H_{16}N_4$; a reagent for nitrate and perchlorate determinations.

nitrosamine   The series of organic compounds with the $=N-N=O$ group; strong carcinogens in animals.

nitrous oxide N_2O; a colorless gas used as an anesthetic; laughing gas.

NLM   National Library of Medicine.

NLMA   National Labor Management Association; National Lumber Manufacturers Association.

nm   Nanometer, a unit of measure equal to 10^{-9} meters.

NMR   Nuclear magnetic resonance; now called MRI in medical uses.

NMSHA   National Mine Safety and Health Academy.

No   Nobelium; number; north.

NO   Normally open (electrical contact or mechanical position); nitric oxide.

NOAA   National Oceanic and Atmospheric Administration.

NOAEL   No observed adverse effect level.

Nobel, Alfred Bernhard   Swedish chemist (1833–1896) who invented a manageable form of nitroglycerin called dynamite and smokeless gunpowder; founded the Nobel Prizes in Physics, Chemistry, Medicine or Physiology, Literature, and Peace (Economics was added later).

nobelium (No)   A radioactive transuranic element, atomic number 102, atomic weight 251–259, depending on the isotope.

noble   A term meaning generally nonreactive.

noble-gas   Elements displaying chemical stability characterized by a filled outer shell of electrons; also known formerly (and incorrectly) as inert gases.

NOC   Not otherwise classified.

node   A small mass of tissue from which other growth may occur; the intersection of two points or lines.

nodule   A small mass of tissue that can usually be detected by touch.

NOEL   No observable effect level.

noise   Unwanted and/or interfering sounds or signals.

noise-induced hearing loss   Slow, progressive inner ear hearing loss resulting from continuous noise exposure over a long time period.

noncombustible   Will not burn.

noncompliance   The state of not being in compliance with regulations and/or laws.

nonconformance   A deficiency in characteristic, documentation, or procedure that renders the quality of an item or activity unacceptable or indeterminate; examples include physical defects, test failures, unacceptable documentation, or deviation from prescribed specifications, drawings, processing, inspection, or test procedures.

nondestructive measurement/assay (NDA)   A process involving no chemical or physical change in the material involved.

nonferrous metal   Describes a metal without iron.

nonflammable   Describes a liquid with a flash point above 100°F; will not easily burn.

nonfriable asbestos   Asbestos-containing material that cannot be crumbled by hand. Examples include transite, vinyl asbestos floor tile, or premolded asbestos pipe insulation in good repair. These materials are hazardous only when made friable by cutting or sanding.

nonionizing radiation   Electromagnetic radiation, such as ultraviolet, infrared, microwave, and radio frequency, that does not cause ionization.

nonpolar   Describes a substance with no permanent electric moment and characterized by a low dielectric constant. Such substances do not ionize or only weakly ionize in solution, e.g., aliphatic and aromatic hydrocarbons.

nonrecoverable residues   Materials equal to or less than the economic discard limit (i.e., waste).

nonvolatile   Will not evaporate under ordinary atmospheric conditions.

No observable adverse effect level (NOAEL)   The highest observed concentration at which a chemical does not cause a noticeable adverse effect compared to control animals.

No observable effect level (NOEL)   The highest observed concentration at which a chemical does not cause a noticeable effect compared to control animals.

normal conditions   Conditions with a temperature of 25°C, pressure of 1 atmosphere.

normal distribution   A statistically symmetrical distribution of values or measurements that approximate a bell-shaped curve.

normal temperature and pressure (NTP)   25°C and 1 atmosphere pressure.

Northrop, John Howard   American chemist (1891–1987) who shared the 1946 Nobel Prize in Chemistry for isolation and crystallization of enzymes.

NOS   Not otherwise specified; a term frequently used in shipping but also applied elsewhere.

Not classifiable as to its carcinogenicity to humans   A definition used by IARC; designated as Group 3, as follows: Agents are placed in this category when they do not fall into any other group.

Not classified as a human carcinogen   A definition used by the ACGIH, designated as A4, as follows: There are inadequate data on which to classify the agent in terms of its carcinogenicity in humans and/or animals.

notice of intended change   A term used by the ACGIH for substances for which a change is anticipated one year after notice is first listed for change.

Not suspected as a human carcinogen   A definition used by the ACGIH, designated as A5, as follows: The agent is not suspected to be a human carcinogen on the basis of properly conducted epidemiologic studies in humans. These studies have sufficiently long follow-up, reliable exposure histories, sufficiently high dose, and adequate statistical power to conclude that exposure to the agent does not convey a significant risk of cancer to humans. Evidence suggesting a lack of carcinogenicity in experimental animals will be considered if it is supported by other relevant data.

Novocain™   A widely used anesthetic.

NO_x   Nitrogen oxides, gaseous products of combustion responsible for the brown color in air pollution; irritants to mucous membranes.

noxious   Hazardous to health.

Np   Neptunium.

NPAA   Noise Pollution and Abatement Act.

NPCA   National Paint & Coatings Association.

NPD   Nitrogen phosphorus detector.

NPDES   National Pollutant Discharge Elimination System.

NPDWS   National Primary Drinking Water Standards.

NPEA   National Petroleum Engineers Association.

NPGA   National Propane Gas Association.

NPL   National Priorities List.

NPRA   National Petroleum Refiners Association.

NPRM   Notice of proposed rulemaking.

NQR   Nuclear quadrupole resonance; a type of NMR that uses natural electric fields around crystals instead of magnets.

NRC   Nuclear Regulatory Commission; National Response Center; National Research Council.

NRCC   National Registry of Certified Chemists.

NRT   National Response Team.

NSA   National Standards Association; National Slag Association.

NSC   National Safety Council.

NSF   National Science Foundation; National Sanitation Foundation.

NSMA   National Safety Management Association.

NSPB   National Society for the Prevention of Blindness.

NSPE   National Society of Professional Engineers.

NSWMA   National Solid Waste Management Association.

NTIS   National Technical Information Service.

NTP   National Toxicology Program; normal temperature and pressure.

NTS   Not-to-scale; Nevada Test Site.

NTSB   National Transportation Safety Board.

nuclear disintegration   *See* decay, radioactive.

nuclear energy   The energy liberated by a nuclear fission or fusion reaction or by radioactive decay.

nuclear force   A powerful short-ranged force that holds together the particles inside an atomic nucleus.

nuclear magnetic resonance (NMR)   An analytical chemistry technique using high-field electromagnets; useful for determining chemical structures; also known as magnetic resonance imaging (MRI) when applied as an aid to diagnosing human medical conditions.

nuclear reaction   Any change effected in the nucleus of an atom.

nuclear reactor   A device to sustain a slow fission reaction and produce heat; used to generate electricity.

Nuclear Regulatory Commission (NRC)   One White Flint North, 11555 Rockville Pike, Rockville MD 20852-2738.

nucleon   Common name for a constituent particle of the atomic nucleus; applied to protons and neutrons but may include any other particles found to exist in the nucleus.

nucleoside   A purine or pyridine base linked to ribose or deoxyribose that lacks the phosphate residue.

nucleotide   Phosphate esters of nucleosides; the precursors of nucleic acids are monoesters, but di- and triphosphates also are known.

nucleus (atomic nucleus); nuclei (plural)   The small, central, positively charged atomic region that carries essentially all the mass. Except for the hydrogen nucleus, which has a single proton, all atomic nuclei contain both protons and neutrons. The number of protons determines the total positive charge or atomic number, which is the same for all the atomic nuclei for the same chemical element. The total number of neutrons and protons is called the mass number. *See* isotope.

nuclide   A general term referring to all known isotopes, both stable (279) and unstable (about 5000), of the chemical elements.

nuisance dust   A particulate that does not induce toxic or adverse health effects or disease in the lung when exposures are under reasonable control.

Nujol   A specific type of mineral oil used to prepare mulls for infrared analysis.

null point   The distance from the source of contamination from which the initial energy or velocity is dissipated, enabling the material to be captured by a hood.

nutrient   Any compound or element essential for an organism to live; a food.

nycto-   A prefix referring to night or darkness.

nylon   A generic term for many forms of a widely used synthetic family of polymers frequently used in fibers.

nystagmus   Spastic, involuntary motion of the eyeballs.

O

Ω Uppercase Greek letter omega; ohm.

o   *Ortho.*

O   Oxygen.

O$_2$ The oxygen molecule.

O$_3$ Ozone.

O&M   Operation and Maintenance.

Oak Ridge National Laboratory (ORNL)   DOE, PO Box 2008, Oak Ridge TN 37831; The US federal laboratory in Oak Ridge, Tennessee, that came into prominence because of its work on atomic energy during World War II.

OBA   Operating basis accident.

objective evidence   A quantitatively or qualitatively documented fact or record pertaining to the quality of an item or activity, based on verified observations, measurements, or tests.

observation   An opinion regarding a deficient condition, procedure, or specification, not covered by a specific requirement, that can be improved.

OC   On center.

OCAW   Oil, Chemical and Atomic Workers (International Union).

occipital   Pertaining to the back of the head.

occlude   To block, close, cover, or shut.

occupational disease   A malady occurring as a result of working conditions or exposure to undesired agents.

occupational exposure limit (OEL)   An upper value for the amount of exposure to a specific chemical, physical or biological agent that a worker should be subjected to in the workplace; often varies between countries, agencies and organizations.

Occupational Health Nursing (OHN)   The profession of providing specific nursing health care to workers.

occupational medicine   The study of disease and its treatment and prevention in the workplace.

Occupational Safety and Health Act (OSH Act)   The 1970 US law establishing workplace safety and health regulations.

Occupational Safety and Health Administration (OSHA)   US Dept. of Labor, 200 Constitution Ave., NW, Washington DC 20210.

Occupational Safety and Health Review Commission (OSHRC)   1825 K St., NW, Washington DC 20006; An independent group set up to review and evaluate any actions taken by the US OSHA.

occupational therapist   A health care professional trained to administer treatment for the relief and elimination of discomfort or pain as a result of an injury or illness in the workplace.

octane number   The number giving the relative antiknocking properties of commercial gasoline. *n*-heptane is 0 and 2,2,4-trimethylpentane is 100 on this arbitrary scale.

octave   The interval between two sounds with a frequency ratio of two.

octave band   An arbitrary frequency range in which the top value is always twice the bottom value.

ocul-, oculo-   A prefix meaning pertaining to the eye.

OD   Optical density; outside diameter; overdose (usually of a drug).

odor   A characteristic of a substance pertaining to smell.

odorant   A substance with a strong (usually unpleasant) odor added to odor-free material to warn (or charm) people.

odor threshold   The minimum concentration of a substance whose odor can be detected by the average individual.

Oe   oersted; a unit of magnetic intensity in a vacuum equal to one dyne; named after the Danish physicist Hans Christian Oersted.

OEG   Occupational exposure guideline.

OEL   Occupational exposure limit.

OEP   Office of Emergency Preparedness.

Oersted, Hans Christian   Danish physicist (1771–1851) who discovered the magnetic effect of an electric current.

Office of Emergency Preparedness (OEP)   US PHS, Office of Emergency Preparedness, National Disaster Medical System, 12300 Twinbrook Parkway, Suite 360, Rockville MD 20850.

Office of Hazardous Materials Transportation (OHMT)   DOT, 400 7th St., SW, Washington DC 20590.

Office of Research and Development (ORD)   401 M St., SW, Washington DC 20460.

Office of Solid Waste and Emergency Response (OSWER)   US EPA, Ariel Rios (5101), 1200 Pennsylvania Ave., NW, Washington DC 20460.

Office of Technology Assessment (OTA)   600 Pennsylvania Ave., SW, Washington DC 20510.

ohm (Ω)   A unit of electrical resistance equal to that of a conductor in which a current of one ampere is produced by a potential difference of one volt.

Ohm, Georg Simon   German physicist (1787–1854) who did pioneering research in electricity, specifically resistance.

ohmmeter   A device used to measure electrical resistance.

Ohm's law   States that electrical voltage is equal to the current or flow in amperes times the resistance in ohms.

OHMT   Office of Hazardous Materials Transportation.

oil   A generic term for any of a number of animal, mineral, vegetable, or synthetic substances; slippery liquids at room temperature with varying viscosities; generally combustible and insoluble in water, used in lubricants and fuels.

Oil, Chemical and Atomic Workers International Union (OCAW)   255 Union Blvd., Lakewood CO 80228.

oil of vitriol   Sulfuric acid.

ointment   An oily petrolatum semisolid salve used to protect wounds, medicate dermatitis, or act as a skin barrier.

OJT   On-the-job training.

Olah, George Andrew   Hungarian-born American chemist (1927–) awarded the 1994 Nobel Prize in Chemistry for work with carbocations (positively charged hydrocarbon molecules), which led to the development of new fuels and ways to raise the octane number of gasoline.

olefin   A class of unsaturated aliphatic hydrocarbons with one or more double bonds.

oleic acid   A long-chain unsaturated organic fatty acid found in many animal and plant oils.

oleum   The Latin word for oil; sometimes still used to denote fuming sulfuric acid.

olfactory   Pertaining to smell.

olfactory fatigue   The phenomenon in which a certain duration or concentration of an odoriferous substance desensitizes the olfactory sense to the point of nondetection; thus a particularly hazardous exposure for agents that produce the effect, e.g., H_2S.

oligo-   Prefix meaning few or little.

oliguria   Formation of diminished amounts of urine.

OMB   Office of Management and Budget.

On-the-job training (OJT)   Hands-on training conducted and evaluated in the work environment by qualified individuals.

oncogen   Any substance that will cause tumors in animals.

oncology   The study of the cause, development, characteristics, and treatment of tumors.

on-line   Describes a computer system that communicates directly and immediately with files.

Onsager, Lars   Norwegian-born American chemist (1903–1976) awarded the 1968 Nobel Prize in Chemistry for theories on electrolyte conduction and dielectrics.

OP   Order point.

opaque   A nontransparent substance with a high optical density.

op. cit.   Abbreviation for the Latin phrase *opere citato*; "in work cited."

operable   Term describing a component or system capable of performing as intended.

operational readiness review   The review of a proposed facility or process system, conducted before startup, to evaluate the capability of the equipment, personnel, and management control systems used to fulfill the system's functional and safety objectives.

operational requirements   The end results that a project should achieve, including needs for operations, maintenance, safety, security, safeguards, quality assurance, and utility requirements.

operational safety analysis (OSA)   A written safety review that outlines the safety hazards involved in an operation, the controls of the hazards, and the responsible personnel.

operational safety requirements (OSRs)   The requirements that define the conditions, safe boundaries, management, or administrative controls required to assure the safe operations of a facility. The purpose is to ensure that the operational status and safety systems of a facility remain consistent with the assumption and provisions of the safety analysis report. OSRs include safety limits, administrative safety controls, limiting conditions for operation, surveillance requirements, design features, and administrative controls.

ophthalmologist   A physician specializing in diseases of the eye.

opium   A mixture of alkaloids; source of morphine; a narcotic.

Oppenheimer, Julius Robert   Controversial American physicist (1904–1967) who helped develop classical quantum physics, worked on the US Manhattan Project, which made the first atomic bomb; directed the Los Alamos National Laboratory; advisor to the US AEC; awarded the US Enrico Fermi medal.

opportunistic infection   An infection caused by a microorganism that is not normally a problem but becomes one under certain circumstances and conditions.

optical density (OD)   The amount of light transmission through or absorption by a substance.

optical isomer   A chemical isomer that rotates light in different directions; also called stereoisomer, diastereoisomer, and enantiomer.

optical microscope   A microscope that relies on visible light, mirrors, and lenses for its magnification.

optical rotation   A change in the direction or rotation of light as it passes through an optically active chemical substance, i.e., one with an asymmetric carbon atom.

optimal   Ideal; the desired value, product, or endpoint.

ORD   Office of research and development.

organ   A group of tissues having a specific function.

organic   Chemical compounds containing carbon; generally associated with living organisms.

organism   A living being.

organoleptic   Relating to a sensory organ, e.g., taste.

orientation   Training that provides familiarization with a subject.

orifice   An opening; often one that serves as an entrance into or out of a living body.

orifice meter   A device used to determine flow rate based on the pressure differential on two sides of a restriction in a duct.

ORM   Operations risk management; other regulated material.

ornithology   The study of birds.

ORNL   Oak Ridge National Laboratory.

ORR   Operational readiness review.

ortho-   A prefix meaning straight ahead, normal, or correct; In chemistry, used in naming substituted aromatic compounds with functional groups adjacent to one another.

Os   Osmium.

OSA   Operational safety analysis.

oscillation   Variation, often over time, with respect to a specific standard or reference.

OSHA   Occupational Safety and Health Administration.

OSHA log   The OSHA Form 200 used as a log to report occupational injuries and illnesses.

OSHRC   Occupational Safety and Health Review Commission.

-osis   Suffix meaning a condition or a process, usually one that is abnormal.

osmium (Os)   A hard, white, toxic, irritating group VIII metallic element, atomic number 76, atomic weight 190.2; used as a hardener and a catalyst and in instruments.

osmosis   The passage of fluid through a membrane.

OSS   Office of Safeguards & Security; off-site shipments.

ossicle   Any one of three bones in the inner ear.

osteo-   Prefix meaning pertaining to bone.

Ostwald, Wilhelm   German chemist (1853–1932) awarded the 1909 Nobel Prize in Chemistry for catalyst studies. Regarded as the founder of modern physical chemistry.

OSW   Office of Solid Waste.

OSWER   Office of Solid Waste and Emergency Response.

ot-, oto-   A prefix pertaining to the ear.

OTA   Office of Technology Assessment.

otologist   A physician specializing in ear diseases.

-ous   A suffix indicating that the key element in the compound exhibits the lower oxidation state.

oven   An enclosed device used for heating.

overload   An excess that produces a stress, strain, or outage (e.g., electrical) on a living or mechanical system.

Ox   Total oxidants.

oxidase   Any oxidizing enzyme.

oxidation   A chemical reaction in which electrons are transferred. Originally defined as any reaction in which oxygen combined with another substance.

oxidation number   The number of electrons that must be added or subtracted from an atom in a formula to convert it to the elemental (neutral) form.

oxidizer   An oxidizing agent; a substance that oxidizes another substance.

oxidizing agent   A chemical that is reduced or gains electrons, e.g., peroxides, chlorates, permanganates; the giving off of oxygen.

oxygen (O)   A gaseous group VI element, atomic number 8, atomic weight 16.00. Molecular oxygen (O_2) is a colorless, odorless gas essential for respiration and practically all combustion.

oxygen deficiency   Describes an atmosphere with less than the amount of oxygen normally present in air, i.e., 21%.

ozone (O_3)   A reactive form of oxygen; pungent and causes irritation to eyes and lungs and is toxic at levels greater than 0.1 ppm; necessary in the stratosphere to shield the surface from UV radiation.

ozonator   A device that produces ozone in the laboratory and therefore requires venting of its effluents to the outside.

P

p   Page; pico- (prefix $= 10^{-12}$)

p   *Para.*

P   Phosphorus; peta- (prefix $= 10^{15}$); poise; uppercase Greek letter rho.

Pa   Protactinium, pascal.

PA   Public address; plant air; project administrator.

PABA   *p*-Aminobenzoic acid.

pachy-   A prefix for heavy, large, thick.

Pacific Northwest National Laboratory   *See* Batelle Pacific Northwest National Laboratories.

packed tower   An air pollution control device in which air effluent is forced through a column filled with absorbent material that is sprayed with water to dissolve or react with the undesired contaminant.

paclitaxel   taxol; a compound used to treat ovarian cancer.

PAH   Polycyclic aromatic hydrocarbon; polynuclear aromatic hydrocarbon.

pair production   The conversion of gamma radiation into a pair of particles, i.e., an electron and a positron.

palladium (Pd)   A relatively nonreactive (noble) group VIII metallic element; atomic number 46, atomic weight 106.4; used for alloys, jewelry, and electronics.

palpate   To feel; to examine by touching.

palpitation   Rapid heartbeat recognized as such by the individual.

papain   A naturally occurring enzyme used as a meat tenderizer.

papilloma   A small growth or tumor of the skin or mucous membrane.

PAN   Peroxyacetyl nitrate.

pandemic   A widespread disease affecting many people.

pantothenic acid $C_9H_{17}NO_5$; a B-complex vitamin.

PAPR   Powered air-purifying respirator.

para-   A prefix meaning opposite, near, abnormal; e.g., in chemistry used in naming ring compounds with two functional groups located opposite from one another on the ring.

Paracelsus, Phillippus Aureolus   Swiss alchemist and physician (1493–1541) who investigated diseases of miners; taught that diseases were specific and could be treated and cured by specific remedies; used opium, sulfur, iron, and mercury as treatments; emphasized value of observation and experience.

paraffin   A saturated hydrocarbon, i.e., all carbons are attached to each other by single bonds.

paraformaldehyde   A toxic white solid used as a fungicide, bactericide, and disinfectant.

paralysis   Loss of body movement, sensation, or mental ability.

parameter   A variable or constant in a mathematical expression.

paranoia   A psychosis characterized by delusions of persecution.

paraplegia   Paralysis of the lower body including both legs.

paraquat   A herbicide; toxic by ingestion, inhalation, and skin absorption.

parasite   An organism living on or within another that obtains nourishment from the host.

parathion   An insecticide; a deep brown to yellow liquid, toxic by ingestion, inhalation, and skin absorption; a cholinesterase inhibitor.

paregoric   A derivative of opium used for digestive disorders.

parent   A father or mother; a radionuclide that on radioactive decay or disintegration produces a specific nuclide (the daughter).

parenteral   Any route of entry other than the alimentary canal, e.g., intravenous.

paresis   Partial paralysis.

paresthesia   Numbness or heightened sensitivity.

parietal   Pertaining to the wall of a cavity.

PARMA   Public Agency Risk Managers Association.

paroxysm   A sudden periodic attack or exacerbation of disease symptom(s).

partial pressure   The pressure exerted by one component of a gaseous mixture. *See* Dalton's law.

particle   A minute constituent of matter with a measurable mass, such as a neutron, proton, or meson.

particle accelerator   A device used to accelerate atomic particles, e.g., cyclotron.

particle size distribution   The statistical distribution of sizes in a group of particles.

particulate   A very finely divided solid.

partition coefficient   A measure of the amount of distribution of a solute in a solvent at equilibrium; sometimes used to predict how a substance will be distributed and absorbed in the human body.

pascal (Pa)   The SI unit of pressure defined as the pressure resulting from a force of one newton acting over an area of one square meter; named after the French scientist.

Pascal, Blaise   French mathematician and physicist (1623–1662) who invented a calculating machine, a barometer, a hydraulic press, and the syringe.

passive dosimeter   A sample collection device based on gaseous diffusion that usually gives immediate results, results after a minimum of manipulations, or measurements that can be directly read.

passive systems   Systems containing no moving parts, such as concrete pads, support brackets, conduit, walls, etc.

Pasteur, Louis   French chemist (1822–1895) noted for heat-treating food to kill or inactivate toxic microorganisms; also popularized inoculations.

pasteurization   A process of heating a substance to just below its boiling point to kill human pathogens present.

PAT   Proficiency analytical testing.

path-   A prefix meaning disease or suffering.

pathogen   A microorganism or other substance that causes disease.

pathology   The study of disease and the changes produced by disease.

Pauli, Wolfgang   Austrian-born American physicist (1900–1958) who studied nuclear physics and postulated the existence of the neutrino; awarded the 1945 Nobel Prize in Physics.

Pauli exclusion principle   States that it is not possible for two atomic particles to be in and occupy the same place at the same time.

Pauling, Linus Carl   American chemist (1901–1994) awarded the 1954 Nobel Prize in Chemistry for crystallography and bonding theories; an advocate of vitamin C and orthomolecular medicine. Also awarded the Nobel Peace Prize in 1962.

PAW   Plasma arc welding.

Pb   *Plumbum* (Latin); lead.

PC   Personal computer; politically correct; production control; programmable controller.

PCA   Portland Cement Association.

PCB   Polychlorinated biphenyl; printed circuit board.

PCC   Poison control center.

PCP   Pentachlorophenol; phencyclidine hydrochloride, a veterinary medicine used as an illegal hallucinogen; also called angel dust.

PCV   Pressure control valve.

PCW   Process cooling water.

Pd   Palladium.

PDAS   Process data acquisition system.

PDR   *Physicians' Desk Reference*; preliminary design review; property disposal report.

PE   Professional engineer; project engineer; product engineer; program engineer.

peat   Partially carbonized organic matter, e.g., moss; found in bogs; used as fertilizer and fuel.

pectin   A high-molecular-weight colloidal substance found in ripe fruit; used to jell foods and pharmaceuticals.

Pedersen, Charles John   Korean-born American (1904–1989) awarded the 1987 Nobel Prize in Chemistry for studies of the mechanisms of molecular recognition.

pediculosis   Infestation with lice.

PEL   Permissible exposure limit.

pellagra   A chronic disease characterized by skin lesions, digestive disorders, and mental deterioration, caused by deficiency of the vitamin niacin.

penetrating radiation   Radiation that can travel long distances and penetrate the body, impart some of its energy, and continue at a lower energy.

pentachlorophenol (PCP)   A fungicide, bactericide, and herbicide and TCDD precursor.

peptide   A compound of two or more amino acids bound together to form a polymer; essential for proteins to function.

per-   A prefix meaning highest oxidation state or complete substitution, e.g. perchlorate.

peracetic acid CH_3COOOH; An irritating, noxious, strong oxidizing chemical used as a biocide.

perchloroethylene $Cl_2C{=}CCl_2$; A colorless, nonflammable solvent; used in dry cleaning and to remove tars and other organic matter.

percutaneous   Through the skin.

performance-based training   Formal, systematic training based on tasks, knowledge, and skills required for competent job performance.

performance test   A system or component test to verify that required performance characteristics can be achieved; used to detect abnormal performance characteristics and to determine effects of maintenance and operating activities on equipment performance. Also an equipment performance check to determine whether the equipment is operating to specifications.

peri-   A Greek-derived prefix meaning near, around, enclosing, or about.

periodic table (periodic chart)   An arrangement classifying chemical elements in order of increasing atomic number in which elements of similar properties are placed under each other in groups. Chemical and physical properties vary within each group but group elements generally have similarities in chemical behavior.

peripheral hardware   Material used with a glove, boot, or bag, i.e., outer retaining ring, inner ring, shielded port cover.

peripheral neuropathy   Deterioration of the nervous system of the extremities, e.g., arms, legs, feet.

Perkin, Sir William Henry   English chemist (1838–1907) credited with making the first synthetic dyestuff.

Permalloy™   A corrosion-resistant alloy containing 78.5% nickel and 21.5% iron.

permeable   Allowing penetration through a barrier; used to describe penetration of a chemical through a membrane or specific types of glove or clothing.

permissible dose   The limit or level of tolerable dose.

permissible exposure limit (PEL)   The legal exposure level published by US OSHA (29CFR.1910.1000 Subpart Z) as the standard.

permit   A written authorization allowing the bearer to work in a certain area or location doing a specific type of work, e.g., confined space, lock-out/tag-out permit.

peroxidase   Any oxidizing enzyme.

peroxide   —O—O— functional group; usually reactive and unstable.

personal hygiene   An individual's responsibility to ensure personal and environmental cleanliness, minimize the spread of infection, and help ensure sterility.

personal protective equipment (PPE)   Equipment or clothing, such as respirators, gloves, etc., used for protection against environmental hazards.

personnel monitoring   The determination of the degree of exposure to individuals, using wipes, air sampling, survey meters, dosimeters, etc.

Perutz, Max Ferdinand   Austrian biologist (1914–) who shared the 1962 Nobel Prize in Chemistry for work on crystalline protein structures such as hemoglobin.

pest   Any type of animal or plant life that is injurious to health or the environment.

pesticide   In general, any of a group of agents used to destroy pests; it may broadly include insecticides, fungicides, rodenticides, herbicides, fumigants, repellents, etc.

PET   Positron emission tomography.

Petit, Alexis Thérèse   French physicist (1791–1820) associated with the Dulong and Petit's law that for all elements the product of the specific heat and the atomic weight is the same.

Petri, Julius R.   German bacteriologist (1852–1921) who invented the dish used in microbiology named after him.

petri dish   A shallow, cylindrical dish with a lid; used to culture microorganisms.

petrochemical   A chemical substance produced from petroleum products.

petrolatum   A semisolid mixture of petroleum hydrocarbons.

pewter   A silver-gray alloy of tin, antimony, copper, and lead, once widely used for kitchen utensils.

peyote   A hallucinogenic substance derived from mescal from a cactus plant.

pH   A mathematical expression of the molar concentration of hydrogen ions in solution i.e., $pH = -\log[H^+]$. Water solutions are acidic below pH 7 and basic above pH 7; the lower the pH, the more acidic the solution, and vice versa.

PHA   Pulse height analyzer; preliminary hazards analysis, process hazard analysis.

phagocyte   A cell that surrounds and consumes foreign bodies.

pharmaceutical   Term describing drugs, chemicals, and medicines produced and marketed by companies for human or animal consumption.

pharmacognosy   The branch of pharmacology concerned with natural drugs.

pharmacology   The study of drugs and their use and interaction.

Pharmacopeia, US   An authorized book on drugs and their standards published every five years.

pharyngeal   Pertaining to the pharynx (throat); the area between the mouth and the esophagus.

phase   A state of matter, e.g., gas, liquid, or solid.

phase rule　The rule formulated by J. Willard Gibbs in 1877 stating that $F = C - P + 2$, where F = the number of degrees of freedom, C = the components, and P = the phases. The rule defines systems such as water and demonstrates how an equilibrium state can exist between water vapor, ice, and liquid water.

phenol　C_6H_5OH; toxic by ingestion, inhalation, and skin absorption; a general disinfectant.

phenol coefficient　A measure of the effectiveness of a germicide.

phenolphthalein　An acid-base indicator; used as a laxative before it was found to be carcinogenic.

phenyl　C_6H_5-group.

pheromone　An odiferous chemical released by animals; used in traps to eliminate these pests.

φ　Lower case Greek letter phi.

phlebo-　Pertaining to the veins.

phlegm　The secretion of mucus.

phosgene　$COCl_2$; toxic by inhalation; gas formed in many building fires; used as poison gas in World War I.

phosphenes　The EMF-caused light flashes sensed within the eye.

phosphine　PH_3; a colorless, highly toxic gas, spontaneously combustible in air, with a garlic-like odor used as a fumigant and in the electronics industry.

phosphor　Material capable of absorbing electromagnetic energy (e.g., UV, visible, IR) and then emitting a portion of it as UV, visible, IR, etc. radiation. Also broadly includes fluorescent and luminescent materials.

phosphorescence　Emission of radiation a fairly long time after the energy is initially absorbed.

phosphorus (P)　A toxic, solid, nonmetallic group VA element, atomic number 15, atomic weight 30.97; can spontaneously combust in air.

photochemistry　The part of chemistry that studies chemical changes caused by radiant energy interacting on substances.

photochromism　When a transparent material, exposed to light, darkens.

photoelectric effect　The emission of an electron from an atom bombarded by incident light or gamma rays.

photoionization   The process by which a substance is ionized when subjected to bombardment by a photon, which ejects an electron.

photoionization detector   A type of detector often used in gas chromatography for the detection of aromatics; frequently used in portable instruments.

photokeratitis   Injury to the eye by UV-B or UV-C radiation; also called welder's flash.

photolysis   Chemical decomposition initiated by light.

Photomultiplier (PM)   A vacuum tube that multiplies the electron input.

photon   A quantum (or packet) of emitted electromagnetic energy, i.e., gamma rays, X rays.

photophobia   Intolerance to light.

photosynthesis   The synthesis of carbohydrates by green plants in the presence of sunlight due to the absorption of light by chlorophyll; also releases oxygen.

phototropic   Describes an organism that obtains, or a chemical reaction that occurs because of, energy from light.

PHS   Public Health Service.

phyll-,phyllo-   A prefix meaning leaflike.

physical hazard   A chemical able to cause or promote a fire, explosion, or uncontrolled chemical reaction because of its being flammable, combustible, pyrophoric, explosive, oxidizable, unstable, reactive, or a compressed gas.

Physicians' Desk Reference **(PDR)**   An annual publication useful to the medical profession, which describes drugs and their proper use, actions, interactions, and hazards.

physiological salt solution   Isotonic salt solution; NaCl and water (0.9%) in concentrations identical to those in the human body.

physiology   The study of the functioning of living organisms.

phyte-, -phyto   Pertaining to plants.

phytochemistry   Plant growth chemistry.

π Lower case Greek letter pi; an amount equal to about 3.14159.

PI   Principal investigator.

PI&S   Product integrity & surveillance.

pi bonds   Covalent bonds envisioned as electron density above and below the plane of organic molecules with double or triple bonds.

pickling   The process of removing a coating of scale, oxide, or tarnish from metals by immersion in an acid bath to obtain a chemically clean surface.

pico- (p) 10^{-12}; a prefix that divides a basic unit by one trillion.

picric acid $C_6H_2(NO_2)_3OH$; a yellow crystalline compound; very shock- and heat-sensitive explosive.

pig   A container (usually lead) used to ship or store radioactive materials; container walls are designed to protect the surrounding environment and people from radiation.

pigment   A finely divided, insoluble substance that gives a particular color to the substance to which it is added.

pile   A nuclear reactor.

pilot plant   A small-scale simulation of a larger industrial process used to scale up chemical reactions and industrial processes.

PIN   Product identification number; personal identification number.

pinch point   The area in a moving system or apparatus where a body part or item of clothing or jewelry can get caught or entangled and cause harm to the worker.

pine oil   A colorless extract of pine wood used as an odorant, disinfectant, fragrance, and preservative.

pitch   A black or dark substance obtained as a residue in the separation of organic materials, especially tars; any of varied bituminous substances or resins from trees (conifers).

pitchblende   A naturally occurring mixture of uranium ores.

Pitot, Henri   French hydraulic and civil engineer (1695–1771) who invented the pitot tube used to measure the relative velocity of a fluid past the orifice of the tube.

pitot tube   A system of two concentric tubes used to measure total and static pressure in an air stream; used to measure the relative velocity of fluids past the orifice of the tube.

pK   The negative log of the equilibrium constant K.

placebo   An inactive substance given as a blank or control in drug studies.

placenta   An organ that connects the mother to the fetus for nourishment but through which undesirable agents can pass.

plaintiff   The initiator of a lawsuit.

plan-of-the-day (POD)   A plan made after a meeting held each working day to discuss the current activity schedule, update the activity schedule, and set priorities. Attendees include representatives from production, safety, quality assurance, construction, maintenance, and safeguards and others deemed necessary by the appropriate manager.

Planck, Max Karl Ernst Ludwig   German physicist (1858–1947) who worked on thermodynamics and black body radiation and introduced the quantum theory; awarded the 1918 Nobel Prize for Physics.

Planck's constant (h) 6.626×10^{-19} JHz; the constant that multiplied times the electromagnetic frequency equals the energy of that photon.

plant safety program   A program established to implement the guidelines, policies, and requirements that ensure safe operation of the facility.

plantwide systems   Systems that provide services to several buildings and areas; including plant power, utilities, fire protection, and alarms.

plasma   The component of blood in which the cells and platelets are suspended; a state of matter.

plasma arc welding (PAW)   The use of an arc formed as a jet at a small opening through which a high-pressure stream of air, argon, helium, nitrogen, hydrogen, or a mixture flows, for joining metals; hazards may be present as a result of the gases and emissions from the substances being welded; hence, proper ventilation is essential.

plasmid   A fragment of genetic material that exists outside the chromosomes.

plastic   A synthetic polymer, which is usually considered to be either thermoset (permanently hard) or thermoplastic (capable of resoftening).

platelet   The minute substance present in blood and necessary for coagulation.

platinum (Pt)   A silvery-white group VIII noble metal element, atomic number 78, atomic weight 195.09; flammable when powdered; used for jewelry, surgery, and catalysts.

platinum black   A fine black powder of platinum used as a gas absorbent and catalyst; also called platinum sponge.

PLC   Programmable logic controller.

pleio-, pleo-, plio-   A prefix meaning more.

plenum   A chamber or enclosure that provides and distributes uniform air pressure for gases flowing though it to ventilation ducts.

pleura   The membrane surrounding the lungs and thoracic cavity.

pleurisy   Inflammation of the pleura.

Pliny the Elder (Gaius Plinius Secundus)   Roman scholar (23–79) who wrote an encyclopedia on natural history; used animal bladders as filters against inhaling lead dust and fumes, observed the eruption of Mt. Vesuvius and was killed by its fumes.

plumbism   Lead poisoning.

plume   Visible discharge, effluent, or release.

plutonium (Pu)   A transuranic fissionable metallic element, atomic number 94, atomic weight 239.11; used for nuclear power plants and nuclear weapons.

Pm   Promethium.

PM   Photomultiplier tube; project manager.

PM-10   Particulate matter ten micrometers or less in aerodynamic diameter.

PMCC   Pensky–Martens closed cup.

PML   Probable maximum loss; physical metrology laboratory.

PMS   Preventive maintenance system; performance measurement system; premenstrual syndrome.

PM$_x$ Particulate matter x micrometers or less in aerodynamic diameter.

PNA   Polynuclear aromatic hydrocarbon; also known as PAH.

pneo-   A prefix pertaining to breath.

pneumo-   A prefix meaning pertaining to air, the lungs, or breathing.

pneumoconiosis   Scarred lungs resulting from the prolonged inhalation of dusts and particles; usually resulting in impaired breathing.

pneumonitis   Inflammation of the lungs.

pneumonoultramicroscopicvolcanicsiliconiosis   Believed to be the longest word in the English language; a lung disease caused by exposure to very small particles of silica from volcanic eruptions.

Po   Polonium.

PO   Purchase order; production order; post office.

POC   Products of combustion; purgeable organic compounds.

pocket dosimeter   A small ionization detection instrument that indicates radiation exposure directly or indirectly.

podiatry   The medical study of foot disease.

POE   Point of exposure.

POGO   Privately-owned, government-operated.

pOH   A mathematical expression used to express the molar concentration of hydroxide ion $[OH^-]$ in solution, i.e., $pOH = -\log [OH^-]$. Water solutions are basic above pH 7 and acidic below pH 7; the higher the pH, the more basic the solution, and vice versa.

point source   A source small enough in size compared to the distance between the source and the receiver to be considered as a point not an object with dimensions which would complicate calculations.

poison   A substance that causes illness or death when taken into the body in relatively small quantities.

Poisson, Siméon Denis   French mathematician (1781–1840) who applied mathematics to physics, especially electricity and magnetism; authored works on Fourier series, calculus of variation, and probability.

Poisson distribution   The probability distribution used to describe unlikely occurrences in a large number of events.

Polanyi, John Charles   German-born Canadian (1929–) who shared the 1986 Nobel Prize in Chemistry for studies of the dynamics of elementary chemical processes; his work led to the development of lasers.

polar   A description of the extent of permanent separation of positive and negative charges in a molecule.

polarimeter   A device used to measure optical rotation of chemicals.

polarization   The separation of charge; the uniform variation of light rotation.

polarography   An electroanalytical chemistry method.

polar solvents   Solvents that have a high dielectric constant, e.g., alcohols, ketones.

pollen   Airborne microspores of plants responsible for certain types of allergies, e.g., hay fever.

pollution   Introduction or presence in the environment of substances that are not naturally present.

polonium (Po)   A radioactive metallic element, atomic number 84, atomic weight 192–218; used as a source of α particles and neutrons.

poly-   A prefix meaning many.

polybrominatedbiphenyl (PBB)   Carcinogenic, toxic, multibrominated chemicals consisting of two joined benzene rings; once commonly used in plastics and as fire retardants.

polycarbonate   A hard, nontoxic, resistant thermoplastic polymer.

polychlorinated biphenyl (PCB)   Persistent, toxic, multichlorinated chemical consisting of two joined benzene rings; once used as fire retardants, lubricants and plasticizers; causes chloracne.

polycyclic   Organic compounds made up of more than two ring structures.

polycyclic aromatic hydrocarbon (PAH)   Highly toxic, multi-ring aromatic hydrocarbon chemicals, some of which are carcinogenic, e.g., benzo(*a*)pyrene; found in cigarette smoke, etc.

polyester fiber   Generic name of synthetic fibers used in tires, fire hoses, and other fabrics.

polyester film   Very strong, thin synthetic film used in electronics and tapes.

polyethylene   A polymer used in plastics and rubber.

polymer   A substance composed of very large molecules formed by joining together many small similar molecules.

polymerization   A reaction in which a large number of relatively simple molecules are combined, usually through linking or cross-linking, into a macromolecule.

polymorphism   *See* allotrope.

polyp   A growth in the mucous membranes.

polypropylene   A polymeric material used in appliances, films, surgical casts, etc.

polystyrene   A polymer used in packaging, insulation, and molding.

polyurethane   A polymer used in sealant, caulking, adhesives, and foams.

polyvinyl chloride (PVC)   ($—CH_2CHCl—)_n$; a polymer used for pipe material and many construction applications.

Pople, John Anthony   English-born American chemist and mathematician (1925–) awarded the 1998 Nobel Prize in Chemistry for developing computational methods in quantum chemistry; a pioneer in nuclear magnetic resonance spectrometry (NMR).

porosity   The degree to which space comprises a solid substance; a measure of the ease with which a fluid can pass through a solid.

porphyrin   A group or class of naturally occurring physiologically active nitrogen-containing compounds comprised of pyrrole rings.

Porter, George   English chemist (1920–) who shared the 1967 Nobel Prize in Chemistry for studies of fast chemical reactions and photosynthesis.

Portland cement   A particular type of hydraulic calcium silicate cement containing specific amounts of specific additional materials such as lime, alumina, silica, iron oxide, etc.

Portland Cement Association (PCA)   5420 Old Orchard Rd., Skokie IL 60077.

positive pressure check   A qualitative respirator fit in which the wearer covers the exhalation valve, gently exhales, listens and feels for outward air leakage around the facepiece seal, and holds the test for 5–10 seconds.

positron   A subatomic particle equal in mass, but opposite in charge, to the electron.

positron emission tomography (PET)   The use of the internal distribution of positron-emitting isotopes to produce sectional images of the body.

possibly carcinogenic to humans   A definition used by IARC and designated as 2B as follows: This category generally includes agents for which there is limited evidence in humans in the absence of sufficient evidence in experimental animals. It may also be used when there is inadequate evidence of carcinogenicity in humans or when human data are nonexistent but there is sufficient evidence of carcinogenicity in experimental animals. In some instances, agents may be included for which there is inadequate evidence or no data in humans but limited evidence of carcinogenicity in experimental animals together with supporting evidence from other relevant data.

potable water   Water fit for human consumption, i.e., free from biological, chemical, or physical contamination.

potassium (K)   A soft, white, highly reactive, water-unstable group IA metallic element, atomic number 19, atomic weight 39.10; flammable in moist air; may form explosive peroxides. Symbol from *kalium* (Latin).

potential energy (PE)   Energy not yet in use but available for use; energy of position.

potentiator   A substance that enhances the taste of the food to which it is added; a substance that increases the toxic effect of one agent in a mixture so that the combined effect is greater than the sum of the components.

potentiometer   A device that measures or controls the electrical voltage in a circuit.

pother   A cloud of smoke or dust.

Pott, Sir Percivall   English physician (1714–1788) who established the occupational relationship between scrotal cancer in chimney sweeps and poor personal hygiene; introduced techniques to make surgery less painful; Pott's fracture and Pott's disease are named after him.

POTW   Publicly owned treatment works.

poultice   A thick paste used as a medicinal salve for sores or wounds.

pour point   The lowest temperature at which a liquid will flow from an inverted container.

power   The time rate at which work is done, usually expressed in units of watts (J/sec) or horsepower.

powered air-purifying respirator (PAPR)   A battery powered respirator that filters and purifies the air breathed by the wearer.

ppb, PPB   Parts per billion.

PPC   Personal protective clothing.

PPE   Personal protective equipment.

ppm, PPM   Parts per million.

ppt, PPT   Parts per trillion; precipitate, prepared.

PQE   Procurement quality engineering.

Pr   Praseodymium.

PR   Purchase request; procurement request.

PR   Rated power.

praseodymium (Pr)   A lanthanide group IIIB metallic element, atomic number 59, atomic weight 140.91; used in lasers and welding.

precautionary clothing and equipment   Company-issued clothing and equipment that workers may be required to wear in controlled areas, includes coveralls, safety shoes, underwear, etc. The clothing is a precaution to avoid contaminating personal clothing or a worker's skin if there is an inadvertent release of contamination. It is not intended to substitute for or to be used as anticontamination clothing.

precipitation   The separation of an insoluble compound from a solution as a result of a chemical reaction; results when the solubility product for relevant ions is exceeded; naturally occurring moisture deposition from the atmosphere.

precipitator   An air pollution control device that removes contaminant particles by passing them through an electric field to ionize them and then collecting them on oppositely charged collection plates.

precision   A generic concept used to describe the dispersion of repeated measurements with respect to a measure of central tendency, usually the mean; sometimes measured by repeatability and reproducibility. Repeatability refers to the within-group dispersion to measurement, whereas reproducibility refers to the between-group dispersion; term often accompanying accuracy.

precursor   A starting compound or agent that is necessary to form another compound or substance.

prednisone $C_{21}H_{26}O_5$; an analog of cortisone used to treat inflammation of arthritis, skin, and eyes.

Pregl, Fritz   Austrian chemist and physician (1869–1930) awarded the 1923 Nobel Prize in chemistry for his work in developing microanalytical techniques.

preliminary hazard analysis (PHA)   A systems safety technique used in the early design stages to estimate potential hazards.

Prelog, Vladimir   Swiss chemist born in the former Yugoslavia (1906–1998) who shared the 1975 Nobel Prize in Chemistry for synthesis and stereochemistry of organic compounds.

presbycusis   Normal hearing loss due to aging.

preservatives   Agents added to retard spoilage.

pressure   The force applied to or distributed over an area. *See* pascal.

pressure demand respirator   A type of respirator in which inward leakage is unacceptable and that is always under positive pressure during inhalation and expiration.

preventive maintenance   Predictive, periodic, or planned maintenance actions performed to prevent equipment breakdown or malfunction.

Priestley, Joseph   English chemist and physicist (1733–1804) who emigrated to America; discovered oxygen and nitrous oxide.

Prigogine, Ilya   Russian-born Belgian chemist (1917–) awarded the 1977 Nobel Prize in Chemistry for nonequilibrium thermodynamics studies; applied his work to living systems; helped establish chaos theory.

prion   An infectious particle; small proteinaceous material.

probability   The likelihood or chance that an event will take place.

probably carcinogenic to humans   A definition used by IARC and designated as Group 2A, to mean: This category includes agents for which there is limited evidence of carcinogenicity in humans and sufficient evidence of carcinogenicity in experimental animals. On occasion, IARC may classify an agent in this category solely on the basis of limited evidence of carcinogenicity in humans or of sufficient evidence of carcinogenicity in experimental animals in view of supporting evidence from other relevant data.

probably not carcinogenic to humans   A definition used by the IARC and designated as Group 4, to mean: For agents in this category, there is evidence suggesting lack of carcinogenicity in humans together with evidence suggesting lack of carcinogenicity in experimental animals. In some circumstances, agents for which there is inadequate evidence of or no data on carcinogenicity in humans but evidence suggesting lack of carcinogenicity in experimental animals, consistently and strongly supported by a broad range of other relevant data, may be classified in this group.

process capability   The limits within which a tool or process operates, based on minimum variability determined by prevailing circumstances.

process hazard analysis (PHA)   A systems safety technique that evaluates the hazards and consequences of an accident in the event one occurred, e.g., in a chemical industrial process.

process safety management (PSM)   The systems safety process of evaluating the entire system of a chemical industry process, which includes identification, evaluation, prevention, consequences, and emergency actions.

product   Manufactured material not meeting the definitions of waste, scrap, or residue that includes vendor material that must be repackaged, either for use on site or for shipment.

product engineer (PE)   The individual in program management responsible for coordinating technical efforts and scheduling for an identifiable product.

product liability   The manufacturer's liability for injury and/or illness as a result of the use of a product.

product safety management (PSM)   The systems safety technique that evaluates the testing and design of products to minimize injury and/or illness from their use.

Proficiency Analytical Testing (PAT)   An analytical method that complies with the NIOSH criteria for the PAT program.

progesterone $C_{21}H_{30}O_2$; female sex hormone; a carcinogen.

prognosis   A prediction of effects, results, and outcome.

program manager (PM)   Person responsible for meeting the requirements and objectives of specific programs who determines and defines the scope of work, approves budgets, provides program plans, and schedules and monitors the performance of functional organizations.

program planning & support (PP&S)   The group within program management consisting of product definition and configuration management, systems engineering, technical writing, program operations planning, and packaging program; provides required technical and administrative support to programs and provides systems analysis and other technical, administrative, and management support for program management.

prokaryote   Single-celled organisms with cell walls, mitochondria, and a nucleus.

proliferation   The reproduction or multiplication of a substance or form, e.g., biological cells.

promethium (Pm)   A radioactive, lanthanide metallic element, atomic number 61, atomic weight 141 to 154; used in the semiconductor industry and as a source of X rays.

prompt critical   An uncontrolled condition in which a reactor period is determined by prompt neutrons and the reactor flux increases extremely rapidly.

promulgate   Make known; pass or enact into law.

propagation of flame   The spread of fire through a substance.

propagation of variance   Determination of the uncertainty value of a given quantity using mathematical formulas for the combination of errors.

prophylaxis   Preventative measures.

proportional counter   An electronic detection system that receives pulses proportional to the number of ions formed in a gas-filled chamber by ionizing radiation.

prospective study   An technique used to study the relationship between the characteristics of a disease and its occurrence.

prosthesis   An artificial replacement for a body part.

prostration   Exhaustion to the point of collapse.

protactinium (Pa)   A highly toxic, radioactive, actinide metallic element, atomic number 91, atomic weight 231.04.

protease   An enzyme that breaks peptide links in protein.

protection factor (PF)   The degree of protection provided by the proper fit and use of respiratory protective equipment. Determined by calculating the ratio of the concentration of a substance outside to the concentration inside the respirator.

protective atmosphere   A highly controlled gaseous area immediately surrounding and encompassing a work area such as welding. Control parameters may include composition, temperature, pressure, flow rate, etc.

protective cream   A barrier used to prevent contact between the skin and a hazardous substance.

protein   A naturally occurring, complex, large molecule containing carbon, hydrogen, oxygen, nitrogen, and sometimes sulfur and phosphorus that are connected by peptide linkages; found in all living cells and biological fluids.

protocol   An agreed-upon way or method of performing a procedure, task, operation, or experiment.

proton   A fundamental particle of matter in the atomic nucleus, with a single positive charge and a mass approximately 1847 times that of the electron. The atomic number of an atom equals the number of protons in the nucleus. *See* atomic number.

protoplasm   The contents of a cell, nucleus and cytoplasm.

Prout, William   British chemist and physiologist (1785–1850) who discovered hydrochloric acid in the stomach,

proximal   Nearby; close; nearest the center.

proximate cause   The immediate cause of an accident, injury or illness, without which it would not occur.

PRP   Potentially responsible party.

pruritis   Severe itching.

PRV   Pressure relief valve.

PSAR   Preliminary safety analysis report.

PSC   Personnel status change; project status control.

psi   Pounds per square inch.

ϕ   Lower case Greek letter psi.

psoriasis   A chronic skin infection.

PSV   Pressure safety valve.

psychosomatic   Describes disorders attributed to the emotional state of an individual; usually with real physiological manifestations.

psychotropic   Affecting the behavior or mental activity of psychologically disturbed people.

psychrometer   An instrument used to measure relative humidity, consisting of a wet and a dry bulb thermometer.

Pt   Platinum.

PT   Plant training; liquid penetrant testing; process tool; physical therapist.

ptomaine   A group of very toxic compounds.

ptosis   Drooping or dropping of a part, e.g., the eyelid.

PTS   Pesticides and Toxic Substances Branch (US EPA).

Pu   Plutonium.

PU&D   Property utilization and disposal.

Public Agency Risk Managers Association (PARMA)   5750 Almaden Expressway, San Jose CA 95118.

Public Health Service (PHS)   200 Independence Ave., SW, Washington DC 20201; a uniformed branch of the US government founded to protect the health of the public. Headed by the US Surgeon General.

publicly owned treatment works (PTOW)   A state or municipally owned facility for waste treatment, recycling, or reclamation.

PUC   Public Utility Commission.

pulmonary   Pertaining to the lungs.

pulmonary edema   Lung edema.

pulmonary function tests   Any of several tests of the lung and respiratory system used to determine lung disease and disorder and also whether a person can wear a respirator.

pulse   The rate or rhythm of the heart; any regular beating.

pulsed laser   A class of lasers in which the emission operation occurs in one or more short flashes of short duration.

pumice   A naturally occurring silicate of volcanic ash or lava; used for abrasion.

pupil   The opening in the iris through which light enters the eye.

purge   To evacuate.

PURPA   Public Utilities Regulatory Policies Act of 1978

purulent   Full of pus.

pus   Yellowish liquid produced in inflammation.

pustule   A small skin bump filled with lymph or pus.

PVC   Polyvinyl chloride.

PW   Process waste.

pyelogram   An X ray of the ureter and renal pelvis.

pyknometer   A calibrated container for determining the specific gravity of a liquid.

pyridine $N(CH)_4CH$; a highly flammable, explosive, toxic, colorless to pale yellow, volatile liquid with a foul odor used in pharmaceuticals and as a reagent.

pyro-   A prefix meaning formed by heat.

pyrogen   A substance that causes elevation in human temperature.

pyroligneous acid   The reddish-brown acidic liquid obtained from wood fires or distillation; a mixture of acetic and other acids, alcohols, tars, etc. Also known as wood vinegar.

pyrolysis   The breakdown of a substance into other substances by heat.

pyrophoric   Having the characteristic of spontaneously igniting when exposed to the air.

pyrotechnics   Pertaining to fireworks.

Q

Q   Volume, as in ventilation Q = va, volume equals rate of flow (velocity) multiplied by the area; coulomb.

QA   Quality assurance.

QAR   Quality assurance record.

QC   Quality control.

Q.E.D.   Abbreviation for the Latin phrase *quod erat demonstrandum*, "which was to be proved."

QF   Quality factor, also neutron quality factor.

Q fever   A disease marked by fever and pain caused by rickettsia *Coxiella burnetti*; human infection occurs by contact with sheep, cattle, and their tissues and fluids, as well as infected humans. Certain laboratory workers and people around such animals are at risk. Also known as nine mile fever and query fever.

QL   Quality level; quality laboratory.

Q-switched laser   A high-powered, short-duration pulsed laser.

quadrant   One quarter of a circle, 90°; shaped like one quarter of a circle.

quadratic equation   A second-degree mathematical expression, $ax^2 + bx + c = 0$, from which the roots of the equation may be determined.

qualification   The combination of an individual's experience, attributes, and technical, academic, and supervisory knowledge and skills developed through training, education, and demonstrated on-the-job performance; refers to educational, experiential, training, and/or special requirements necessary to perform assigned responsibilities.

Qualified instructors   Instructors who are certified to teach, have attended on-the-job training courses for their area of instruction, are competent in the area, and are approved by management to provide the training.

qualitative analysis   An estimate or approximation; pertaining to or a description of the character or nature.

qualitative fit test   A nonprecise estimate of the ability of a respirator to adequately protect the wearer; often performed with volatile, odiferous substances like isoamyl acetate (banana oil) to determine whether the wearer can smell the substance.

quality assurance (QA)   The overall program or system for measuring and maintaining a predetermined standard or level of excellence or performance or a product.

quality assurance record (QAR)   A document that furnishes evidence of the quality of items and/or activities affecting quality.

quality control (QC)   A component of a quality assurance program that helps ensure the level, grade, or acceptability of an individual component, aspect, or product of the system.

quality evidence   Written information indicating the extent of conformance to quality specifications or drawing requirements that may be based on inspection, data, or physical and chemical tests.

quality factor (QF)   The principal factor by which the absorbed dose is multiplied to obtain a quantity that expresses, on a common scale for all ionizing radiation, the biological damage to exposed individuals. The term is used because some types of radiation, such as alpha particles, are more biologically damaging than other types (also known as Q factor).

quantify   To measure and express mathematically.

quantitative   Describes the ability to be mathematically measured and have a value determined.

quantitative fit test   A mathematical expression of the ability of a respirator to properly fit the wearer. Determined by measuring the amount of a test substance outside (C_0) and inside (C_i) the respirator facepiece and expressed as a percentage called the fit factor, i.e., $FF = C_0/C_i \times 100$.

quantum   The unit of electromagnetic energy; a noncontinuous, discrete packet or bundle; usually applied to energy.

quantum mechanics   The study of a theory of atomic and molecular energy based on the theory that energy occurs in discrete packets or bundles. A system devised, in part, by Max Planck.

quantum state   A discrete energy level.

quarantine   A time of enforced isolation and restriction. Usually done to ensure containment and/or to limit the spread of contamination or disease.

quark   A subatomic particle theorized as one of the smallest fundamental units of matter.

quarry   An open excavation in the earth used for mining or the extraction of stone or other materials.

quartz   Crystallized silicon dioxide or silica; used in electronics.

quaternary ammonium compound   A type of charged or ionized nitrogen-containing organic compound; very useful for cleaning and disinfecting and as detergents.

quench   To extinguish or put out; also to harden, e.g., by plunging a hot metal into a cold liquid like water or an oil.

quicklime   Calcium oxide; CaO.

quicksilver   Mercury metal.

quiescent   Inactive, still, motionless.

quinine $C_{20}H_{24}N_2O_2 \cdot 3H_2O$, a bitter, irritating (if pure) alkaloid used medicinally for its antimalarial properties and as a flavorant.

q.v.   Abbreviation for the Latin phrase *quod vide*, "which see."

R

ρ Lower case Greek letter rho.

R   The gas constant, $= 0.08206$ L-atm/mole degree; roentgen; resistance; the representation of an organic group in a formula.

Ra   Radium.

RA   Remedial action.

RAC   Recombinant advisory committee.

racemic   The mixture of two optically active isomers that cancel out one another; equal mix of d- and l-forms.

rachi-, rachio-   Prefix pertaining to the spine.

RACT   Reasonably achievable control technology, reasonable available control technology.

R&D   Research and development.

rad   A unit of absorbed dose of ionizing radiation (100 erg/g). 1 gray = 100 rads = 1 J/kg); radiation absorbed dose; radian (plane angle).

radiant temperature   The temperature resulting from direct absorption of radiant heat energy and not due to contact, i.e., touch.

radiation   Indicates alpha, beta, gamma, X ray, and neutron types of ionizing radiation; electromagnetic energy traveling through matter or space.

radiation area   An area, accessible to personnel, where the radiation level is such that a major portion of the body could receive a dose in excess of 5 millirem/hour or a dose in excess of 100 millirem over 5 consecutive days.

radiation, nuclear   Particles (alpha, beta, neutrons) or photons (gamma) emitted from the nucleus of an unstable (radioactive) atom as a result of radioactive decay.

radiation shielding   Reduction of radiation by interposing a shield of absorbing material between the source and the individual, work area, or radiation-sensitive device.

radiation source   Usually a person-made, sealed source of radioactive material used in teletherapy, radiography, as a power source for batteries, or in various types of industrial gauges. Machines such as accelerators, X-ray units, and radioisotope generators and natural radionuclides may be considered sources.

radiation standards   Exposure standards; concentration guides; safe handling rules; transportation regulations, regulations for industrial radiation control and control of radioactive material by legislative means.

radiation warning symbol   The official symbol, a magenta trefoil on a yellow background, that must be displayed where prescribed quantities of radioactive materials are present or where certain doses of radiation could be received; uses are prescribed by law.

radiator   An emitter of energy, e.g., light or heat.

radical   A group of atoms that behaves as a single atom in chemical reactions; e.g., carbonate, nitrate, and sulfate.

radioactive series   A succession of nuclides, each of which is transformed by radioactive disintegration into the next, until a stable nuclide results. The first member is called the parent, the intermediate members are called daughters, and the final stable member is called the end product.

radioactive sources   An electroplated, sealed solid gaseous or liquid radioactive material used for chemical tracers, radiation comparison measurements, radiation instrumentation calibration, irradiation, assay, or nondestructive testing.

radioactivity   The spontaneous emission of radiation, generally alpha or beta particles, often accompanied by gamma rays, from the nucleus of an atom.

radio frequency (RF)   That part of the electromagnetic spectrum used in communication; 10 kHz to 100 gHz.

radiogenic   That which is the product of radioactive decay.

Radioimmunoassay (RIA)   A sensitive method to determine the concentration of substances, e.g., hormones in blood plasma.

radioisotope   An unstable isotope that decays or disintegrates spontaneously, emitting radiation.

radiological area   An area within a controlled area in which an individual can receive a dose equivalent greater than 5 mrem in one hour at 30 cm from the radiation source or any surface through which radiation penetrates or where airborne radioactive concentrations greater than 1/10 of the derived air concentrations are present (or are likely to be) or where surface contamination levels are greater than ten times those specified in Attachment II of DOE Order 5480.11.

radiological control area (RCA)   An area to which access is controlled to protect individuals from exposure to known or suspected radiation and radioactive materials. *See* radiological area.

radiological incident   Release of radioactive material to areas where radioactive materials are not normally found. Includes personnel contamination, positive wound counts, inhalation, and exposure above the control levels.

radiological survey   A survey of radioactivity in a given location to assure that levels are below regulatory standards or institutional requirements. Generally they are regularly scheduled but they may be performed if a special circumstance arises.

radiology   The diagnostic and therapeutic applications of radiant energy, including X rays and radioisotopes.

radionuclide   *See* radioisotope.

radiosensitivity   The susceptibility of cells, tissues, organs, organisms, etc. to injury from ionizing radiation.

radium (Ra)   A white, highly toxic, luminescent, radioactive group IIA element, atomic number 88, atomic weight 226.03, discovered by the Curies in 1898; used in medicine.

radius   A straight line from the center of a circle or sphere to the circumference; the bone on the thumb side of the human forearm.

radon (Rn)   A colorless, radioactive group VIII noble gas element, atomic number 86, atomic weight 222; used for cancer treatment and as a tracer.

raffinate   The undissolved portion of an oil in solvent refining.

rale   An abnormal breath sound emanating from the chest, detectable with a stethoscope.

RAM   Random access memory; reliability, availability, maintainability; responsibility assignment matrix.

Raman, Sir Chandrasekhara Venkata   Indian physicist (1888–1970) awarded the 1930 Nobel Prize in Physics for his discoveries relating to the scattering of light; the Raman effect is named for him.

Raman spectroscopy   An analytical spectroscopic technique that now incorporates lasers used as a complementary tool to infrared spectroscopy for fingerprinting certain compounds.

Ramazzini, Bernardino   Italian physician (1633–1714) and pioneer of occupational health. In *De Morbis Artificum Diatriba* (Diseases of Workers) he wrote about occupational diseases and environmental hazards such as exposure to lead by potters and painters and made observations on epidemics.

Ramsey, Sir William   British chemist (1852–1916) born in Scotland who was awarded the 1904 Nobel Prize in Chemistry for his discovery of the noble gases.

random error   Error that can not be attributed to a specific cause and is believed to follow a normal or gaussian distribution.

random noise   An instantaneous sound or signal that occurs as a nonroutine function of time.

Raney nickel   A gray powder that ignites spontaneously in air and is used as a catalyst.

range   The difference between smallest and largest values; how far radiation will penetrate a given material.

Rankine, William John Macquorn   British engineer (1820–1872) born in Scotland who studied steam engines, machinery, and shipbuilding; helped develop thermodynamics and the theories of elasticity and waves.

Rankine temperature   An absolute temperature scale based on degrees Fahrenheit ($°F + 460 = °R$)

Raoult, Franois Marie   French chemist (1830–1910) who studied solution chemistry and vapor pressure. Raoult's law is named for him.

Raoult's law   Relates the vapor pressure of a solution to the number of solute species dissolved in it.

rapeseed oil   A yellow-brown, viscous vegetable oil from which canola oil is obtained. Produced mainly in Canada.

rare earth   Any of the 15 elements in group IIIB of the periodic table (atomic numbers 57 to 71) that are chemically related.

rare gas   Any of the six gases occupying column VIII of the periodic table; also called inert (obsolete) or noble gases.

rash   Abnormal or unusual reddening of the skin.

Rayleigh, Lord John William Strutt   English physicist (1842–1919) awarded the 1904 Nobel Prize in Physics for investigations of gases and the discovery of argon.

Raynaud, Maurice   French physician (1834–1881) who described the disease named after him in his 1862 MD thesis entitled, "Local Asphyxia and Symmetrical Gangrene of the Extremities."

Raynaud's syndrome   Pain, paleness, cold, numbness, and tingling of the fingers and toes caused by repeated exposure to vibration.

Rb   Rubidium.

RBC   Red blood cell(s).

RBE   Relative biological effectiveness.

RCA   Radiological control area.

RCRA   The 1976 US EPA Resource Conservation & Recovery Act and subsequent amendments as codified in Title 40 CFR, Parts 260–270. These regulations provide for the protection of human health and the environment through proper management and minimization of hazardous wastes.

RCRA-regulated   Describes waste materials containing hazardous constituents regulated by US EPA under regulations 40 CFR, Part 260–270.

RDA   Recommended dietary allowance; established by the Food and Nutrition Board of the National Academy of Sciences.

R&D   Research and development.

Re   Rhenium; Reynolds number.

reaction   A process involving a chemical or nuclear change occurring by combination, replacement, or decomposition.

reaction order   A description indicating the rate and mechanism by which a chemical reaction occurs.

reactivity   A description of chemical change, e.g., burning, combining violently with air or water.

reagent   A substance used in a reaction to detect, analyze, or measure another substance.

real-time   Term describing the performance of a computer during the actual time the physical process occurs.

reasonably anticipated to be carcinogens   A definition used in the *NTP Annual Report on Carcinogens* to mean the following   A. There is limited evidence of carcinogenicity from studies in humans, which indicates that casual interpretation is credible, but that alternative explanations, such as chance, bias or confounding, could not adequately be excluded. or B. There is sufficient evidence of carcinogenicity from studies in experimental animals which indicates that there is an increased incidence of malignant tumors: (a) in multiple species or strains, or (b) in multiple experiments (preferable with different routes of administration or using different dose levels), or (c) to an unusual degree with regard to incidence, site or type of tumor, or age at onset. Additional evidence may be provided by data concerning dose-response effects, as well as information on mutagenicity or chemical structure.

recognition   The distinguishing of a hazard; one of the four basic principles of industrial hygiene, i.e., anticipation, recognition, evaluation, and control.

recoil energy   The energy emitted, produced, or shared when an object hits, interacts, or collides with another; the energy emitted when a nucleus undergoes fission or radioactive decay.

recombinant DNA   The formation of new DNA from existing DNA from other sources.

recommended exposure limit (REL)   NIOSH-recommended occupational exposure limit that over a working lifetime will protect the health and safety of the worker.

recordable injury   An injury occurring in the workplace that must be reported under OSH Act of 1970.

recrudescence   Relapse; return of symptoms after a period of remission.

red blood cell (RBC)   Erythrocyte; a blood cell that contains hemoglobin.

Red Dye No. 2   A type of dye reported to be carcinogenic.

red lead   Red marine paint containing iron oxide.

red tide   A discoloration of sea water due to the presence of a high concentration of undesired microorganisms, which cause a substance toxic to humans to be produced by shellfish.

redox   Oxidation-reduction occurring simultaneously; as in a redox reaction.

reducing agent   A substance that combines with oxygen or loses electrons.

reduction   Addition of one or more electrons to an atom or ion; opposite of oxidation.

reflux   Prolonged heating that allows the vapors to condense and return to the reaction vessel containing the liquid being heated; a technique used to assist the reaction between combined reagents.

refractory   Term describing a ceramic material of low thermal conductivity able to withstand very high temperatures without change, commonly used to line furnaces.

refrigerant   A substance that by changing physical state lowers the temperature of its environment as a result of its latent heat.

regulatory standard   Laws and standards promulgated by a governing body, e.g., federal, state or local, to which compliance is mandatory.

REL   Recommended exposure limit as determined by NIOSH.

relative error   The ratio of the absolute error to the exact value.

relative humidity (r.h.)   The ratio of the quantity of water vapor in air versus the value for saturated air at a specific temperature.

reliability   The measure of sustained performance over a time period.

relief/safety/vacuum breaker reseat pressure setpoint   Value of the pressure/ vacuum required to restore a valve to its original relief/safety/vacuum breaker lift pressure setpoint; the value of the pressure/vacuum required to actuate a valve or rupture disk.

rem   Roentgen-equivalent-man.

render   To separate fat from other animal tissue.

renal   Pertaining to the kidney.

rennin   A digestive enzyme secreted by the stomach to curdle milk, also called chymosin.

REP   Roentgen equivalent physical.

repeatability   The ability to obtain duplicate, triplicate, etc. results that are within statistical acceptability; precision.

repetitive strain injury   An injury or illness due to repeated exposure to the same conditions and physical body movements; also known as repetitive motion injury.

replicate   An independently obtained duplication of the desired value within acceptable statistical limits.

representative sample   A portion or aliquot of a particular lot or batch of a manufactured or synthesized substance that is assumed to be indicative of any part of that particular production run and used for analysis to show that the quality or properties of the run conform to required standards.

reproducibility   A value for an analytical determination or experiment that can be produced again in another experiment or determination.

reproductive toxicity   An effect that is harmful to a species' ability to reproduce a healthy offspring; different from developmental toxicity.

resin   A solid or semisolid amorphous mixture of organic compounds without a definite melting point and incapable of crystallizing.

resist   A substance that prevents fixing of a dye or dissolving of a material; used for making patterns or designs.

resistance (R)   Opposition to air flow through canisters, chemical cartridges, or particulate filters in respiratory protection devices; electrically, volts per ampere.

resonance   A mathematically based concept used in chemistry to describe the true structure that cannot be otherwise accurately depicted. Used in spectroscopy to mean a condition in which the energy of incident radiation equals that of the absorbing entity. Used in sound to mean that an object or air volume strengthens a sound at a particular frequency.

respirable   The potential to be inhaled and deposited in the lower respiratory tract.

respirable size particle   Particle of a size that permits deep lung penetration on inhalation, usually considered to be less than 5 micrometers in diameter.

respirator   A device to protect the wearer from inhalation of hazardous substances that may or may not provide oxygen or air and may cover all or specific portions of the head and face.

respiratory system   The nose, mouth, nasal passages, pharynx, larynx, trachea, bronchi, bronchioles, air sacs (alveoli), lungs, and muscles of respiration.

restricted area   A controlled-access area to protect individuals from exposure to harm.

restricted-use pesticides   Any pesticide or pesticide use classified for restricted use by the US EPA.

restricted work envelope   The portion of the work envelope in robotics restricted by limiting devices that the robot cannot exceed.

reticle   The scale or grid on the eyepiece of a microscope.

retina   The inner part of the eye that receives and transmits the image formed by the lens.

retort   A type of laboratory glassware with a round bottom and an extended sidearm; used to distill volatile materials.

return air   Air that is removed from an area and returned to it as supply air; sometimes, but not necessarily, after treatment for removal of contamination.

rev   Revision; revolution.

review   A formal, systematic inspection of a program or project.

Reynolds, Osborne   English engineer, physicist, and educator (1842–1912) known for his work in hydraulics and heat transfer and the parameters (Reynolds stress and Reynolds number) named after him.

Reynolds number   The Reynolds stress in fluids with turbulent motion and the Reynolds number used for modeling in fluid flow describe turbulence; values above 2500 indicate turbulence; lower values denote smooth flow.

rf, RF   Radio frequency.

RFD   Reference dose.

RFETS   Rocky Flats Environmental Technology Site.

RFP   Request for proposals; Rocky Flats Plant (obsolete; *See* RFETS)

Rh   Rhodium.

r.h.   Relative humidity.

rhenium (Re)   A silver-white group VIIB metallic element; atomic number 75, atomic weight 186.21; flammable if powdered; used for alloys and high-temperature components and as a catalyst.

rheology   The study of deformation and flow of materials with respect to stress, strain, and time.

rheometer   A device used to measure viscosity and elasticity of resins and polymers.

rhesus factor (Rh factor)   A substance in red blood cells; 85% of the American population test positive (Rh positive) for it.

Rh Factor   *See* rhesus factor.

rhin-, rhino-   Pertaining to the nose.

rhodium (Rh)   A white group VIII metallic element, atomic number 45, atomic weight 102.91; flammable if powdered; used for high-temperature alloys, jewelry, and coatings.

rhodopsin   A red light-sensitive pigment in the human eye.

RIA   Radioimmunoassay.

riboflavin $C_{17}H_{20}N_4O_6$; vitamin B_2.

ribonucleic acid (RNA)   A universal constituent of all living cells made up of amino acids that are needed in protein synthesis.

ribosome   The smallest structure in a cell.

Richards, Theodore William   The first American (1868–1928) to win a Nobel Prize in Chemistry in 1914 for work in the determination of the atomic weights of about 60 elements indicating the existence of isotopes.

right-to-know law   The common name for the 1983 OSHA Hazard Communication Standard.

RIMS   Risk and Insurance Management Society.

risk   A measure of the probability and the consequence of the hazards of an activity or condition.

Risk and Insurance Management Society (RIMS)   205 E. 42nd St., New York NY 10017.

risk assessment   An evaluation of the probability, extent, and impact an undesirable event could have on the health and welfare of humans and the environment.

risk factor   A parameter or characteristic included in the risk assessment process.

risk management   The process of managing and controlling risks to obtain a desired level.

risk perception   The risk associated with a particular event by an individual or group that is not necessarily the correct assessment.

RM   Raw material; resources management.

rms   Root mean square.

Rn   Radon.

RN   Registered Nurse.

RNA   Ribonucleic acid.

roasting   Heating in the presence of air or oxygen.

Robinson, Sir Robert   English chemist (1886–1975) awarded the 1947 Nobel Prize in chemistry for work on plants with biochemical significance.

Rochelle salt $KNaC_4H_4O_6 \cdot 4H_2O$; potassium sodium tartrate, used for mirrors and electronics and as a laxative.

rock salt   Large crystals of common salt, sodium chloride.

Rocky Flats Environmental Technology Site (RFETS)   PO Box 464, Golden CO 80402-0464.

rodenticide   A pesticide designed to kill rodents.

roentgen (R)   A unit of exposure to ionizing radiation. Defined as the amount of gamma or X rays required to produce ions carrying 1 electrostatic unit of electrical charge in 1 cubic centimeter of dry air under standard conditions. Named after Wilhelm K. Roentgen, a Nobel Prize-winning German physicist who discovered X rays in 1895.

Roentgen, Wilhelm Konrad   A German physicist (1845–1923) awarded the first Nobel Prize in Physics in 1901 for the discovery of X rays in 1895.

roentgen-equivalent-man   A unit of absorbed radiation dose in biological matter; equal to the absorbed dose in rads, multiplied by a quality factor to express the relative biological effectiveness of the radiation. *See* quality factor.

roi   Return on investment; region of interest.

rolfing   Deep massage.

ROM   Read-only memory; range of motion.

room temperature   68–77°F or 20–25°C.

root mean square   The square root of the arithmetic mean of the squares of a set of numbers; used to determine the deviation of a value from a desired value.

rosacea   Pustules or hyperplasia of the sebaceous glands, usually on the face.

rosin   The resin of pine trees used to produce turpentine and other products; used for the manufacture of soaps, gums, and varnishes.

rotary kiln   A revolving type of cylinder used in furnaces to heat materials.

rotometer   An instrument to measure the flow rate of fluids, i.e., gases and liquids.

rouge   A red pigment, often containing finely ground iron oxide; used in cosmetics and as a polishing agent.

route of entry   The path by which a biological, chemical, or physical agent may enter the body; most often inhalation, ingestion, and dermal.

Rowland, Frank Sherwood   American chemist (1927–) who shared the 1995 Nobel Prize in Chemistry for work in atmospheric chemistry.

royal jelly   A waxy mixture of compounds secreted by honeybees.

RPG   Radiation protection guide.

RQ   Reportable quantity.

RRT   Regional response team.

R_t   Reliability.

RS   Registered Sanitarian.

RSD   Relative standard deviation.

RSO   Radiological safety officer.

RT   Radiographic testing.

RTECS   *Registry of Toxic Effects of Chemical Substances*, published by NIOSH.

RTK   Right to know.

RTP   Request for technical proposals.

RTR   Real-time radiography.

Ru   Ruthenium.

rubidium (Rb)   A soft, silvery white group IA alkali metal element, atomic number 37, atomic weight 85.47; ignites in air and reacts violently with water; used as a catalyst and in photocells.

rubber   An amorphous, pliable solid; a generic term including many types of natural (polyisoprene) and synthetic rubber.

ruby   Synthetic rubies are composed of aluminum oxide and are used for lasers; natural rubies are red, precious gems.

rupture disk   A thin piece of metal or other material that breaks when subjected to a certain pressure.

rust   Ferric oxide, Fe_2O_3; the reddish corrosion product of iron and oxygen; more generally a mixture of oxides.

ruthenium (Ru)   A hard, white group VIII metallic element, atomic number 44, atomic weight 101.07; used in alloys and to harden jewelry.

Rutherford, Sir Ernest (First Baron Rutherford of Nelson)   New Zealand-born physicist (1871–1937) who lived in England; awarded the 1908 Nobel Prize in Chemistry for establishing the theory of a nuclear atom.

rutherfordium (Rf)   A transuranic element, atomic number 104, atomic weight 261.

Ruzicka, Leopold Stephen   Croatian-born chemist (1887–1976) who moved to Switzerland; shared the 1939 Nobel Prize in Chemistry for organic syntheses of ringed molecules, terpenes (found in the essential oils of many plants), and sex hormones.

S

σ   Lower case Greek letter sigma.

s   Second.

S   Sulfur; south; entropy; siemens.

Σ   Upper case Greek letter sigma.

S&S   Safeguards and Security.

SA   Safety analysis; statistical applications.

SAAM   Selective alpha air monitor; an instrument that continuously monitors air for alpha-emitting radioactive material. The instruments are located throughout areas where radioactive materials exist and are set to alarm when airborne radioactive contamination exceeds a predetermined level.

SAB   Science Advisory Board.

sabulous   Gritty, sandy.

saccharides   Carbohydrates, including sugars.

saccharin $C_7H_5NO_3S$; a white powder artificial sweetener; National Academy of Science classified as a weak carcinogen, hence its use is banned in some countries but not the US. Delisted in the NTP 9th Report On Carcinogens as "reasonably anticipated to be a human carcinogen."

sacculus   Having the shape of a sack, pouch, or bag.

sacro-   Prefix referring to the sacrum.

sacrum   The triangular bone between the fifth lumbar vertebra and the coccyx; the base of the vertebral column.

SAE   Society of Automotive Engineers. The initials are often applied to specifications and tests for motor oil and fuel.

safe   Essentially without risk, harm, or danger.

safeguards   A system of physical protection, material accounting, and material control measures to deter, prevent, detect, and respond to unauthorized possession, use, or sabotage of special material.

safety analysis   The process to systematically identify operations and hazards of a facility or process; describe and analyze the measures to eliminate, control, or mitigate identified hazards; and analyze and evaluate potential accidents and their associated risks.

safety analysis report (SAR)   A report describing all aspects of a nuclear facility, including the findings of the safety analysis process for the facility and/or its operations. SAR documents describe a facility and its design features related to safety; operations conducted at the facility, operational safety requirements (OSRs), support functions and their impact, safety analyses performed, and the findings of the safety analysis and assess the risk to the public, employees, facility, and environment resulting from normal operations, operational accidents, and natural events. Preliminary SARs are prepared in the design phase of new facilities, and a final SAR is prepared and approved before starting operations.

safety can   An approved container, not more than five gallons, with a spring-closing lid and spout cover and flash arrestor; designed to safely relieve internal pressure when subjected to fire or heat exposure.

safety engineering   Application of engineering principles to safety-related issues or concerns.

Safety Equipment Institute (SEI)   1307 Dolly Madison Blvd., Suite 3A, McLean VA 22101.

safety glass   Shatterproof glass usually composed of multilayered glass bonded or compressed so that shattering is not possible.

safety limit (SL)   Limits on process variables necessary to protect the physical barriers guarding against uncontrolled release of hazardous materials.

safety (scram) systems   Systems designed to initiate automatic reactor protection by reducing reactivity to a safe level, thereby shutting down the reactor.

salicylate $C_7H_5O_3{}^-$; the anion of compounds used to reduce pain and elevated temperature.

sallow   Having a sickly, yellowish complexion.

Salmonella   A genus of Gram-negative, rodlike bacteria.

salt   A compound formed by the neutralization of an acid with a metal or base; e.g., table salt, sodium chloride; potassium nitrate.

samarium (Sm)   A hard, brittle, lanthanide group IIIB metallic element, atomic number 62, atomic weight 150.35; used in lasers and metallurgy.

sample   Part of a group or larger set; a process used to obtain a product on which a chemical, biological, physical, or statistical analysis is performed to determine whether the product meets or exceeds limitations and criteria.

sample medium   A material on which a substance is absorbed, adsorbed, or collected.

sample volume   The amount of fluid (e.g., air) pulled through a pump and onto a medium to collect a representative sample for analysis.

sampling strategy   A plan devised to obtain valid, representative sets of samples for analysis to determine whether contaminants are present.

sand   Particles of silicon dioxide ranging in size from 1/16 to 2 mm.

Sandia National Laboratories Albuquerque (SNLA)   DOE, PO Box 5800, 7011 East Ave., Albuquerque NM 87185-5800.

Sandia National Laboratories Livermore California (SNLL)   DOE, PO Box 969, Livermore CA 94551.

Sanger, Frederick   English biochemist (1918–) awarded the 1958 Nobel Prize in Chemistry for work on protein structures, particularly insulin.

sanitize   To reduce contaminants to safe levels as determined by public health requirements.

saponification   The heating of an ester with aqueous alkali to form an alcohol and the sodium salt of the acid called a soap.

sapro-   A prefix meaning putrid or rotten.

SAR   Safety analysis report; simultaneous activity request; structure activity relationship.

SARA   The EPA Superfund Amendment and Reauthorization Act (Emergency Planning & Community Right To Know Act).

sarco-   A prefix meaning flesh.

sarcoid   Resembling flesh.

sarcoma   A malignant tumor arising from connective tissue.

sarin $[(CH_3)_2CHO](CH_3)FPO$; A cholinesterase inhibitor, highly toxic by inhalation and skin absorption.

SAS   Society for Applied Spectroscopy.

saturated compound   An organic compound without double or triple bonds.

saturation   A solution that contains the maximum quantity of a dissolved substance at a specific temperature; the atomic state in which all bonds are attached to other atoms, e.g., the group of organic compounds referred to as paraffins.

saturnine   Related to or produced by lead.

saturnism   Lead poisoning.

Savannah River Site (SRS)   DOE, PO Box A, Aiken SC 29802.

Saybolt universal second (SUS)   A unit of viscosity defined as the number of seconds required for an oil heated to 130°F (lighter oils) and 210°F (heavier oils) to flow through a standard orifice and fill a 60-ml flask.

Sb   *Stibium* (Latin); antimony.

Sc   Scandium.

SCA   Single-channel analyzer.

scale   A weighing apparatus; deposition of minerals; markings or graduations.

scale-up   The process of taking a smaller operation (pilot) or reaction to one of a larger scale (plant).

scandium (Sc)   A silver-white, lightweight group IIIB metallic element, atomic number 21 atomic weight 44.96; used as a tracer and in the semiconductor field.

scanning electron microscope (SEM)   A microscope that produces a three-dimensional image by passing a focused beam of electrons across the surface of the object being studied.

scaphoid   Boat shaped.

scato-   A prefix pertaining to fecal matter.

scavenger   A substance added to remove impurities or to react with oxygen or nitrogen.

SCBA   Self-contained breathing apparatus.

SCE   Sister chromatid exchange.

SCEM   Scanning electron microscope.

SCG   Storage compatibility group (chemicals that can be stored together safely).

Schawlow, Arthur Lawrence   American physicist (1921–1999) who shared the 1981 Nobel Prize in Physics for contributions to laser spectroscopy.

Scheele, Carl Wilhelm   German-born Swedish chemist (1742–1786) who discovered oxygen, chlorine, glycerine, hydrogen sulfide, and several types of acid.

schisto-   A prefix meaning split.

Schrödinger, Erwin   Austrian physicist (1887–1961) who shared the 1933 Nobel Prize in Physics for the discovery of new productive forms of atomic theory, e.g., wave mechanics.

scientific notation   A method of representing very large or very small numbers as the products of a nonexponential term and an exponential term of the form $n \times 10^m$ where n is a number between 1 and 10 and m is a whole number that may be either positive or negative.

scintillation counter (detector)   An instrument that detects and measures ionizing radiation by counting the light flashes (scintillations) produced by radiation.

scler-   A prefix meaning hardness.

sclera, sclerae   The tough, white fibrous outer covering of the eyeball, continuous from the cornea to the optic nerve sheath.

scleraderma   A skin disease in which the skin becomes thick.

sclerosis   Hardening of tissue.

scopolamine $C_{17}H_{21}NO_4$; "truth serum;" also used for motion sickness.

scotoma   A blind spot in the vision field.

scram   Initiation of an operation to rapidly make a critical system subcritical.

scrap   Material generated during processing that is no longer suitable for use.

screen   A structure of woven material such as wire, cloth, or polymer to separate particles of a specified size (mesh).

scrubber   A device to remove components of gases or vapors by passing them through a liquid in which they are soluble or react.

SCUBA   Self-contained underwater breathing apparatus.

SDWA   Safe Drinking Water Act.

Se   Selenium.

Seaborg, Glen Theodore   American chemist (1912–1999) who shared the 1951 Nobel Prize in Chemistry for discoveries in the transuranium elements.

sealant/encapsulant   A substance that either penetrates or coats another material to form a tough or protective skin.

sealed source   Nuclear material (usually for testing and calibration) that has been packaged to be environmentally and critically safe.

seborrhea   A disease of the sebaceous glands.

sebum   Waxy skin secretions.

sec   Second; section; secant.

secondary radiation   Radiation originating as the result of absorption of other radiation.

secondary standard   A reference standard that can be directly traced to primary standards, that is used for calibration, and whose accuracy can be maintained through reasonable handling and treatment.

sedative   An agent known to induce relaxation and depression of the central nervous system.

sedentary   Sitting; not exercising.

sediment   Solid material that has settled out from being suspended in a liquid as a result of gravity.

seed   Part of a plant that includes the embryo and is sown to grow additional plants.

SEI   Safety Equipment Institute.

select carcinogen   Any agent that is regulated by OSHA as a carcinogen and listed as Known to be Carcinogens by the National Toxicology Program (NTP); listed in Group 1, Carcinogenic to Humans by the International Agency for Research on Cancer (IARC); listed in either Group 2A or 2B by IARC; or listed under the category Reasonably Anticipated to be Carcinogenic by NTP.

selenium (Se)   A nonmetallic group VIA element, atomic number 34, atomic weight 78.96; poisonous to some animals but an essential element for humans; used in electronics and ceramics, as a catalyst, in the rubber industry, and as a food supplement.

self-contained breathing apparatus (SCBA)   An air- or gas-supplying unit that utilizes an air or gas container carried by the user.

SEM   Scanning electron microscope (microscopy).

Semenov (Semyonov), Nikolay Nikolayevich   Russian chemist (1896–1986) who shared the 1956 Nobel Prize in Chemistry for research into the mechanism of chemical reactions.

semiconductor   A substance, used in solid-state physics, with electrical conductivity properties between those of a conductor and a nonconductor.

Semiconductor Industry Association (SIA)   4300 Stevens Creek Blvd., San Jose CA 95129.

sensitivity   An indication of the minimum amount of a substance that can be reliably detected or that will produce an abnormal effect or response (e.g., in biological systems).

sensitization   An immune-response reaction in which further exposure produces an immune or allergic response.

sensitizer   A substance capable of inducing inflammation, an allergic type response, or aggravation of the response. Initial exposure may or may not produce an effect; however, extended or repeated exposure may develop an allergic reaction. *See* allergen.

sensorineural   A type of hearing loss that is the result of inner ear and ear nerve damage.

sentient   Able to perceive sensation.

separation   A device, procedure, method, or process that divides and isolates one component or a series of components from a mixture of components.

sepsis   The presence of disease-causing organisms in blood or tissue.

septa   A dividing wall or partition as in nasal septum.

sequence of operations   Sequentially listed manufacturing operations required to produce specific pieces, parts, or assemblies safely.

sequestrant   A coordination or chelating compound used to remove or deactivate components in a solution or mixture that may interfere with the functions or properties of the solution or mixture.

SERI   Solar Energy Research Institute.

serious chronic effect   A level of prolonged, sustained, or repeated exposure that results in persistent effects such as mortality, disfigurement or deformity, or a major reduction in functional capacity.

serum   Blood plasma without fibrinogen.

SES   Standards Engineers Society.

SETA   Seta flash closed tester.

Seveso   The site in Italy of a large dioxin spill.

sewage   Waste liquids or solids disposed via drains.

sex chromosomes   Female X and male Y.

Sexually Transmitted Disease (STD)   A term that includes more than venereal diseases.

SG   Specific gravity; strain gauge; safeguards.

shall   The word used to denote a requirement or regulation as opposed to a recommendation.

sharp   An item capable of puncturing a fabric or skin, e.g., needle, scalpel.

Shaver's disease   A type of fibrous pneumoconiosis associated with aluminum workers breathing fumes of aluminum oxide and silica.

shielding (shield)   A material or obstruction that serves to protect personnel or material from adverse effects.

Shockley, William Bradford   English-born American physicist (1910–1989) who shared the 1956 Nobel Prize in Physics for inventing transistors.

shoe cover (booty)   A covering to protect shoes from contact with contamination.

short-term exposure limit (STEL)   A 15-minute time-weighted average exposure to airborne contaminants recommended by the ACGIH®, which should not be exceeded at any time during a work shift. *See* threshold limit value (TLV®).

should   The word used for a recommendation as opposed to a requirement. *See* shall.

Si   Silicon.

SI   Le Système International d'Unités; the international set of rules for units of measurement, generally metric.

SIA   Semiconductor Industry Association.

sick building syndrome   A poorly defined medical disorder affecting the inhabitants of buildings; caused by structures tightly constructed and well insulated and the emission of off-gases and vapors from building materials.

side-chain   A sequence of carbon atoms in organic compounds extending in a straight line.

sidero-   A prefix relating to iron or steel.

siderosis   A benign pneumoconiosis caused by the inhalation of iron oxide particles or dust; associated with welders, grinders, polishers, and boiler and metal workers exposed to these materials.

sieve   A structure of woven material such as wire, cloth, or polymer to separate particles of a specified size (mesh). *See* screen.

sievert (Sv)   An International System of Units (SI) designation of dose equivalent; 1 Sv = 100 rem.; J/kg.

significant   Important, meaningful, and representative of a larger set.

significant figures   The digits of a number in a decimal form that are needed to represent accuracy.

silanes   Compounds of silicon and hydrogen; analagous to alkanes.

silica   Silicon dioxide, e.g., sand, quartz.

silica gel   An amorphous silica containing adsorbent used as a drying agent.

silicate   Any compound containing silicon, oxygen, and a metal with or without hydrogen.

silicic acid   Hydrated silica.

silicon (Si)   A nonmetallic group IVA element, atomic number 14, atomic weight 28.09; widely used in semiconductors.

silicone   A group of organosiloxane polymers with alternating silicon and oxygen atoms.

silicosis   A fibrotic disease of the lungs causing shortness of breath due to prolonged inhalation of certain types of silica dusts.

silo filler's disease   A lung disease caused by inhalation of nitrogen oxides given off in silos during fermentation.

silver (Ag)   A precious group IB metallic element, atomic number 47, atomic weight 107.87; a soft ductile valuable metal widely used in jewelry, coins, etc.; the symbol is derived from the Latin word *argentum.*

silver solder   Solder that appears silver and contains cadmium.

sinestro-   A prefix meaning left.

sinter   A cohesive mass obtained by heating without melting, e.g., sinter glass crucible.

sinusitis   An inflammation of a sinus, usually nasal.

SIO   Signal input/output.

sister chromatid exchange (SCE)   The exchange of genetic material between chromosomes during cell division; used as a test for mutagenicity.

skatole C_9H_9N; a chemical derived from feces.

skin   A notation (e.g., on a label or MSDS or ACGIH-TLV® designation) indicating possible significant contribution to overall exposure by absorption through the skin, mucous membranes, and eyes by direct or airborne contact.

skin absorption   The dermal absorption of chemical substances. A skin notation used with TLV® or PEL exposure data indicates that the substance may be absorbed by the skin, eyes, or mucous membranes.

slag   The combination of flux and impurities that rise to the surface in smelting and refining; the residue from coal gasification.

slaked lime   Calcium hydroxide, hydrated lime.

slimicide   A chemical that destroys aqueous slime-producing fungi and bacteria.

slot velocity   The linear flow rate through openings in a slot-type hood. *See* face velocity, capture velocity.

sludge   A thick mass of waste sediment, e.g., from boilers, borings, or sewage treatment.

slurry   A pourable mixture of solid and liquid.

Sm   Samarium.

Smalley, Richard Erret   American chemist (1943–) who shared the 1996 Nobel Prize in Chemistry for the discovery of fullerenes, closed spheroidal aromatic molecules with an even number of carbon atoms.

small-quantity generator   A facility defined by EPA as one that produces more than 100 kg, but less than 1000 kg, of waste per month.

SME   Society of Manufacturing Engineers; subject matter expert.

smear   A procedure in which a piece of dry filter paper is wiped across a surface and measured to determine whether the surface is contaminated.

smectic   A layered molecular structure.

smelting   A heat treatment process of ore to remove the metallic portion.

Smith, Michael   Canadian chemist (1932–) who shared the 1993 Nobel Prize in Chemistry for developing site-directed mutagenesis.

smog   A hazy appearing mixture of fog and smoke found in polluted urban areas that is irritating to the eyes and mucous membranes; caused by the effect of sun on industrial and automobile emissions.

smoke   Aerosol particles and droplets, often carbon or soot, generated by the incomplete combustion of organic material.

smoke test   A qualitative test using an irritant smoke to determine the relative fit of a respirator.

SMSA   Standard Metropolitan Statistical Area (BLS).

Sn   *Stannum* (Latin); tin.

SNLA   Sandia National Laboratories Albuquerque.

SNLL   Sandia National Laboratories Livermore.

snorkel   A device that permits the exhaust of stale air and the intake of fresh air; originally used in submarines; a flexible local ventilation exhaust duct; elephant trunk. *See* canopy hood, laboratory hood.

SO testing plan   A document that provides information on system operating testing setup, conduct, criteria, etc.

soap   The water-soluble metal salt reaction product of a fatty acid and an alkali.

Society for Applied Spectroscopy (SAS)   201 B Broadway St., Frederick MD 21701-6501.

Society of Automotive Engineers (SAE)   400 Commonwealth Dr., Warendale PA 15096.

Society of Manufacturing Engineers (SOM)   PO Box 930, Dearborn MI 48121.

Society of Plastics Engineers (SPE)   14 Fairfield Dr., Brookfield Center CT 06805.

Society of Risk Analysis (SRA)   1313 Dolly Madison Blvd., McLean VA 22101.

Society of Risk Management Consultants (SRMC)   3255 Fritchie Dr., Baton Rouge LA 70809.

Society of Toxicology (SOT)   1767 Business Center Dr., Reston VA 22190.

SOCMA   Synthetic Organic Chemical Manufacturers Association.

soda   Sodium carbonate, Na_2CO_3; any sodium-containing compound.

soda ash   Commercial grade sodium carbonate.

soda lime   A mixture of calcium oxide and sodium or potassium hydroxide.

Soddy, Frederick   English physicist (1877–1965) awarded the 1921 Nobel Prize in Chemistry for studies of the chemistry of radioactive substances.

sodium (Na)   A soft, silvery group IA alkali metal element, atomic number 11, atomic weight 22.99; flammable in air; one of the principal cations in humans; the symbol is derived from the Latin word *natrium*; used as a lab and synthetic reagent, catalyst, and heat transfer agent.

sodium azide   A highly toxic compound used for automobile air bag inflation.

Software Quality Assurance (SQA)   Planned and systematic actions that provide confidence that software will meet specific needs or requirements.

soft water   Water that is relatively free of calcium or other precipitating cations that form insoluble compounds.

solar cell   A battery or photovoltaic cell that converts the radiant energy of the sun to electrical energy by use of semiconductors.

solar energy   Energy radiated from the sun.

solder   A low-melting alloy, usually of lead and tin; used to join solids, usually metals, at relatively low temperatures, i.e., below 450°C.

solid   The most dense form of matter, in which the atoms or molecules are closest together.

solid waste management unit (SWMU)   An inactive waste disposal area defined in the EPA Resource Conservation and Recovery Act (RCRA) that represents known and unknown hazards to human health and the environment.

solubility   A measure of the amount of a substance that will dissolve in a given amount of water or other solvent.

solubility in water   A term expressing the percentage of a material, by weight, that will dissolve in water at ambient temperature. Negligible, $<0.1\%$; slight, 0.1 to 1.0%; moderate 1 to 1.0%; appreciable $>1.0\%$; complete, soluble in all proportions.

soluble   Readily dissolved in a specified fluid.

solute   The substance that is dissolved in a solvent to produce a solution.

solution   The homogeneous mixture formed by combining a gaseous, liquid, or solid substance (the solute) with a liquid (the solvent), in which the molecules or ions of the solute are uniformly dispersed throughout the solvent.

solvent   A material that dissolves other materials (solute) to form a solution; sometimes used loosely to refer to organic solvents.

solvent extraction   Removal of a substance from a solution by mixing it with a second liquid (not miscible with the first) in which the substance is more soluble.

soma-, somat-, somato-   Pertaining to the body.

somatic effects of radiation   Radiation effects limited to the body of the exposed individual, as distinguished from genetic effects, which may also affect subsequent unexposed generations.

sonogram   A two dimensional picture obtained through utilization of the echo pattern produced from high-frequency sound waves being reflected off internal organs.

soot   Fine particles, usually black, formed by combustion (complete or incomplete), which consist mainly of carbon.

SOP   Standard operating procedure; specified operating power.

soporific   Sleep inducing.

sorbant   A substance with a large capacity to absorb gases and vapors.

SOT Society of Toxicology.

sound Changes in air pressure detected by the human ear or by another device if the frequency is above or below human detection.

sound absorption Changes in sound pressure when it passes through a medium other than air.

sound level meter A device that measures sound pressure levels, usually in decibels.

sound pressure level (SPL) The level of a sound in decibels is 20 times the log to the base 10 of the ratio of this sound to the stated reference pressure.

source A radiation-producing substance in a piece of equipment designed to emit a specific amount of radiation for research, industrial, or medical use.

SO_x Oxides of sulfur; common air pollutants.

Soxhlet extractor A laboratory device used to extract alcohol or ether soluble substances from other materials.

spanish fly Derived from the European blister beetle; widely believed to be an aphrodisiac, but it is not.

sparge To flush air or some gas into a liquid to agitate.

spasm An involuntary, convulsive muscular contraction.

SPCC Spill prevention control and counter measure.

SPE Society of Plastics Engineers.

special process A test or process that requires special qualifications, controls, equipment, procedures, and/or personnel to ensure conformance to specifications; a process, dependent on the control procedures or operator skill, or both, in which the specified quality cannot be readily determined by inspection or test.

specific activity The total radioactivity of an atom or molecule per unit mass, e.g., Ci/g.

specification A detailed description of product design and performance.

specific gravity The ratio of the mass of a unit volume of a substance to the mass of the same volume of a standard substance (usually water) at a standard temperature.

specificity The degree which a device or method can accurately and precisely measure or detect a substance.

spectrograph   The recording, graph, plot, or print-out of absorbed or emitted electromagnetic radiation.

spectrophotometry   The detection, measurement, and analysis of the absorption or emission of electromagnetic radiation, commonly in the IR, visible, or UV region of the spectrum, in which the radiation detection is done using a photoelectric cell.

spectroscope   An instrument used to detect, measure, and analyze absorbed or emitted electromagnetic radiation.

spectroscopy   The study of the detection, measurement, and analysis of absorbed and emitted electromagnetic radiation.

spectrum   The full range of electromagnetic energy; the electromagnetic radiation emitted by a substance characterized by a particular wavelength or frequency by which it can be identified.

SPEGL   Short-term public emergency guidance level.

spent (depleted) fuel   Nuclear reactor fuel that has been used to the extent it can no longer effectively sustain a chain reaction.

spermato-   A prefix pertaining to sperm.

sphincter   A muscle that surrounds an orifice and functions to close it.

sphygmomanometer   A device used to measure blood pressure.

spirometer   An instrument used to measure the volume of air entering and leaving the lungs.

spontaneous fission   Fission occurring without an external stimulus.

spontaneous ignition   The automatic generation of fire when a substance reaches its ignition temperature, in the absence of spark and without the application of external heat.

spore   A dormant form of certain microorganisms, e.g., molds, bacteria, etc.; can be revived to an active form by certain stimuli.

spot test   Micro- or semi-microanalytical detection tests for chemical compounds, ions, groups, or classes using sensitive reactions.

SPR   Simplified Practice Recommendations (of the US Department of Commerce)

spray paint booth   A power-ventilated structure exhausted to the outside or suitably filtered; used to enclose paint spraying operations, contain spray vapors, gases, particulates, and other aerosols and residue and to direct them to an exhaust system away from the operator.

sputum   Fluid excreted from the upper respiratory system.

SQG   Small-quantity generator.

squamous   A scaly or platelike cell structure.

squamous cell carcinoma   A malignant tumor with a scaly or platelike structure.

sr   Steradian (solid angle)

Sr   Strontium.

SRA   Society of Risk Analysis.

SRM   Standard Reference Material.

SRMC   Society of Risk Management Consultants.

SRP   Savannah River Plant.

SRS   Savannah River Site.

SS   Stainless steel; safeguards systems; safe and secure; shift superintendent; safety stock; special source; Social Security.

SSHO   Site safety and health officer.

SSS   Systems Safety Society.

SST   Safe, secure transport; safe, secure trailer.

stability   An expression of the ability of a material to remain unchanged. A material is considered stable if it remains in the same form under expected and reasonable conditions of storage or use.

stable   Unchanged for a specified time period.

stable isotope   A nuclide that does not undergo radioactive decay.

stack   A vertical device that discharges air or other material.

stack velocity   The speed at which the effluents from a stack are discharged into the atmosphere.

stagnant   Lack of motion or movement.

stain   A dye used to color microorganisms to aid in their visual detection and observation.

stainless steel (SS)   A steel alloy containing 10–52% chromium; there are many different types and grades.

standard base cost   The cost of labor and materials used in computing yield cost.

standard conditions   The conditions of one atmosphere of pressure and 0°C.

standard deviation   The square root of the variance used to measure deviation from the mean.

Standard Industrial Classification code (SIC)   A system of identification of employment categories according to major type of activity.

Standard operating conditions (SOP)   A written description of a process or operation that describes precisely how the process is to be performed so that, theoretically, a new person can independently determine how to perform the operation.

Standard reference material (SRM)   An item or substance that comes from, and is certified by, the National Institute of Standards and Technology (NIST) and is verified as to its purity or conformance to accepted standards and tolerances.

Standard temperature and pressure (STP)   0°C and 760 mm of mercury.

Standards Engineers Society (SES)   13340 SW 96th Ave., Miami FL 33176.

Stanley, Wendall Meredith   American biochemist (1904–1971) who shared the 1946 Nobel Prize in Chemistry for virus and enzyme protein research.

stannosis   Pneumoconiosis caused by inhalation of tin-containing dust.

Staphylococcus   A genus of Gram-positive micrococcus; a generic term for any pathogenic micrococcus.

starch   A carbohydrate macromolecule with a unique repeating structural unit.

static   Without change or movement.

static pressure   The potential pressure exerted by fluids in all directions. For fluids in motion, it is measured at right angles to the direction of flow.

stationary source   A nonmoving emitter of pollution.

statistical sampling   A statistical technique used to select segments to be sampled from a total sample population; methods include probability, random; systematic; stratified; and cluster sampling.

statute   A law.

Staudinger, Hermann   German chemist (1881–1965) awarded the 1953 Nobel Prize in Chemistry for his discoveries in the field of macromolecular chemistry that led to the development of plastics.

stay time   The period when personnel may remain in a restricted area before accumulating permissible exposure.

steam   Water at 100°C and 760 mm of pressure, with a latent heat of vaporization of 540 calories per gram.

steel   An iron alloy with 0.02–1.5% carbon; There are many types of steel, e.g., stainless steel.

Stein, William Howard   American biochemist (1911–1980) who shared the 1972 Nobel Prize in Chemistry for contributions to the understanding of the connection between chemical structure and catalytic activity of the active center of ribonuclease.

STEL   Short-term exposure limit.

STEM   Scanning transmission electron microscope (microscopy).

stem cell   A cell that gives rise to a specific type of cell.

steno-   A prefix meaning narrow or short.

step-off pad   The area just outside of a contaminated area that is the area designated for donning and removing shoe covers and to survey personnel and materials to prevent the spread of contamination.

stereochemistry   A sub-branch of chemistry that studies the three-dimensional spatial configurations of molecules.

steric hindrance   An arrangement of atoms that interfere, restrict, or retard the movement of atoms within the same molecule because of their spatial orientation.

sterile   The absence of all life on or in an object; especially microorganisms.

sterility   An inability to reproduce.

sterilization   The killing of all life; to render sterile.

sterling silver   An alloy of silver legally defined as containing 925 parts of silver and 75 parts of copper.

steroid   A group of polycyclic compounds containing a 17-carbon atom fused ring system related to terpenes, including vitamins, cholesterol, certain hormones, bile acids, and other specific compounds.

STEV   Short-term exposure value.

stibialism   Antimony poisoning.

stibium (Sb)   Antimony.

still   An apparatus to convert solids or liquids to vapors that are condensed again into solids or liquids of higher purity.

still air   Air with a velocity of 25 fpm or less.

stink damp   A mining term used to denote presence of hydrogen sulfide.

stochastic effect   A biological effect on a population due to radiation exposure, when no threshold and increasing effects with increasing dose are assumed.

Stoddard solvent   An ASTN-defined petroleum distillate solvent used extensively for dry cleaning with a flash point of 100°F and an autoignition temperature of 450°F that is clear and free from suspended material, undissolved water, and objectionable odor.

stoichiometry   The quantitative aspects of chemistry; a term used particularly in balancing equations.

Stokes, Sir George Gabriel   Irish physicist and mathematician (1819–1903) who used spectroscopy to determine the chemical compositions of the sun and stars; studied diffraction and formulated Stokes' law for the force opposing a small sphere in its passage through a viscous fluid.

Stokes' law   An equation that relates terminal, settling velocity of a spherical particle in a fluid, such as air, of known conditions to its diameter.

stom-   Pertaining to the mouth.

stomatitis   Inflammation of the mucous membrane of the mouth.

STP   Standard temperature and pressure; sewage treatment plant.

straight chain   The part of organic compounds in which the carbon atoms are aligned in a straight line.

Streptococcus   A Gram-positive, chainlike genus of bacteria.

stress   A physical, chemical, biological, or emotional factor that causes bodily or a mental tension and may be a factor causing disease or fatigue.

strip mine   The method of removal of an ore or mineral from the earth after removal of the overlying layers of earth and rock.

strontium (Sr)   A soft, silvery group IIA metallic element, atomic number 38, atomic weight 87.62; easily oxidized and used in many alloys and electronics.

strontium 90   A radioactive isotope of strontium, with atomic weight 90, present in radioactive fallout from nuclear explosions; hazardous because it is assimilated and deposited in the bones.

strychnine $C_{21}H_{22}N_2O_2$; a bitter-tasting white powder alkaloid; toxic by ingestion and inhalation.

stupor Partial or nearly complete unconsciousness.

Styrofoam™ A light-weight polymer, widely used for insulation, packing, and manufacturing.

SU Standard unit.

subacute toxicity A 5- to 14-day toxicity study in animals.

subchronic toxicity An animal toxicity study longer than 2 weeks and shorter than 6 months.

subclinical The presence of a mild form of a disease without any clinical signs, symptoms, or indications.

subcritical mass The amount of fissionable material insufficient in quantity or of improper geometry to sustain a fissionable chain reaction.

subcutaneous Beneath the skin.

Subject matter expert (SME) A person with technical expertise and knowledge in a specific area.

sublimation The process in which a substance is transformed directly from a solid to a gaseous state, or from a gaseous to a solid state, without becoming a liquid.

subpart Z The OSHA Z list-Toxic and Hazardous Substances, 29 CFR 1910.1000-1500.

substantial dividing wall An interior wall designed to prevent simultaneous detonation of explosives on opposite sides of the wall. The walls are often made of 12-inch-thick reinforced concrete with reinforcing rods.

substitution The replacement of a more toxic or hazardous substance or process by one that is less so; replacing ions in chemical reactions.

substrate A solid surface that is coated or on which another material is deposited; a substance acted upon by an enzyme.

sucrose $C_{12}H_{22}O_{11}$; table sugar; a sweetener for drink and food.

sugar A sweet-tasting carbohydrate composed of one or more saccharose entities.

sulfa drug A chemical compound containing sulfur and nitrogen; highly effective in controlling certain specific types of bacteria.

sulfur (S)   A pale yellow, nonmetallic group VIA element, atomic number 16, atomic weight 32.06; a common element in nature, both free and in compounds; used in black gunpowder, to vulcanize rubber, and in insecticides, pharmaceuticals, and a variety of commercial products; "brimstone."

sulfur dioxide SO_2; A toxic, colorless, irritating gas with a sharp pungent odor; a constituent of smog.

Sumner, James Batcheller   American biochemist (1887–1955) who shared the 1946 Nobel Prize in Chemistry for his discovery that enzymes can be crystallized and confirmed the existence of protein.

sump   A device or container in which waste liquids are collected and removed by pumping.

superconductivity   A condition, usually near absolute zero, in which compounds or substances lose electrical resistance and magnetic permeability and have infinite electrical conductivity.

supercooled   Describes a liquid cooled below its freezing point that is still liquid.

supercritical fluid   A dense fluid maintained above critical temperature under pressure, with useful properties.

supernatant   The clear liquid layer above a precipitate.

supplied air respirator   A device that provides a constant flow of breathing air through hoses from a compressor or tanks.

supra-   Prefix meaning above, upon.

surface   The place, area, or location of contact between two different states or physical forms of matter.

surface tension   The net molecular attractive force exerted from below the surface on those molecules at the surface or air interface resulting from the higher molecular concentration of a liquid compared to a gas.

surfactant   Surface-active agent, detergent, wetting agent, or emulsifier; it reduces the surface tension of water, allowing that water to more easily penetrate another liquid or porous solid.

surveillance   An investigation that determines by direct observations that activities are performed in accordance with procedures, drawings, and specifications; The consistent measurement of attributes and identification of priorities requiring corrective actions.

survey meter   A portable radiation detection instrument specially adapted to establish the existence and amount of ionizing radiation present. *See* counter.

SUS   Saybolt universal seconds.

Suspect Human Carcinogen   Definition used by the ACGIH® as follows   The agent is carcinogenic in experimental animals at dose levels, by route(s) of administration, at sites(s), of histologic type(s), or by mechanism(s) that are considered relevant to worker exposure. Available epidemiological studies are conflicting or insufficient to confirm an increased risk of cancer in exposed humans.

suspension   The uniform dispersal of small, fine particles in a liquid or gas.

Sv   Sievert (J/kg)

Svedberg, Theodor   Swedish chemist (1884–1971) awarded the 1926 Nobel Prize in Chemistry for studies of colloids.

SWDA   Solid Waste Disposal Act.

sweat   The liquid emitted at the skin surface in animals and humans during perspiration.

SWIMS   Solid waste information management system.

swinging vane velometer   A device that measures air velocity by the deflection of a piece of metal or other thin plate, as measured against a calibrated scale.

swipe   A procedure to sample surface material; done by wiping a piece of soft filter paper over a representative surface area, usually 100 cm^2.

SWMU   Solid waste management unit.

SWRF   Stored waste retrieval facility.

symbiosis   When two different organisms live in close association to their mutual benefit.

symmetry   The property of having the same, identical size, shape, proportion, and position on both sides of a given, stated, or defined dividing line, plane, or otherwise described boundary, limitation, or condition.

symptom   Evidence of a disease, disorder, condition, or feeling.

synalgia   The condition in which one feels pain in a location some distance from the site of its origin; referred pain.

syncope   Fainting.

syndrome   A collection of signs and symptoms, often of a disease or disorder.

synergism   A cooperative interaction in which the result or effect is greater than the sum of the separate effects.

Synge, Richard Laurence Millington   Irish physicist (1914–) who shared the 1952 Nobel Prize in Chemistry for his studies of the physical chemistry of proteins; devised gas chromatography.

synonym   A word with the same meaning as another word; a different name for the same thing.

synthesis   Creation of a substance by causing one or more substances to react using physical, chemical, or nuclear interactions.

synthetic   Man-made as opposed to naturally occurring.

Synthetic Organic Chemical Manufacturers Association (SOCMA)   1330 Connecticut Ave., NW, Suite 300, Washington DC 20036.

SYS   System, systemic effects.

system oxygen analyzer   An instrument with an oxygen sensor for controlling nitrogen flow and alarming at 5% oxygen by volume.

systematic error   A consistent error that can be attributed to a specific cause.

systemic   Describes an effect spread throughout the body, affecting all body systems and organs, not localized in one spot or area. *See* skin absorption.

systemic effects   Effects on the metabolism and excretory functions.

systems engineering   The program management group that provides technical, administrative, and management support in the plant.

systems safety   The evaluation of all or part of a system or process for safety over a specified time period or life cyclic, taking corrective action to correct flaws, instituting procedures in the event of a systems failure, evaluating risk, and determining acceptability.

Systems Safety Society (SSS)   Technology Trading Park, 5 Escort Dr., Sterling VA 21210.

syzygiology   The study of interrelationships as contrasted to study of isolated functions.

T

τ   Lower case Greek letter tau.

θ   Lower case Greek letter theta.

T   Temperature; tritium; ton; tera, one trillion, (prefix $= 10^{12}$); tesla (Wb/m^2); uppercase Greek letter tau.

T-M   Time and materials.

2,4,5-T   2,4,5-Trichlorophenoxyacetic acid.

TA   Teaching assistant; travel authorization (form).

Ta   Tantalum.

TAC   Toxic air contaminant.

tachy-   A prefix meaning swift.

tachycardia   Excessively rapid heartbeat.

tachypnea   Excessively rapid respiration.

tacky   Sticky, adhesive.

tactile   Determined by touch.

tag   A prominent warning notice to be securely attached, which forbids the operation of a device; (as in Tag closed cup (TCC), Tag open cup, Tag open tester): abbreviation for Tagliabue; a type of apparatus for determining flash points. The Tag closed tester is intended for testing liquids with a viscosity of less than 45 SUS at 100°F. (38°C) and a flash point below 200°F (93.4°C). The Tag open tester is intended for testing liquids with low flash points in open tanks.

tailing, tails   Chromatographic peak broadening due to impurities or incorrect solvent choice; mining refuse. *See* mill tailings.

talc   A naturally occurring hydrous magnesium silicate; Mohs hardness of 1 to 1.5; toxic by inhalation.

tallow C_{16} to C_{18} animal fats.

tamper   Material used to direct the force of an explosion, by itself resisting motion.

tamper-indicating device (TID)   A device that may be used on containers and areas that, because of their uniqueness in design or structure, reveal violations of their containment integrity; includes seals, mechanisms, containers, and enclosures.

tannic acid $C_{76}H_{52}O_{46}$; a toxic, natural substance found in tree bark; used to cure and preserve animal hides and skins.

tanning   The curing process for preservation of animal hides and skins that uses chemicals, e.g., tannic acid, chromium, aluminum and zinc sulfates, and can cause occupational diseases such as dermatitis.

tantalum (Ta)   A hard, heavy group VB metallic element, atomic number 73, atomic weight 180.95; used for capacitors, electronics, and alloys; flammable when powdered.

TAPPI   Technical Association of Pulp and Paper Industry.

tar   A dark, oily semisolid mixture produced from the destructive distillation of organic material, e.g., coal, cigarettes, petroleum; a general term for a petroleum exudate.

tare   The weight removed or deducted when weighing a substance, due to the container.

target organ   That organ in the body most affected by a particular exposure, e.g., to a hazardous chemical.

Taube, Henry   Canadian born chemist (1915–) awarded the 1983 Nobel Prize in Chemistry for studies of electron transfer reactions.

Tb   Terbium; biological half-life.

TB   Tuberculosis.

TBD   To be determined.

TC   Thermocouple.

Tc   Technetium.

TCC   Tag closed cup.

TCDD   Tetrachlorodibenzodioxin; dioxin.

TCDF   Terchlorodibenzofuran.

TCE   Trichloroethylene.

T-cells   A class of lymphocytes derived from the thymus that control cell-mediated immune reactions by secreting lymphokine hormones. Three types are known: helper cells, which help the production of antibody-forming cells; killer cells, which are formed after interaction with an antigen on foreign cells; suppressor cells, which suppress production of antibody-forming cells.

TC_{Lo}   Toxic concentration low. The lowest concentration of a substance in air to which humans or animals have been exposed for any given period of time that has produced any toxic effect in humans or produced a tumorigenic or reproductive effect in animals or humans.

TCLP   Toxicity characteristic leaching procedure.

TCRI   Toxic chemical release inventory.

TD   Toxic dose.

TDL   Toxic dose, lethal; total dust loading.

TD_{Lo}   The lowest dose of a substance in noninhalation studies at which an adverse health effect is noted.

TDS   Total dissolved solids.

TE   Test engineer; totally enclosed; technical evaluation.

Te   Tellurium.

TEC   Total estimated cost.

technetium (Tc)   A radioactive group VIIB element, atomic number 43, atomic weight 98.91; used as a tracer.

Technical Association of Pulp and Paper Industries (TAPPI)   PO Box 105113, Atlanta GA 31348.

technical justification   A statement of the basis for a "repair" or "use-as-is" disposition in engineering terminology; may include comments to the effect that the item's sizing calculations show it can perform the needed function, but with a reduced design factor or that the item still complies with applicable codes and standards.

technical reviewer   An independent, outside observer trained in a specific technical field who reviews the operations or proposed project plans of another organization or group.

technical specialist   An individual assigned to the quality or safety audit team who has applicable technical experience or expertise to assist in the investigation and evaluation of the organization or activity being audited.

TEFC   Totally enclosed fan cooled.

T_{eff} Effective half-life.

Teflon™   Tetrafluoroethylene polymer; a widely used DuPont polymer.

TEL   Tetraethyl lead.

telecon approval   Document approvals received through telephone communication.

tellurium (Te)   A silvery-white, solid, toxic, nonmetallic group VIA element, atomic number 52, atomic weight 127.60; used in alloys and vulcanizing rubber and in glass and ceramics.

TEM   Transmission electron microscope (microscopy).

temper   A process to relieve stress in a substance such as glass or metal; anneal.

Temperature (T)   The thermal state of a substance and its ability to transfer heat as measured against a reference scale, e.g., Celsius, Fahrenheit, etc.

temporal effect   Effect over time.

temporary threshold shift   Recoverable hearing loss due to noise.

tendon   The tissue that connects bone to muscle.

tendonitis   Inflammation of a tendon(s).

tennis elbow   Inflammation of the lateral elbow tendon; also called epicondylitis.

tenosynovitis   A repetitive motion injury characterized by swelling and inflammation of the tendon sheath.

tenth thickness   The thickness of a given material that will decrease the amount (or dose) of radiation to one-tenth of the amount incident upon it. Two-tenth thicknesses will reduce the dose received by a factor of 10×10; i.e., 100, etc. *See* shielding.

TENV   Totally enclosed nonventilated.

tepid   Slightly warm.

TER   Teratogen.

tera- (T)   A prefix $= 10^{12}$.

teratogen (TER)   An agent that elicits permanent structural or functional malformation in the embryo or fetus exposed during development, usually at doses that do not harm the mother; derived from Greek word meaning "monster."

Teratology Society   9650 Rockville Pike, Bethesda MD 20814.

terbium (Tb)   A lanthanide group IIIB metallic element, atomic number 65, atomic weight 158.93; used for doping electronic devices.

terminal velocity   The final rate of fall of a particle in a fluid due to gravity.

terpine $C_{10}H_{16}$; a naturally occurring unsaturated hydrocarbon present in most plant essential oils and resins.

terrestrial radiation   The portion of natural radiation (background) that is emitted by naturally occurring radioactive materials in the earth.

tesla (T)   The unit of magnetic flux density; Wb/m^2.

Tesla coil   A resonant air transformer used to check glass vacuum lines for leaks.

Tesla, Nikola   Croatian-born American electrical engineer (1856–1943) who worked on electromagnetic motors that became the basis for alternating-current machinery; high-frequency electricity; developed the Tesla coil, a resonant air-core transformer, and hydroelectricity, wireless communication, solar power, and the forerunner of radar.

testosterone   An androgen produced in the testes; chief testicular hormone in men.

tetanus   An acute disease caused by a bacteria that generally enters the body via a wound; characterized by muscle spasms and rigidity; "lockjaw."

tetracycline   Any one of the tetracycline group; broad-spectrum antibacterial drug.

tetraethyl lead (TEL)   An antiknock gasoline additive no longer used in the US because of environmental lead concerns.

TFX   Toxic effects.

TGA   Thermogravimetric analysis.

Th   Thorium.

thallium (Tl)   A toxic, blue-white, solid, heavy group IIIA metallic element, atomic number 81, atomic weight 204.37; used as a rodenticide and in alloys.

THC   Total hydrocarbons.

therm   A heat unit = 100,000 Btu.

thermal decomposition   The chemical breakdown of a material as a result of exposure to heat.

thermal (slow) neutron   A neutron in thermal equilibrium with its surrounding medium; a neutron that has been slowed down by a moderator.

thermodynamics   The mathematical study of the relationship between energy, heat, work, temperature, equilibrium, and chemicals.

thermogravimetric analysis (TGA)   An analytical technique utilizing weight change as a function of temperature or time.

thermoluminescent dosimeter (TLD)   A dosimeter utilizing one or more phosphors that, when heated, produce light in proportion to their absorbed radiation dose.

thermonuclear reaction   The technical name for "fusion"; "thermo" because very high temperatures are necessary, and "nuclear" because forces in the nucleus of an atom are involved.

thermoplastic   A plastic capable of being reshaped after reapplication of heat.

thermoset plastic   A plastic that is permanently hard and shaped and that cannot be reshaped through heating.

thia-, thio-   Containing sulfur, usually replacing oxygen.

-thiol   A suffix for compounds with the —SH group.

thixotropy   A property of some gels to change from a solid to a liquid on shaking or stirring but return to a solid on sitting.

Thomson, Sir Joseph John   British physicist (1856–1940) noted for his theory of atomic structure, which earned him the first Nobel Prize in Physics in 1901.

thoracic   Pertaining to the chest or thorax.

thorium (Th)   A soft, silvery, radioactive actinide group IIIB metallic element, atomic number 90, atomic weight 232.04; used in nuclear fuel, X-ray targets, and sunlamps.

threshold   The level at which first effects are observed.

threshold limit value (TLV®)   The level of exposure to airborne concentrations of contaminants, at or below which it is believed that nearly all workers may be repeatedly exposed, day after day, without adverse effect. The TLV® was developed by the ACGIH®, which publishes these recommended exposure limits for chemical and physical hazards. Because of wide variation in individual susceptibility, however, a small percentage of workers may experience discomfort from some substances at or below the TLV®. A smaller percentage may be affected more seriously by aggravation of a preexisting condition or by development of an occupational illness. There are three categories of TLV®: time-weighted average limit, short-term exposure limit, and ceiling limit. TLV® ceiling is the concentration limit that should not be exceeded even instantaneously. For some substances, such as irritant gases, even short excursions above the TLV may not be permissible. In these cases, a ceiling limit is assigned. TLV®-STEL is the short-term exposure limit. This is a 15-minute time-weighted average exposure that should not be exceeded at any time during a work shift, even if the 8-hour time-weighted average is below the TLV. No more than four excursions per day up to the STEL are permitted, and there must be at least 60 minutes between successive exposures at the STEL. TLV®-TWA is the time-weighted average concentration limit for a normal 8-hour workday or 40-hour workweek. Nearly all workers can be repeatedly exposed at this limit, day after day, without adverse effect.

thrombosis   The presence of a blood clot in the vascular system.

thulium (Tm)   A rare earth group IIIB metallic element, atomic number 69, atomic weight 168.93; used as an X-ray source.

Ti   Titanium.

time-response relationship   Monitoring some toxic effect as a function of time. Similar to a dose-response curve, the information is medically useful.

time-weighted average (TWA)   The average exposure to a hazardous chemical over a specified time period (usually an 8-hour work period). *See* threshold limit value (TLV).

tin (Sn)   A silver-white, ductile group IVA metallic element, atomic number 50, atomic weight 118.69; widely used commercially, e.g., in alloys and to coat other metals; symbol derived from the Latin word *stannum*.

tincture   An alcoholic solution of a chemical or drug.

tinnitus   A ringing sound in the ears.

TIRC   Toxicology information response center.

Tiselius, Arne Wilhelm Kaurin   Swedish biochemist (1902–1971) awarded the 1948 Nobel Prize in Chemistry for his research with electrophoresis and adsorption separations.

titanium (Ti)   A strong, corrosion-resistant, low-density metallic element, atomic number 22, atomic weight 47.90; widely used in aircraft and other alloys.

Tl   Thallium.

TLC   Thin-layer chromatography; total lung capacity.

TLCC   Total life cycle costs.

TLD   Thermoluminescent dosimeter.

TLV®   Threshold limit value.

Tm   Thulium.

TNT   Trinitrotoluene, $CH_3C_6H_2(NO_2)_3$; the explosive compound found in dynamite and other high explosives.

TOC   Total organic carbon.

Tocopherol   Vitamin E.

Todd, Sir Alexander Robertus   Scottish chemist (1907–1997) awarded the 1957 Nobel Prize in Chemistry for his studies on nucleotides, co-enzymes, vitamins Bv-1 and E.

tolerance   The ability of an organism to resist; a limit of inaccuracy, imperfection, or impurity.

topical   Applied to the skin.

topography   Surface configuration.

torr   A unit of pressure equal to one mm of mercury (1333.2 bars). *See* pressure.

Torricelli, Evangelista   Italian mathematician and physicist (1608–1647) who invented the mercury barometer.

tort   A wrongful act, damage, or injury done willfully, negligently, or in circumstances involving strict liability.

total pressure   The sum of the velocity and static pressure equal to the total energy present in a ventilation system.

Townes, Charles Hard   American physicist (1915–) who shared the 1964 Nobel Prize in Physics for work in quantum electronics that led to masers and lasers.

toxemia   Poison that acts through blood.

toxic   Descriptive of any substance that can produce injury or illness to humans through ingestion, inhalation, or absorption through any body surface.

toxic chemical (material)   A substance shown to be lethal to laboratory animals under the following conditions. (If data on human experience indicate results different from those obtained on animals, the human data take precedence.) (1) A chemical that has a median lethal dose (LD_{50}) of more than 50 mg/kg but not more than 500 mg/kg of body weight when administered orally to albino rats weighing between 200 and 300 g each. (This dosage is defined as highly toxic.) (2) A chemical that has a LD_{50} of more than 20 mg/kg but not more than 1000 mg/kg of body weight when administered by continuous dermal contact for 24 hours or less with the bare skin of albino rabbits weighing between 2 and 3 kg each. (3) A chemical that has a median lethal concentration (LC_{50}) in air of more than 200 ppm but not more than 2000 ppm by volume or less of gas or vapor, or 2 mg/l but not more than 20 mg/l of mist, fume, or dust, when administered by continuous inhalation for 1 hour or less to albino rats weighing between 200 and 300 g each, provided such concentration is likely to be encountered by a person when the chemical is used in any reasonably foreseeable manner. *See* acute toxicity scale, highly toxic poison.

toxic dose (TD_{Lo})   The lowest toxic dose (TD) of a chemical known to cause signs of toxicity in laboratory animals (or in humans). The toxic dose is expressed in terms of milligrams per kilogram (mg/kg) of body weight of the test animal. *See* acute toxicity scale, LC_{50}, LC_{Lo}, LD_{50}, LD_{Lo}.

toxicity   An inherent property of a chemical agent, referring to the harmful effect it can have on a biological mechanism and the conditions under which this effect can occur. Compare with hazard.

toxicology   The study of the nature, effects, and detection of poisons in living organisms; Also, substances that are otherwise harmless but prove toxic under particular conditions. The basic assumption of toxicology is that there is a relationship among the dose (amount), the concentration at the affected site, and the resulting effects.

Toxicology Information Response Center (TIRC)   ORNL, 1060 Commerce Park, MS 6480, Oak Ridge TN 37830.

Toxic Substance Control Act (TSCA)   US legislation passed in 1976; legal basis for establishing and enforcing regulations by the EPA.

Toxic Substance List   An annual listing prepared by NIOSH listing toxic substances.

toxic tort   A legal suit brought about when there is exposure to a hazardous agent.

toxin   A poisonous substance.

TOXLINE   On-line toxicology information from the National Library of Medicine (NLM).

TPQ   Threshold planning quantity.

TQ   Threshold quantity.

trace element   Any element deemed essential to the organism in very low concentrations.

tracer   A substance used or added to another substance to follow its path, e.g., radioactive tracer.

trachea   Windpipe or tube that conducts air to the lungs.

trade name   The commercial name by which a chemical is known.

training   A job-specific activity that enhances or provides knowledge and skills needed to perform in the present job.

training plan   A plan that describes course management, organization, course loading and scheduling requirements, trainee management and evaluation guidelines, instructor qualifications and responsibilities, course facility and equipment requirements, test administration guidelines, training record requirements, and a course curriculum outline.

transient   Of short duration.

transmutation   The process of changing atoms of one element into atoms of another by nuclear reaction; that which the alchemist once sought to do.

transport velocity   The force required to move a specific type of particle in an air stream.

Transportation Research Board (TRB)   2101 Constitution Ave., NW, Washington DC 20418.

Transportation Safety Institute (TSI)   DOT, 6500 S. MacArthur Blvd., Oklahoma City OK 73125.

transuranic element   Any element beyond uranium in the periodic table, i.e., with an atomic number greater than 92; all transuranic elements are radioactive and produced artificially; e.g., curium, lawrencium, and plutonium.

trauma   A wound or injury.

TRB   Transportation Research Board.

treatment   A method or process designed to remove (waste); a process designed to reduce, remove, or change (biology).

Treatment, storage, and disposal facilities (TSDF)   Facilities designated by RCRA as requiring a permit to treat, store, and/or dispose of waste designated as hazardous by EPA.

tremor   Shaking or movement.

trend analysis   The systematic evaluation of data to monitor and identify changes in equipment or activity performance.

TRI   Toxics release inventory.

triage   The classification of wounded or sick during a crisis, war, or emergency to help ensure timely treatment and efficient use of resources.

trichloroethylene $CCl_2{=}CHCl$; A colorless, volatile liquid used as an analgesic and anesthetic agent (with N_2O); formerly widely used in degreasing operations.

2,4,5-trichlorophenoxyacetic acid (2,4,5-T) $C_6H_2Cl_3OCH_2CO_2H$; a toxic herbicide that often contains dioxin contaminant.

trigger finger   An occupational repetitive motion disease causing swelling of the tendon in the index finger.

trinitrotoluene (TNT)   A toxic, yellow, solid, flammable substance detonated by violent shaking, shock.

triple point   The temperature and pressure at which all three phases of a substance are in equilibrium.

tritium (T) ^3H; A radioactive isotope of hydrogen containing one proton and two neutrons; chemically identical to natural hydrogen, tritium can easily be taken into the body by any inhalation, ingestion, or absorption path. It decays by beta emission with a radioactive half-life of about $12\frac{1}{2}$ years.

trophology   The study of nutrition.

tropism   The involuntary reaction of an organism toward or away from some stimulus, e.g., heat or light.

Trouton's rule   States that the entropy of vaporization for many substances at their normal boiling points is approximately 21 cal/deg mole.

TRU   Transuranium or transuranic.

TRUPACT   Transuranic package transporter; a type B container used to transport drums or boxes of transuranic waste.

truth serum   Scopolamine, which may also be used to treat motion sickness.

trypsin   A pancreatic enzyme that produces amino acids from albuminoid foodstuffs.

tryptophan   A colorless, solid essential amino acid; formerly used to treat insomnia, but in late 1989 the FDA discontinued the use of L-tryptophan because of illnesses, e.g., muscle pain, weakness, fever, etc., reported.

TS   Technical security.

TSA   Technical safety appraisal.

TSCA   Toxic Substances Control Act.

TSCA-regulated   Describes waste materials that contain hazardous constituents regulated by the US EPA pursuant to 40 CFR, Parts 702–799; materials may or may not be contaminated with radioactive materials.

TSD   Treatment, storage, and disposal.

TSD/DOE   Transportation Safeguards Division/Department of Energy.

TSDF   Treatment, storage, and disposal facility.

TSI   Transportation Safety Institute.

TSO   Time-sharing option.

TSP   Total suspended particulates; trisodium phosphate.

TSS   Total suspended solids.

TTU   Transportable treatment unit.

tularemia   A bacterial infection of rabbits and other rodents capable of transmission to humans.

tumor   An abnormal mass of noninflammatory tissue that persists and grows independently and has no physiologic use. Tumors may be benign (not likely to recur after total removal) or cancerous (likely to re-form and eventually cause death).

tungsten (W)   A hard, brittle group VIB metallic element, atomic number

74, atomic weight 183.85; used for electric light filaments; symbol is derived from German name *wolfram*.

turbid   Cloudy.

turbulent   Nonlinear, nonsmooth as applied to air flow.

turgor   The normal pressure or tension in a cell.

Turnbull's Blue $[Fe_3Fe(CN)_6]_2$; An iron ferricyanide compound used in blueprints.

turning vanes   Devices placed in ducts at turns and elbows to help direct air flow and reduce turbulence.

turnover checklist   A standard mechanism used by operating or support personnel as an aid in verifying and recording procedural and technical requirements; provides a convenient method of denoting equipment in service, limiting conditions of operation status, surveillances in progress, and other documents oncoming personnel should review to ensure a complete transfer of building status information.

TW   Tower water; technical writer; technical writing.

TWA   Time-weighted average.

TWX   Teletype transmission.

Tygon^{TM}   A series of vinyl polymers used as lining, tubing, etc., that are corrosion resistant.

Tylenol^{TM}   *p*-Acetylaminophenol; acetaminophen; a headache remedy.

Tyndall, John   Irish physicist (1820–1893) who did research on heat radiation, acoustics, and light scattering.

Tyndall effect   The phenomenon of light scattering when passed through a fluid suspension, creating a visible cone.

Typhoid Mary   Mary Mallon, a New York City cook determined to be a carrier of typhoid in 1907.

U

U   Uranium.

UAQI   Uniform air quality index.

UAW   United Auto Workers.

UBC   Uniform Building Code.

UCL   Upper confidence limit; upper control limit.

UEL   Upper explosive limit.

UFA   United Farmworkers of America.

UFC   Uniform Fire Code.

UFL   Upper flammability limit.

UGST   Underground storage tank.

uhf   Ultrahigh frequency.

UL   Upper limit; Underwriters Laboratories Incorporated.

UL approved   Designation that product meets the standards and requirements of the Underwriters Laboratories, a private testing organization.

ulcer   Skin or mucous membrane surface lesion, usually with inflammation; infection occurs. Also called ulcus.

ULD   Upper level discriminator.

ultracentrifuge   A very high-speed centrifuge often used for separation of materials in biology and biochemistry.

ultrasonic   Sound frequencies greater than 20,000 hertz, i.e., above audible range.

ultraviolet (UV)   Electromagnetic radiation of a wavelength between the shortest visible violet and low-energy X rays, i.e., 200–400 millimicrons.

UMW   United Mine Workers.

unattended laboratory operation   A laboratory procedure or operation at which there is no person present who is knowledgeable regarding the operation and emergency shutdown procedures; An unattended laboratory operation can be one that is carried on automatically for long periods of time or overnight. In this case, special planning to circumvent potential problems is required; it can also consist of informal absences for lunch, telephone calls, etc., without coverage by a knowledgeable person. This also requires precautions to forestall problems.

uncertainty principle   The quantum mechanical theory proposed by Heisenberg that if the speed of an atomic particle is precisely known, its location is unknown and vice versa.

underground storage tank (UST, UGST)   According to US EPA, any tank, container, or pipe used for the accumulation of a regulated substance whose volume is 10% or more below ground.

Underwriters Laboratories, Inc. (UL)   333 Pfingsten Rd., Northbrook IL 60062.

Uniform Hazardous Waste Manifest   A multi-copy shipping document that must accompany hazardous waste shipments off-site from any generating source.

United Auto Workers (UAW)   8000 E. Jefferson Ave., Detroit MI 48214.

United Farmworkers of America (UFW)   PO Box 62, Keene CA 93570.

United Mine Workers of America (UMW)   900 15th St., NW, Washington DC 20005.

United States Coast Guard (USCG)   DOT, 2100 Second St., SW, Washington DC 20593.

United States Code (USC)   The code of laws, i.e., statutes of the US federal government.

United States Department of Agriculture (USDA)   *See* Department of Agriculture.

United States Metric Association (USMA)   10245 Andasol Ave., Northridge CA 91325.

United States Pharmacopeia (USP)   12601 Twinbrook Parkway, Rockville MD 20852; A private organization that establishes drug and pharmaceutical standards.

United States Recommended Dietary Allowance (USRDA)   The amount of a specific food or nutrition specified by the FDA as the basis for receiving proper nutrition and which must be specified on labels.

United States Transuranic and Uranium Registries (USTUR)   Washington State University, at Tri-Cities, 2710 University Dr., Richland WA 99352.

unit operation   A specific industrial chemical process based on a physical change that is used in manufacturing of chemicals, e.g., filtration, distillation.

unit process   A specific type of chemical reaction used in industry for the manufacturing of chemicals, e.g., oxidation.

universal donor   A person with blood type O positive.

universal precautions   A CDC guideline for control of blood-borne pathogens; established for handling human blood or certain body fluids.

unrestricted area   The area outside the owner-controlled portion of a facility (usually the site boundary).

unsaturated   Describes an organic compound with one or more carbon-to-carbon double or triple bonds.

unscheduled scram   Any unplanned termination of an experiment caused by actuation of the safety system, operator error, equipment malfunction, or a manual scram in response to conditions that would adversely affect safe operation, not including scrams during testing or check-out operations. *See* scram.

unstable   Subject to unwanted change, e.g., oxidation, degradation.

upper confidence limit (UCL)   The highest expected value for a parameter within statistical levels of confidence.

upper explosive limit (UEL)   The maximum proportion of vapor or gas in air above which propagation of flame does not occur, expressed as a volume percentage. *See* flammable limits.

upper flammability limit (UFL)   *See* upper explosive limit.

upper respiratory tract   The nose, mouth, and throat.

UPS   Uninterruptible power supply.

UPW   Uniform present worth.

uranium (U)   A naturally radioactive, dense, silvery, actinide metallic element, atomic number 92, atomic weight 238.03; used as a power source and to generate electricity.

urea CH_4N_2O; A crystalline solid in blood, urine, and lymph.

urethro-   A prefix relating to the urethra.

Urey, Harold Clayton   American chemist (1894–1981) awarded the 1934 Nobel Prize in Chemistry for his discovery of the heavy isotopes of hydrogen and oxygen.

URI   Upper respiratory infection.

uro-   A prefix relating to urine.

urology   The study of the urinary tract.

urticaria   Nettle rash; hives; elevated, itching, white patches.

USC   United States Code.

USCG   The United States Coast Guard.

USCS   United States Commercial Standard.

USDA   United States Department of Agriculture.

use-as-is   A disposition permitted for a nonconforming item when it can be established that the item under consideration is satisfactory for its intended use and that the item will continue to meet all functional requirements, including performance, maintainability, fit, and safety.

user   In robotics, a person who uses robots and who is responsible for the personnel associated with the robot operation; anyone properly trained to handle radioactive sources; any employee or group who has primary authority over a piece of equipment and who is accountable for its proper use.

USGS   United States Geological Survey.

USMA   United States Metric Association.

USP   United States Pharmacopeia.

USRDA   United States Recommended Dietary Allowance.

UST   Underground storage tank.

USTR   United States Transuranic Registry; former name of what is now the United States Transuranic and Uranium Registries (USTUR).

USTUR   United States Transuranic and Uranium Registries.

UT   Ultrasonic testing.

UTS   Ultimate tensile strength.

UV   Ultraviolet.

UV-A   The ultraviolet region between 315 and 400nm; called the black light UV region. Exposure in this region can cause eye cataracts.

UV-B   The ultraviolet region between 280 and 315 nm; called the actinic UV region. Exposure in this region can cause cataracts and corneal injuries called welder's flash.

UV-C   The ultraviolet region between 100 and 280 nm; called the actinic, far UV, or vacuum UV region. Exposure in this region can cause corneal injuries called welder's flash.

V

V   Vanadium; volt; volume.

VA   Veterans Administration; now called Department of Veterans Affairs.

vaccine   A substance composed of living or dead microorganisms used to produce antibodies in humans and help prevent the contraction of a particular disease through immunization.

vacuum   The condition when the pressure in a container is less than atmospheric pressure.

valence   *See* oxidation number.

validation   The act of giving official sanction or status to an item, process, service, or document.

vanadium (V)   A silvery-white, ductile group VB transition metal element, atomic number 23, atomic weight 50.94; used for alloys, as a catalyst, in making rubber, and as an X-ray target.

van der Waals, Johannes Diderik   Dutch physicist (1837–1923) awarded the 1910 Nobel Prize in Physics for applying the ideal gas laws to real gases and deriving the van der Waals equation of state.

van der Waals forces   Weak forces that hold atoms and molecules together.

van't Hoff, Jacobus Henricus   Dutch chemist (1852–1911) awarded the first Nobel Prize in Chemistry in 1901 for his discovery of the laws of chemical dynamics and osmotic pressure.

vapor   The gaseous phase that evolves from the normally liquid or solid state of a substance; vapors are generated by substances that can exist as liquids or solids at standard temperature and pressure (0°C and 1 atm pressure). With few exceptions, vapors are more dense than air. An example would be water vapor. *See* sublimation.

vapor density   The weight of a vapor or gas compared to the weight of an equal volume of air, expressed as a ratio.

vapor pressure (VP)   A physical property of substances that provides a relative index of material volatility. Vapor pressure is dependent only on the temperature and the liquid itself.

variable   A symbol, as in a mathematical equation, used to represent a particular value; a value that changes.

variance   A measure of the dispersion of a set of results; approval to do something normally or formerly not allowed.

Vaseline^{TM} Trademark for petrolatum or petroleum jelly.

vaso-   A prefix relating to a vessel, e.g., a blood vessel.

vasoactive   That which affects blood vessel size; that which causes vasoconstriction or vasodilation.

vat dye   A water-insoluble, easily reduced, coal tar fabric dye, e.g., indigo.

VAX   Virtual address extension.

vdf   Video frequency.

VDT   Video display terminal.

VE   Visible emissions.

vector   A living animal carrier and/or transmitter of infectious microorganisms; a mathematical quantity that expresses quantity and direction.

vehicle   A substance or medium, with no therapeutic value, used to transmit, contain, or introduce another substance, e.g., the vehicle used for dosing animals.

vein   A blood vessel; a vessel that carries blood toward the heart; a body of rock containing a high concentration of a particular substance or mineral.

Velcro™   Trademarked fabric fastener.

velocity   A vector quantity expressing movement per unit time, e.g., miles per hour, feet per minute.

velocity pressure (VP)   The pressure that is created by a moving fluid.

velometer   A device for measuring air velocity.

vena contracta   The point of maximum reduction in the diameter of a stream of air in a duct caused by its movement.

venereal   Relating to or resulting from sexual intercourse.

venom   A toxic substance secreted by an animal.

ventilation   An engineering method used to control potential airborne health hazards. Ventilation causes fresh air to circulate, remove, and replace contaminated air. There are two general types of ventilation; local and general or dilution.

ventral, ventro-   Relating to the stomach.

Venturi, Giovanni Battista   Italian physicist (1746–1822) who did research on fluids; discovered the Venturi effect, i.e., the decrease in pressure of a fluid in a pipe when the diameter is reduced by a gradual taper, which led to the carburetor and flow-collecting devices.

Venturi collector   A high-velocity collection device used to remove dust by scrubbing it through a fine water mist.

verification   The act of reviewing, inspecting, testing, checking, auditing, or otherwise determining and documenting whether items, processes, services, or documents conform to specific requirements; an act of ensuring that a condition of a vital system conforms to specified requirements.

vermicide   A substance that kills intestinal worms.

vermiculite   A lightweight mica-type substance composed of hydrated magnesium iron aluminate frequently used as a packing and absorbing material to minimize breakage because of its ability to expand; also used as an insulator and soil conditioner.

vertical entry   A confined-space entry that is accomplished by lowering from above.

vertigo   A feeling of revolving in space; dizziness, giddiness.

very high frequency (vhf)   Radio frequencies between 30 and 300 megahertz.

very high radiation area   Any area within a controlled area where an individual can receive a dose of 5 rem or greater in one hour at 30 cm from the radiation source or from any surface through which the radiation penetrates.

vesicant   Anything that produces blisters.

vesico-   A prefix relating to the bladder.

Veterans Affairs, Department of (VA)   810 Vermont Ave., NW, Washington DC 20420.

Veterans of Safety (VOS)   203 North Wabash Ave., Chicago IL 60601.

vhf   Very high frequency.

vibr.   Vibration, vibrator.

vibration   Rapid linear movement.

Vibrio   A genus of curved, mobile, Gram-negative bacilli.

vicinal   Adjoining; substituent groups next to one another.

Video display terminal (VDT)   A cathode ray tube used as an input-output device for computers or other devices.

vinyl chloride $CH_2{=}CHCl$; an explosive, toxic, and carcinogenic gas that is easily liquefied; widely used in syntheses and for polymer products.

virology   The study of viruses or viral diseases.

virulence   The relative ability of a microorganism to produce disease or a toxin to kill.

virus   A microorganism composed of a core of a single nucleic acid enclosed by a protein coat and able to replicate only within a living cell.

viscera   Body organs.

viscosity   The temperature-dependent property of a fluid that measures its resistance to flow. Also known as the internal resistance to change exhibited by a liquid or gas resulting from the combined effects of adhesion and cohesion. Unit of measure is commonly the poise.

visible radiation   350 to 800 nm; that part of the electromagnetic spectrum generally perceived by the human eye.

vitrification   The process of changing or making into a glassy substance.

vitriol   Obsolete term for some metal sulfates.

vlf   very low frequency.

VOC   Volatile organic compound.

Volatile organic compound (VOC)   Any hydrocarbon with a vapor pressure >0.1 mmHg, except methane and ethane.

volatility   The tendency or ability of liquid to vaporize. Alcohol and gasoline, which evaporate rapidly, are well-known examples of volatile liquids. *See* vapor pressure.

volt (V)   The international unit of electromotive force equal to the difference in electrical potential between two points in a wire carrying one ampere of constant current with a power of one watt; W/A.

Volta, Alessandro Giuseppe Antonio Anastasio   Italian physicist (1745–1827) who invented a device to generate static electricity, discovered methane gas, and developed the first electric battery.

voltaic cell   An electrolytic cell that generates an electric current.

volume (V)   A measure of the three-dimensional size or space occupied by a body.

volume flow rate (Q)   The amount of a fluid flowing per unit time, e.g., cubic feet per minute, liters per second.

Voluntary Protection Program (VPP)   An OSHA standard used to promote safety in the work place by conformance and compliance to voluntary, internally generated safe work practices and polices.

von Baeyer, Johann Friedrich Wilhelm Adolf   German chemist (1835–1917) awarded the 1905 Nobel Prize in Chemistry for work with dyes, photosynthesis, and aromatic compounds.

VOS   Veterans of Safety.

VP   Vapor pressure; velocity pressure.

VPP   Voluntary Protection Program.

W

W   Tungsten, wolfram (German); watt; west.

W/g   Watts per gram.

W-HA   Walsh–Healy Act (law).

WAC   Waste acceptance criteria.

WAD   Work-authorizing document.

walkdown   A detailed administrative or physical review of each process or support control system.

Walker, John Ernest   English chemist (1941–) who shared the 1997 Nobel Prize in Chemistry for his elucidation of the enzymatic mechanism underlying the synthesis of adenosine triphosphate (ATP).

Wallach, Otto   German chemist (1847–1931) awarded the 1910 Nobel Prize in Chemistry for his work with alicyclic compounds. Also did pioneering work on essential oils, sex hormones, and vitamins.

Walsh-Healy Act (W-HA)   The 1936 US federal law that established minimum occupational health and safety standards for contractors working on federal projects. Considered the forerunner of the OSH ACT.

warfarin   A rodenticide that acts by preventing the coagulation of blood (hence, use as an anticoagulant in medicine) and causing the animal to die by internal hemorrhaging.

WARNING   A labeling term that indicates an intermediate level of hazard, between "Danger" and "Caution"; an energized and/or audible annunciator or light that serves to alert an operator that action must be taken to prevent an alarm condition.

warning limits   Areas designated on gauges or charts indicating that special attention is required; quantity limits for inventory differences that, when exceeded, require investigation and appropriate action; For processing, production, and fabrication operations, warning limits will be established with a 95% confidence level.

warning odor   An odor produced by a substance added to an odorless hazardous gas, e.g., natural gas, so that its presence will be known in case of a leak.

washing soda $Na_2CO_3 \cdot 10 H_2O$. *See* soda ash.

WASP   Waste accountability, shipping and packaging.

waste   Residues that have been determined to be uneconomical to recover.

waste, radioactive   Solid, liquid, and gaseous materials from nuclear operations that are radioactive or become radioactive and for which there is no further use; Wastes are generally classified as high level (having radioactivity concentrations of hundreds of curies per gallon or cubic foot) or low level (in the range of 1 microcurie per gallon or cubic foot). De minimis levels, below which there is no health or safety concern, are being established for the most common isotopes.

water column   A unit of pressure also called inches of water.

water gas   Principally CO and H_2 with some N_2, CO_2, etc.; the flammable and explosive gas produced when steam is passed over incandescent coke.

water gauge (WG)   A device that indicates the level of water in something, e.g., a stream, a boiler or a tank. *See* water column.

water reactive   A characteristic of material, such as chemical waste, that causes it to react with water to generate explosive mixtures or toxic gases; material that contains cyanide or sulfide can release these substances on exposure to acids or bases with a pH between 2.0 and 12.5. Reactive metals (such as the alkali metals) react with water, liberating explosive hydrogen gas. The hydrogen gas is often ignited by the metal, which gets very hot during the reaction. The resulting alkali metal hydroxide is a very corrosive alkali.

Watson, James Dewey   American geneticist (1928–) who shared the 1962 Nobel Prize in Medicine or Physiology with Crick and Wilkins for determining the structure of DNA and RNA.

watt (W)   A unit of electrical power equal to one joule per second.

Watt, James   Scottish engineer (1736–1819) who invented the modern steam engine and the centrifugal governor.

watts/cm²   The unit of power density, used to denote the power or intensity of lasers.

Wavelength (λ)   The length of a wave, usually measured from peak to peak or trough to trough between two waves denoting one full cycle, usually expressed in angstroms, centimeters, meters, micrometers, or nanometers.

Wb   Weber (Vs).

WBC   White blood cell.

WBGT   Wet bulb globe temperature.

WBS   Work breakdown structure.

WC   Water column.

WDS   Wavelength dispersive spectroscopy.

Weber   Magnetic flux; Wb; Vs.

WEEL®   Workplace Environmental Exposure Level.

weight   The force by which a body is attracted to Earth; a measure of the heaviness of an object.

welding   Joining by heat; bringing together.

WERC   Wind Energy Research Center.

Werner, Alfred   Swiss chemist (1866–1919) awarded the 1913 Nobel Prize in Chemistry for developing the coordination theory of atoms in molecules, especially inorganic molecules.

Wet bulb globe temperature (WBGT)   An index of heat stress in humans that includes air temperature, humidity, air velocity, and radiation.

Wet bulb temperature (T_{wb})   The temperature measured as influenced by the exposure to the movement of air across the thermometer.

Wet globe temperature (WGT)   A temperature measurement that combines air temperature, humidity, air movement, and the result of radiant heat.

WG   Water gauge.

Wheatstone bridge   An electrical circuit used to measure electrical resistivity; used in many analytical and other instruments.

WHIMS   Workplace hazardous materials information system.

white blood cell (WBC)   Leukocyte; the cell that fights infections; phagocytes.

white damp   Carbon monoxide, sometimes found in coal mines.

white noise   Acoustical noise whose intensity is the same at all frequencies within a given band; sometimes used to mask disturbing nuisance noise.

white room   *See* clean room.

WHO   World Health Organization.

whole body counter   A device used to identify and measure the radiation in the body (body burden) of humans and animals; uses heavy shielding to minimize the interference of background radiation on ultrasensitive radiation detectors and electronic counting equipment.

Wieland, Heinrich Otto   German chemist (1877–1957) awarded the 1927 Nobel Prize in Chemistry for his work on bile acids, cholesterol, toxins, and oxidation.

Wien, Wilhelm Carl Werner Otto Fritz Franz   German physicist (1864–1928) awarded the 1911 Nobel Prize in Physics for his studies of heat radiation; he stated his displacement law of blackbody radiation spectra at different temperatures in 1893.

Wilkins, Maurice Hugh Frederick   New Zealand-born biophysicist (1916–) living in Great Britain who shared the 1962 Nobel Prize in Medicine or Physiology with Crick and Watson for determining the structure of DNA and RNA.

Wilkinson, Geoffry   British chemist (1921–) who shared the 1973 Nobel Prize in chemistry for his studies of organometallics, particularly the "sandwich" compounds, e.g., ferrocene.

Williams-Steiger Occupational Safety and Health Act   The 1970 US law that regulated occupational health and safety and established the US Occupational Safety and Health Administration (OSHA) in the Department of Labor (DOL) and established the National Institute for Occupational Safety and Health (NIOSH) of the Public Health Service (PHS) and the Centers for Disease Control and Prevention (CDC) of the Department of Health and Human Services (DHHS).

Willstater, Richard Martin   German chemist (1872–1942) awarded the 1915 Nobel Prize in Chemistry for his work with plant pigments, chlorophyll, cocaine, and atropine.

Windaus, Adolf Otto Reinhold   German chemist (1876–1959) awarded the 1928 Nobel Prize in Chemistry for his work with sterols and their connections with vitamins.

WIP   Work in process.

wipe sample (swipe or smear)   A sample made for the purpose of determining the presence of removable contamination on a surface; it is done by wiping, with slight pressure, a piece of soft filter paper over a representative type of surface area. Also known as a "swipe sample"; referred to as "smear" at some facilities.

WIPP   Waste Isolation Pilot Plant; the TRU waste depository located near Carlsbad, NM.

WIPP-WAC   Waste Isolation Pilot Plant; Waste Acceptance Criteria.

Wittig, Georg   German chemist (1897–1987) who shared the 1979 Nobel Prize in Chemistry for his development of phosphorus-containing compounds into important reagents in organic synthesis and pharmaceuticals.

WLM   Working level month(s).

WLN   Wiswesser line notation; a system used to convert chemical structures into letters and diagrams; used for structure activity predictions and by the pharmaceutical industry.

WO   Work order.

WOG   Water, oil, gas.

Wohler, Friedrich   German chemist (1800–1882) who made urea from inorganic ammonium cyanate, thus proving that organic and inorganic chemistry were linked, not separate.

wolfram (W)   Tungsten.

Woodward, Robert Burns   American chemist (1917–1979) awarded the 1965 Nobel Prize in Chemistry for his outstanding work in organic syntheses, e.g., quinine, cholesterol, and chlorophyll.

work   The energy that results when a mass is acted on by a force and caused to move or change.

Work breakdown structure (WBS)   A family tree representation and identification of a program's objectives, including the end objective and successively smaller objectives and supporting work tasks; the subdivision of a project effort.

work envelope   The volume of space enclosing the maximum designed reach or impact of a specific activity or piece of work.

work exposure assessment   An evaluation of the potential hazards associated with an individual's exposure to potentially hazardous workplace conditions and agents and an estimation of the risk.

workers compensation   A requirement under law that payments be made if an employee is injured during work.

working alone   Hazardous or nonhazardous work in locations or at times that cause an employee to be isolated from audio or visual contact for more than one hour; an unsafe situation that should be avoided.

working standards   Process items that have been frequently measured or characterized by a more accurate measurement technique and traceable to a national measurement base; used in performance testing.

Workplace Environmental Exposure Level (WEEL)®   AIHA exposure guidelines to protect workers against exposure to specific hazardous chemicals.

Workplace Hazardous Materials Information System (WHIMS)   A Canadian system that provides information on hazardous materials in the workplace including material safety data sheets (MSDSs).

workstation   The space in which an employee works; term often improperly associated only with a computer.

World Health Organization (WHO)   Avenue Appia 20, 1211 Geneva 27 Switzerland; the United Nations agency that deals with health.

World Wide Web (www)   An international personal computer-linked system that permits rapid, easy access to a wealth of information.

wormwood   A plant whose extract produces absinthe, a toxic, bitter flavoring agent not permitted in the US.

WP   Word processing; work package.

WQCC   Water Quality Control Commission.

WRAP   Waste processing and packaging.

WRC   Water Resources Council.

WSF   Water-soluble fraction.

wt   Weight.

www   World Wide Web.

X

X   Uppercase Greek letter chi.

xanth-, xantho-   A prefix indicative of a yellow color.

X-chromosome   The chromosome that determines female sex characteristics. Females have two X, and males have one X and one Y.

Xe   Xenon.

xeno-   A prefix indicating strange or foreign.

xenobiotic   A nonnatural material in the environment.

xenon (Xe)   A group VIII noble gas element, atomic number 54, atomic weight 131.30; the colorless, odorless gas is used in luminescent tubes and lasers and as an anesthetic.

xerosis   Abnormal skin dryness.

X ray   Highly penetrating electromagnetic radiation of extremely short wavelength ($0.06-120$Å) resulting from transitions of inner orbital electrons as a result of electron bombardment. These rays are sometimes called roentgen rays after their discoverer, W. K. Roentgen.

X-ray diffraction   A method of analysis using X rays; useful for determining the structure of pure solid materials; especially useful for determining crystalline silica in dust.

xylocaine   A local anesthetic.

Y

Y   Yttrium.

Yb   Ytterbium.

Y-chromosome   The chromosome that determines the sex to be male. Males have one X and one Y chromosome.

yeast   A yellowish-white, living, single-cell organism used for fermentation, baking, and brewing.

yellowcake   A synonym for uranium dioxide, it is the initial feed material to the nuclear fuel cycle and the final precipitate produced from milling uranium ore. It is used in crystalline form to pack nuclear fuel rods. The name is derived from its color and texture.

yield   In chemistry, the amount of product in a reaction or process, usually expressed as a percentage of the total starting material. The stress point of a material at which deformation occurs. The total effective energy released in an explosion, usually expressed as equivalent tonnage of trinitrotoluene (TNT). In general, the ratio of the amount of material produced or recovered to the initial or calculated amount.

YTD   Year-to-date.

ytterbium (Yb)   A group IIIB, metallic element used in lasers, atomic number 70, atomic weight 173.04.

yttrium (Y)   A group IIIB metallic element, atomic number 39, atomic weight 88.91; flammable if finely divided; used in alloys, electronics, and the nuclear industry.

Yusho disease   A disease outbreak in the Yusho area of Japan caused by exposure to cooking oil contaminated with chlorinated dibenzofurans. Chloracne is a common symptom of exposure.

Z

Z   Atomic number; the number of protons in the nucleus, characteristic for each element.

Z_{eff} Effective atomic number.

zeolite   A naturally occurring or man-made molecular sieve, filter, or ion exchange matrix composed of hydrated aluminum and sodium and/or calcium silicate.

zero defect   Devoid of defect.

Zero risk level (ZRL)   Devoid of risk.

Zewail, Ahmed   Egyptian-born American chemist (1946–) awarded the 1999 Nobel Prize in Chemistry for his studies of chemical reaction transition states utilizing femtosecond spectroscopy.

ZFF   Zinc fume fever.

Ziegler, Karl   A German chemist (1898–1973) who shared the 1963 Nobel Prize in Chemistry for his development of plastics.

zinc (Zn)   A shiny, lustrous group IIB transition metal element, atomic number 30, atomic weight 65.38; the dust is flammable and explosive; used in alloys, die casting, electronics, and fungicides.

zinc fume fever (ZFF)   Caused by inhalation of zinc oxide fumes and characterized by flulike symptoms, a metallic taste in the mouth, coughing, weakness, fatigue, muscular pain, and nausea, followed by fever and chills; symptoms occur 4–12 hours after exposure.

zinc protoporphyrin   The blood enzyme used to measure zinc exposure.

zirconium (Zr)   A hard, lustrous, gray group IVB metallic element atomic number 40, atomic weight 91.22; the dust is flammable and explosive; used in alloys.

Z list   OSHA's Toxic and Hazardous Substances Tables Z-1, Z-2, and Z-3 of air contaminants, found in 29 CFR 1910.1000. These tables record PELs, TWAs, and ceiling concentrations for the materials listed. Any material listed on these tables is considered to be hazardous.

Zn   Zinc.

zone refining   A purification method using repeated heating and crystallization to move impurities (in the liquid phase) to the end of the object.

zoology   The study of animals.

zoonoses   Diseases of lower animals that can be transmitted to and infect humans.

zoospore   A motile, asexual spore that moves by means of one or more flagella.

Zr   Zirconium.

ZRL   Zero risk level.

Zsigmondy, Richard Adolf   An Austrian chemist (1865–1929) awarded the 1925 Nobel Prize in Chemistry for work on colloids and invention of the ultramicroscope.

zwitterion   A particle that has both a positive and negative charge.

zyg-, zygo-   A prefix meaning pair or united.

zygote   The earliest stage of gestation in the development of an organism. In humans, the zygote stage is generally the first week after fertilization (conception).

-zym, -zymo   A suffix relating to fermentation.

zymase   The enzyme in yeast that converts sugars to alcohol (and CO_2).

Appendix A

Abbreviations and Acronyms

A

α   Lower case Greek letter alpha; alpha particle (radiation).

A   Atomic mass number; ampere; uppercase Greek letter alpha.

Å   Angstrom unit.

a   Atto (10^{-18}).

A&C   Abatement and control.

A&S   Accident and sickness.

A&WMA   Air and Waste Management Association.

A-E   Architect-engineer.

a.u.   Atomic units.

AA   Atomic absorption spectroscopy; activation analysis.

AAALAC   American Association for Accreditation of Laboratory Animal Care.

AAAS   American Association for the Advancement of Science.

AAEE   American Academy of Environmental Engineers.

AAI   Alliance of American Insurers.

AAIH   American Academy of Industrial Hygienists.

AALA   American Academy for Laboratory Accreditation.

AAOHN   American Association of Occupational Health Nurses.

AAOM   American Academy of Occupational Medicine.

AAPCC   American Association of Poison Control Centers.

AARCEP   Alliance for Acid Rain Control and Energy Policy.

AAS   American Astronautical Society.

AASHTO   American Association of State Highway and Transportation Officials.

AATCC   American Association of Textile Chemists and Colorists.

abamp   Absolute ampere.

ABIH   American Board of Industrial Hygiene.

abs.   Absolute.

ABSA   American Biological Safety Association.

Ac   Actinium.

AC   Alternating current.

ACC   American Chemistry Council.

acfm   Actual cubic feet per minute.

ACGIH   American Conference of Governmental and Industrial Hygienists.

ACLAM   American College of Laboratory Animal Medicine.

ACM   Asbestos-containing material.

ACOM   American College of Occupational Medicine.

ACS   American Chemical Society; American Cancer Society.

ACSH   American Council on Science and Health.

ACT   American College of Toxicology.

ACTH   Adrenocorticotropic hormone.

ADA Americans with Disabilities Act; American Dental Association.

ADI Acceptable daily intake.

ADP Adenosine diphosphate; automatic data processing; also known as electronic data processing (EDP).

ADT Automated data terminals.

AEA Atomic Energy Act.

AEC Atomic Energy Commission; original name for what are now the Department of Energy and the Nuclear Regulatory Commission.

AECF Association of Environmental Consulting Firms.

AED Aerodynamic equivalent diameter.

af Audio frequency.

AFFF Aqueous film-forming foam.

AFL-CIO American Federation of Labor-Congress of Industrial Organizations.

AFS American Foundrymen's Society.

AFSHP Association of Federal Safety and Health Professionals.

Ag *Argentum* (Latin); silver.

AGA American Gas Association.

AGSE Association of Groundwater Scientists and Engineers.

AGST Above-ground storage tank(s).

AHA American Hospital Association.

AHERA Asbestos Hazard Emergency Response Act.

AHJ Authority having jurisdiction.

AIA American Institute of Architects; Asbestos Information Association.

AIC American Institute of Chemists.

AICCPU American Institute of Chartered Casualty Property Underwriters.

AIChE American Institute of Chemical Engineers.

AIDS Acquired immunodeficiency syndrome.

AIEEE   American Institute of Electrical and Electronics Engineers.

AIHA   American Industrial Hygiene Association.

AIHF   American Industrial Hygiene Foundation.

AIME   American Institute of Mining, Metallurgical and Petroleum Engineers.

AIP   American Institute of Physics.

AISG   American Insurance Services Group.

AISS   American Iron and Steel Society.

AL   Acceptable level; action level.

Al   Aluminum.

ALAP   As low as practicable. *See* ALARA, which is preferred.

ALARA   As low as reasonably achievable

ALCA   American Leather Chemists' Association.

ALD   Average lethal dose.

ALI   Annual limits/intake.

ALR   Allergenic effects.

Am   Americium.

A/m   Ampere/meter; magnetic field strength.

A/m   Ampere/square meter; current density.

AMA   American Medical Association.

AMS   American Meteorological Society; aerial measuring system.

amu   Atomic mass unit.

ANL   Argonne National Laboratory.

ANS   American Nuclear Society.

ANSI   American National Standards Institute.

AOAC   Association of Official Analytical Chemists.

AOCS   American Oil Chemists Society.

AOM   Annual operating and maintenance cost.

APCA   Air Pollution Control Association.

APCD   Air Pollution Control Division.

APEN   Air pollution emission notice.

APhA   American Pharmaceutical Association.

APHA   American Public Health Association.

API   American Petroleum Institute.

AQRV   Air quality-related values.

AQTX   Aquatic toxicity.

Ar   Argon.

ARAR   Applicable or relevant and appropriate requirement.

ARC   American Red Cross.

ART   Aqueous recycle technologies.

ARU   Acid recovery unit.

As   Arsenic.

AS   Atmosphere, standard.

ASA   American Standards Association; American Supply Association; Acoustical Society of America.

ASCE   American Society of Civil Engineers.

ASCII   American Standard Code for Information Interchange.

A/sec   Ampere/second.

ASHAA   Asbestos in Schools Hazard Abatement Act.

ASHRAE   American Society of Heating, Refrigerating, and Air Conditioning Engineers, Inc.

ASI   Aviation Safety Institute.

ASLAP   American Society of Laboratory Practitioners.

ASM   American Society for Metals International; American Society of Microbiology.

ASME   American Society of Mechanical Engineers.

ASNT   American Society for Nondestructive Testing.

ASSE   American Society of Safety Engineers; American Society of Sanitary Engineers.

AST   Above-ground storage tank.

ASTD   American Society for Training and Development.

ASTM   American Society for Testing Materials.

At   Astatine.

ATAA   Air Transport Association of America.

at %   Atomic percent.

at. no.   Atomic number.

at. wt   Atomic weight.

atm   Atmosphere.

ATMX   Atomic materials rail transport car.

ATP   Adenosine triphosphate.

ATSDR   Agency for Toxic Substances and Disease Registry.

Au   *Aurum* (Latin); gold.

A/V   Ampere per volt; siemens (S).

AVMA   American Veterinary Medical Association.

AVO   Avoid verbal orders.

AVS   American Vacuum Society.

AWG   American wire gauge.

AWMA   American Waste Management Association.

AWQC   Ambient water quality criteria.

AWS   American Welding Society.

awu   Atomic mass unit.

AWWA   American Water Works Association.

AWWARF   American Water Works Association Research Foundation.

B

β   Lower case Greek letter beta, beta particle (radiation).

B   Boron; blower.

b   Barn.

Ba   Barium.

BAC   Blood alcohol concentration; budgeted cost at completion; biologically activated.

BACT   Best available control technology.

BASIC   Beginner's all-purpose symbolic instruction code.

bat   Battery.

BAT   Best available technology.

BATF   US Bureau of Alcohol, Tobacco, and Firearms.

BBL   Body burden level.

bbl   Barrel.

bcc   Body-centered cubic.

BCD   Bar code decal.

BCF   Bioconcentration factor.

BCHM   Board of Certified Hazard Control Management.

BCL   Battelle Columbus Laboratories.

BCM   Blood-clotting mechanism effects.

BCSP   Board of Certified Safety Professionals.

BCSPM   Board of Certified Product Safety Management.

bd ft   Board foot (feet).

BDAT   Best demonstrated available technology.

Be   Beryllium.

Bé   Baumé.

BEA   Building evacuation area.

BEI   Biological exposure index.

BEIR   Biological effects of ionizing radiation.

BEJ   Best expert judgment.

BeV   Billion electron volts.

Bi   Bismuth; biot.

BIA   Brick Institute of America; Bureau of Indian Affairs.

bit/s   Bits per second.

Bk   Berkelium.

BLD   Blood effects.

BLEVE   Boiling liquid expanding vapor explosion.

BLM   Bureau of Land Management.

BLS   Bureau of Labor Statistics.

BM   Bowel movement; building manager.

BNL   Brookhaven National Laboratory.

BNWL   Battelle Pacific Northwest Laboratories.

BOCA   Building Officials & Code Administrators International; Building Officials Conference of America.

BOD   Biochemical oxygen demand.

BOD$_5$ Biochemical oxygen demand, 5-day incubation period.

BOM   Bureau of Mines.

B.P., bp, b.pt.   Boiling point.

Bq   Becquerel.

Br   Bromine.

BSC   Biological safety cabinet.

BSI   British Standards Institute.

Btu   British thermal unit.

BW   Bandwidth.

C

C   Carbon; Celsius; Centigrade; TLV® ceiling limit; coulomb.

Ca   Calcium.

CA   Corrective action.

CAA   Clean Air Act.

CAAA   Clean Air Act Amendments.

CAD/CAM   Computer-aided design/computer-aided modeling.

CAER   Community Awareness and Emergency Response.

CAG   Carcinogen Assessment Group.

CAGI   Compressed Gas Association. Incorporated.

CAIR   Comprehensive assessment information rule.

CAR, CARC   Carcinogen.

CAS   Chemical Abstract Service.

CAS #   Chemical Abstract Service Registry Number.

CBA   Cost-benefit analysis.

CBOD   Carbonaceous biochemical oxygen demand.

CBT   Computer-based training.

CBW   Chemical and biological weapon(s).

cc   Cubic centimeter; carbon copy.

CCHO   Certified Chemical Hygiene Officer.

cd   Candela.

C_D Drag coefficient.

Cd   Cadmium.

CDC   Centers for Disease Control and Prevention.

Ce   Cerium.

cd/m²   Candela/square meter.

cdsr   Candela steradian; lumen (lm).

CEC   Council on European Communities.

CEM   Cost-estimating manager; continuous emission monitoring.

CEP   Capital equipment project.

CEQ   Council on Environmental Quality.

CER   Complete engineering release.

CERCLA   Comprehensive Environmental Response, Compensation and Liability Act.

CESQG   Conditionally exempt small quantity generator.

CET   Corrected effective temperature.

cf   Abbreviation for the Latin word *confer*; compare.

CFC   Control frequency converter; chlorofluorocarbon.

cfm, CFM   Cubic foot (feet) per minute.

CFR   Code of Federal Regulations; cooperative fuel research.

cfs   Cubic feet per second.

cg   Centigram.

CGA   Compressed Gas Association.

CGL   Comprehensive general liability.

cgs   Centimeter-gram-second system of units.

CHEMTREC   Chemical Transportation Emergency Center.

χ Lower case Greek letter chi.

CHMM   Certified hazardous materials manager.

CHO   Chemical hygiene officer.

CHP   Certified health physicist; chemical hygiene plan.

CHRIS   Chemical Response Information System; a reference tabular listing of compatible chemicals.

Ci   Curie.

CIH   Certified industrial hygienist.

CIIT   Chemical Industry Institute of Toxicology.

C/kg   Coulomb/kilogram.

Cl   Chlorine.

Cm   Curium.

CMA   Chemical Manufacturers Association.

C/m²   Coulomb/square meter; electric flux density.

C/m³   Coulomb/cubic meter; electric charge density

CMA   Chemical Manufacturers Association

CMO   Certified manufacturing operation.

CNS   Central nervous system.

CO   Carbon monoxide.

COBOL   Common business-oriented language.

COC   Cleveland open cup; continuity of combustibility.

COCO   Contractor-owned, contractor-operated.

COD   Chemical oxygen demand.

COE   Cost of energy; (US Army) Corps of Engineers.

COHb   Carboxyhemoglobin.

COHN   Certified Occupational Hygiene Nurse.

conc   Concentrated or concentration.

cp   Chemically pure.

CP   Center of pressure.

CPAF   Cost plus award fee.

CPF   Carcinogenic potency factor.

CPFF   Cost plus fixed fee.

cpm   Counts per minute; cycles per minute.

CPM   Critical path method.

CPR   Cardiopulmonary resuscitation.

cps   Counts per second; characters per second.

CPSC   Consumer Products Safety Commission.

CPU   Central processing unit.

Cr   Chromium.

CRM   Certified reference material.

CRT   Cathode ray tube; cargo restraint transporters.

Cs   Cesium.

CS   Cold storage area; Department of Commerce Commerce Standard.

CSA   Campus Safety Association; Canadian Standards Association.

CSL   Chemistry Standards Laboratory.

CSO/ACSSO   Computer security officer/alternate computer system security officer.

CSP   Certified Safety Professional.

CSV   Central storage vault.

C_T   Thrust coefficient.

CTD   Cumulative trauma disorder.

CTS   Carpal tunnel syndrome.

Cu   *Cuprum* (Latin); copper.

CV   Curriculum vitae, cost variance; coefficient of variation.

C/V   Farad.

CW   Cold water; clockwise.

CWA   Clean Water Act.

CWTC   Chemical Waste Transportation Council.

CWU   Chemical Workers Union.

D

δ Lower case Greek letter delta.

Δ Upper case Greek letter delta.

d   Deuteron; day; density; prefix for dextrorotatory.

D   Deuterium.

2, 4-D   Dichlorophenoxyacetic acid.

d/m/l   Disintegrations per minute per liter.

D-test   Destruction testing.

D&D   Decontamination and decommission; decontamination and disposal.

DA   Disbursement authorization; double amplitude; destructive analysis.

DAC   Derived air concentration.

DAS   Data acquisition system; document accountability system.

dB   Decibel.

DBA   Design basis accident.

dBC   Decibels measured on the C scale.

DBCP   1,2-Dibromo-3-chloropropane.

DBE   Design basis earthquake.

DBT   Design basis tornado.

DBW   Design basis wind.

D & C   Drugs and cosmetics.

DC   Direct current; design criteria, duty cycle.

DCG   Derived concentration guide.

DCW   Domestic cold water.

DDE   Dichlorodiphenyldichloroethylene.

DDT   Dichlorodiphenyltricholorethane.

DDVP   Dimethyl dichlorovinyl phosphate; dichlorovos; Vapona.

DEA   Drug Enforcement Agency.

DES   Diethylstilbestrol.

DHHS   Department of Health and Human Services.

DHW   Domestic hot water.

DIN   Do it now.

DL   Detection limit.

dl   Equal rotation.

DMSA   2,3 Dimercaptosuccinic acid.

DMSO   Dimethyl sulfoxide.

DNA   Deoxyribonucleic acid.

DNAPL   Dense non-aqueous phase liquid.

DNI   Do not incorporate.

DO   Dissolved oxygen.

DOC   Department of Commerce.

DOD   Department of Defense.

DODES   Division of Disaster Emergency Services.

DOE   Department of Energy.

DOI   Department of the Interior.

DOJ   Department of Justice.

DOL   Department of Labor.

DOP   Dioctylphthalate.

DOS   Department of State; disk operating system; dosimetry-internal and external.

DOT   Department of Transportation.

DP   Defense programs; data processing; delta pressure; differential pressure.

dpm   Disintegrations per minute.

dps   Disintegrations per second.

DTA   Differential thermal analysis.

DVM   Doctor of Veterinary Medicine; digital voltmeter.

DWS   Drinking water standard.

Dy   Dysprosium.

E

ε Lower case Greek letter epsilon.

E   Antenna; armature; arrester, lightning; binding post; brush, electrical contact; exposure level; contact, electrical; east; uppercase Greek letter epsilon; exa- (10^{10}); environment/environmental (category).

e^- Electron.

E^o Redox or oxidation-reduction potential.

EA   Environmental assessment.

EA&C   Environmental analysis and control.

EACT   Emergency action and coordination team.

EAP   Employee assistance program; emergency action plan.

E&OH   Environmental and occupational health.

EBW   Electron beam welding.

EC   Eddy current; European Community; electron capture.

EC$_{50}$ Effective concentration at 50%.

ECAD   Electronic computer-aided design.

ECC   Estimated cost at completion.

ECD   Electron capture detector.

ECG   Electrocardiogram.

ECL   Exposure control limit.

e/cm^2   Electrons per square centimeter.

E. coli   *Escherichia coli.*

ECS   Emergency control station; engineering computer system; engineering control system.

ED   Effective dose.

ED$_{50}$ Effective dose at 50%.

EDB   Ethylene dibromide.

EDF   Environmental Defense Fund.

EDL   Economic discard limit.

EDP   Electronic data processing.

EDS   Energy-dispersive spectroscopy.

EDTA   Ethylenediaminetetraacetic acid.

EEG   Electroencephalogram; electroencephalography.

EEGL   Emergency response guidance level.

EEI   Edison Electric Institute.

EEL   Environmental exposure limit.

EEOC   Equal Employment Opportunity Commission.

EF   Emission factor.

e.g.   Latin *exempli gratia* ("for example").

ehp   Effective horsepower.

EHS   Extremely hazardous substance.

EI   Emissions inventory.

EI&C   Electrical instrumentation and control.

EIS   Environmental impact statement; effluent information system.

EKG   Electrocardiogram.

EL   Explosive limit; exposure level.

ELF   Extremely low frequency.

ELF-EMF   Extremely low-frequency electromagnetic field.

EM   Equipment management; environmental management; electron microscope.

e-mail   Electronic mail.

emf   Electromotive force.

EMG   Electromyogram.

emi   Electromotive interference.

EML   (DOE) Environmental Measurements Laboratory.

EMR   Electromagnetic radiation.

EMS   Emergency Medical Service.

EMT   Emergency Medical Technician.

emu   Electromagnetic units.

ENCOP   Energy conservation project.

E°   Standard potential.

EOC   Emergency operations center.

EP   Emergency preparedness; extreme pressure.

EPA   Environmental Protection Agency.

EPC   Emergency planning commission.

EPR   Electron paramagnetic resonance; ethylene propylene rubber.

EPRC   Emergency planning review committee.

EPRI   Electric Power Research Institute.

eq   Gram equivalent weight.

Er   Erbium.

ER   Emergency room; electrorefining.

ERDA   Energy Research and Development Administration.

ERP   Emergency response plan.

ERPG   Emergency response planning guidelines.

ERT   Emergency response team.

Es   Einsteinium.

ESD   Electron-stimulated desorption.

ESH   Environmental safety and health.

ESP   Electrostatic precipitator.

ESR   Electron spin resonance.

esu   Electrostatic unit(s).

et   Latin word for "and," as in *et seq.*, and "the following"; shorthand notation for the ethyl group.

ET   Effective temperature; eddy current testing; emission testing.

ETA   Event tree analysis.

EtOH   Ethyl alcohol.

ETS   Environmental tobacco smoke.

ETU   Environmental test unit.

Eu   Europium.

eu   Entropy unit.

eV   Electron volt(s).

exp   Exponential; expanded.

F

f   Femto (10^{-15}).

F   Fluorine; Fahrenheit; farad (C/V).

FAA   Federal Aviation Administration.

FBF   Fluid bed fluorination.

FC   Fail closed.

FCC   Federal Communications Commission; Federal Construction Council.

FD   Fire department.

FDA   Food and Drug Administration.

FD&C color   Food, Drug and Cosmetics color.

FDCA   Food, Drug and Cosmetics Act.

FDR   Final design review.

FE   Facilities engineering.

Fe   *Ferrum* (Latin); iron.

FEA   Finite element analysis.

FEIS   Final environmental impact statement.

FEMA   Federal Emergency Management Agency.

FEP   Field evaluation program.

FERC   Federal Energy Regulatory Commission.

FEV   Forced expiratory volume.

FHWA   Federal Highway Administration.

FI   Facilities inspection.

FID   Flame ionization detector.

FIFO   First-in first-out.

FIFRA   US EPA Federal Insecticide, Fungicide and Rodenticide Act.

Fl. Pt., fl. pt.   Flash point.

FL   Fail last.

FLSA   Fair Labor Standards Act.

FM   Frequency modulation; finished material; facilities manager.

Fm   Fermium.

FMC   Federal Maritime Commission.

FMEA   Failure mode and effects analysis.

FMER   Factory Mutual Engineering Research Organization.

FMS   Flexible manufacturing system.

FMSHRC   Federal Mine Safety and Health Review Commission.

FO   Fail open.

FOD   File or destroy.

FOG   Fat, oil, and grease.

FOI   Freedom of information.

FOIA   Freedom of Information Act.

FOM   Figure of merit.

FONSI   Finding of no significant impact.

fp   Flash point; freezing point.

FPA   Federal Pesticide Act.

FPD   Flame phosphorus detector.

FPM   Fine particle mass.

Fr   Francium.

FR   Federal Register.

FRA   Federal Railroad Administration.

FRMAP   Federal Radiological Monitoring and Assessment Plan.

FRP   Fiberglass-reinforced plastic (or polyester).

FSAR   Final safety analysis report.

FSIS   Food safety and inspection service.

FTA   Fault tree analysis.

FTC   Federal Trade Commission.

FTIR   Fourier transform infrared.

FWPCA   Federal Water Pollution Control Act.

G

γ Lower case Greek letter gamma; gamma ray.

g   Gram; acceleration due to gravity.

G   Amplifier; gauss; giga (prefix = 10^9).

Ga   Gallium.

ga   Gauge, gage (obsolete).

GACT   Generally available control technology.

GC   Gas chromatography.

GC/MS   Gas Chromatography/Mass Spectrometry.

Gd   Gadolinium.

Ge   Germanium.

Ge (Li)   Lithium-drifted germanium detector.

GFCI   Ground fault circuit interrupter.

GFI   Ground fault interrupter.

GI   Gastrointestinal.

GL   General Laboratories.

GLC   Gas liquid chromatography.

GLP   Good laboratory practice(s).

GMP   Good manufacturing practice(s).

GOGO   Government-owned, government-operated.

GOPO   Government-owned, privately-operated.

GPO   Government Printing Office.

GRAS   Generally recognized as safe.

Gy   Gray; J/kg.

H

h Planck's constant (6.613×10^{27} erg seconds).

h   Hecto (prefix = 10^2); height.

H   Hydrogen, henry (Vs/A; Wb/A); enthalpy.

HA   Hazard analysis.

Ha   Hahnium.

³H   Tritium.

HAP   Hazardous air pollutant.

HAZ   hazardous, heat-affected zone.

HAZCOM   Hazard Communication Standard.

HAZMAT   Hazardous materials response team.

HAZOP   Hazard and operability study.

HAZWOPER   Hazardous Materials Waste Operations and Emergency Response.

HBV   Hepatitis B virus.

HC   Hydrocarbon.

HCS   Hazard Communication Standard.

Hct   Hematocrit.

HDPE   High-density polyethylene.

He   Helium.

HEPA   High-efficiency particulate air.

hf   High frequency.

Hf   Hafnium.

HF   Hydrofluoric acid.

HFES   Human Factors and Ergonomics Society.

hfs   Hyperfine structure.

Hg   *Hydrargyrum* (Latin); mercury.

HHS   Health and Human Services.

HIV   Human immunodeficiency virus.

Hivol   High-volume air sampler.

HLW   High-level waste.

HMIS   Hazardous materials identification system.

HMTA   Hazardous Materials Transportation Act.

Ho   Holmium.

hp   Horsepower.

HP   Health physics.

HPGe   High-purity germanium detector.

HPLC   High-performance liquid chromatography; high-pressure liquid chromatography.

HPS   Health Physics Society.

HR   Human resources; hot-rolled.

HRS   Hazard rating system.

HS & E   Health, safety and environment.

HSI   Heat stress index.

HSM   Hospital, surgical, medical (insurance plan).

HSWA   Hazardous and Solid Waste Amendments.

HT   Heat treatment.

hv   High voltage.

HVAC   Heating, ventilating, and air conditioning.

HW   Hot water.

HWTC   Hazardous Waste Treatment Council.

HX   Heat exchanger.

Hz   Hertz (s^{-1}).

HZMD   Hazardous waste management division.

HZTM   Hazardous material response team.

I

I   Iodine; moment of inertia; intermittent.

IACUC   Institutional Animal Care and Use Committee.

IAEA   International Atomic Energy Agency.

IAIABC   International Association of Industrial Accident Boards and Commissions.

IAG   Interagency agreement.

IAM   International Association of Mechanics and Aerospace Workers.

IAP   Indoor air pollution.

IAQ   Indoor air quality.

IARC   International Agency for Research on Cancer.

IATA   International Air Transport Association.

IBC   Institutional Biosafety Committee.

ibid.   Abbreviation for the Latin word *ibidem* meaning in same place.

I & C   Instrumentation and control.

IC   Ion chromatography; inductance-capacitance; integrated circuit; installed cost.

ICAP   Inductively coupled atomic plasma emission spectroscopy.

ICBO   International Conference of Building Officials.

ICC   Interstate Commerce Commission.

ICE   Institute of Civil Engineers.

ICP   Inductively coupled plasma.

ICRP   International Commission of Radiological Protection.

ICRU   International Commission of Radiological Units and Measurements.

ICT   International Critical Tables.

ID   Identification; inside diameter; inventory difference.

IDC   Item description code; initiating device circuits.

IDL   Instrument detection limit.

IDLH   Immediately dangerous to life and health.

IDN   Identification number.

IE   Industrial engineering.

I & E   Inspection and evaluation.

i.e.   Latin *id est*, meaning "that is."

IEEE   Institute of Electrical and Electronics Engineers.

IES   Institute of Environmental Sciences.

IESNA   Illuminating Engineering Society of North America.

IGCI   Industrial Gas Cleaning Institute.

IGT   Institute of Gas Technology.

IH   Industrial hygiene.

IHF   Industrial Health Foundation.

IHMM   Institute of Hazardous Materials Management.

IH&S   Industrial hygiene & safety.

IIE   Institute of Industrial Engineers.

I/M   Inspection & maintenance.

IMDG   International maritime dangerous goods.

IME   Institute of Makers of Explosives.

IMECHE   Institute of Mechanical Engineers.

In   Indium.

INEL   Idaho National Engineering Laboratory.

INMM   Institute of Nuclear Material Management.

insol   Insoluble.

in vitro   Latin meaning literally "in glass."

in vivo   Latin meaning literally "in life."

I/O   Input/Output.

IOC   Initial operational capacity.

IOHA   International Occupational Hygiene Association.

IOM   Institute of Medicine.

IP   Ionization potential; inhalable particles; interproject.

IPS   Interruptible power supply; inside pipe size; in-process.

Ir   Iridium.

IR   Infrared.

IRI   Industrial risk insurers.

IRDS   Primary irritation dose.

ISA   Instrument Society of America.

ISEA   Industrial Safety Equipment Association.

ISO   International Organization for Standardization.

ISTM   International Society for Testing Materials.

ITC   Interagency Testing Committee.

ITE   Institute of Transportation Engineers.

ITSB   International Transportation Safety Board.

IUPAC   International Union of Pure and Applied Chemistry.

IV   Intravenous.

IW   Industrial waste.

J

J   Joule.

J/K   Joule/Kelvin; entropy; heat capacity.

J/kg   Joule/kilogram; gray (Gy); sievert; specific range.

J/kg/K   Joule/kilogram degree Kelvin; specific heat capacity.

J/m³   Joule/cubic meter; energy density.

J/mol   Joule/mole; molar energy.

J/mol K   Joule/mole degree Kelvin; molar entropy; molar heat capacity.

J/s   Joule/second; watt (W).

JA   Job analysis.

JHA   Job hazard analysis.

JSA   Job safety analysis.

JSHA   Job safety and health analysis.

K

k   Kilo-(10^3).

K   *Kalium* (Latin); Potassium; Kelvin; kayser; uppercase Greek letter kappa.

kcal/mole   Kilocalories per mole.

KE   Kinetic energy.

Kev, keV   kilo electron volt; a unit of energy; one Kev equals 1000 electron volts.

kg   Kilogram, i.e., 1000 grams (g).

kg/m^3   Kilogram/cubic meter; mass density.

kgm/s^2   Kilogram meter/second^2; newton (N).

kilo-   1000.

Kr   Krypton.

L

λ   Lower case Greek letter lambda; wavelength.

l   Levo, levorotatory.

L   Liter; lambert.

La   Lanthanum.

LAER   Lowest achievable emission rate.

LANL   Los Alamos National Laboratory; formerly LASL

laser   Light amplification by stimulated emission of radiation.

LASL   Los Alamos Scientific Laboratory.

LC   Liquid chromatography; lethal concentration.

LC_{50}   Lethal concentration at 50%.

lcd   Liquid crystal display; lowest common denominator; least common divisor.

LCL   Lower confidence limit; lower control limit.

LC_{LO}   The lowest confidence limit; lower control limit.

LC_x   Lethal concentration at x%.

LD_{50}   Lethal dose at 50%, usually 14 days.

$LD_{50/30}$ Acute lethal dose at 30 days to 50% of exposed population.

LD_{lo}   Lowest lethal dose.

LD_x   Lethal dose x%, usually 14 days.

LED   Light-emitting diode.

LEL   Lower explosive limit.

LET   Linear energy transfer.

LFL   Lower flammability limit; also known as the lower explosion limit (LEL).

Li   Lithium.

LIA   Lead Industries Association.

LLD   Lower-level discriminator.

LLNL   Lawrence Livermore National Laboratory.

LLW, LLRW   Low-level radioactive waste.

lm   Lumen.

LNG   Liquefied natural gas.

LOC   Level of concern; level of confidence; Library of Congress.

loc. cit.   Latin *loco citato*, "in place cited."

LOD   Limit of detection.

LOE   Level of effort.

LOEL   Lowest observable effect level.

log   Logarithm.

LOX   Liquid oxygen.

LPA   Local planning authority.

LPG   Liquefied petroleum gas.

LQG   Large quantity generator.

Lr   Lawrencium.

LSA   Laser Institute of America.

l/s   lumen per second.

LSD   Lysergic acid diethylamide.

LSO   Laser safety officer.

LTD   Long-term disability; lowest toxic dose.

Lu   Lutetium.

LUST   Leaking underground storage tank.

lx   Lux.

M

μ   Lower case Greek letter mu; micro- (prefix $= 10^{-6}$); one-millionth; dipole moment.

m   Mass; meter; milli- (10^{-3}); meta-; minem (0.06 mL); molal.

m^{-1}   Reciprocal meter(s); wave number.

M   Molar; mega- (prefix $= 10^{6}$); moment; uppercase Greek letter mu; Mach number.

MA   Maritime Administration; aerodynamic moment; mental age.

M&S   Materials and supply.

MAAM   Mobile ambient air monitoring.

MAC   Maximum allowable concentration.

MACT   Maximum achievable control technology.

MAK   Maximum allowable concentration (German PELs).

MC&A   Material control and accountability.

MCA   Maximum credible accident; multichannel analyzer.

MCL   Maximum contaminant level.

MCS   Multiple chemical sensitivity; Metal Casting Society.

Md   Mendelevium.

MD   Medical doctor; medical.

MDA   Minimum detectable amount.

MDC   Minimum detectable concentration.

Me   Methyl.

MED   Minimum erythemal dose.

MEDLARS®   Medical Literature Analysis and Retrieval System.

MEDLINE   MEDLARS on-line.

MEK   Methyl ethyl ketone.

MeV   Million electron volts.

mf   Medium frequency.

MFD   Minimum fatal dose.

mfp   Mean free path.

Mg   Magnesium.

mg   Milligram.

mg/kg   Milligrams administered per kilogram of body weight.

mg/m³   Milligrams per cubic meter.

mH   Millihenry.

mho   Conductivity unit; reciprocal ohm.

MHW   Mixed hazardous waste.

MIG   Metal inert gas.

mil   1/1000 inch thickness.

MIPS   Millions of instructions per second.

ML, ml   Milliliter.

MLD   Median lethal dose.

mmHg   Millimeter of mercury.

Mn   Manganese.

Mo   Molybdenum.

MOA   Memorandum of agreement.

mol   Mole.

mol/m^3   Mole/cubic meter; concentration.

mol. wt   Molecular weight.

MORT   Management oversight risk tree.

MOU   Memorandum of understanding.

mp, MP   Melting point.

MPA   Maximum probable accident; manufacturing project approval.

MPC   Maximum permissible concentration; maintenance publication coordinator.

MPD   Maximum permissible dose.

MPE   Maximum permissible exposure.

MPH   Master of Public Health.

mppcf   Million particles per cubic foot.

MRI   Magnetic resonance instrument; magnetic resonance imaging.

m/s   Meter(s)/second; velocity.

m/s^2   Meter(s)/second^2; acceleration.

m^2/s   Square meter(s)/second; kinematic viscosity.

MS   Mass spectrometry; mass spectrometer.

MSA   Mine safety appliance; Management Science America (computer system).

MS&C   Material scheduling and control.

MSDS   Material Safety Data Sheet.

MSE   Molten salt extraction.

MSG   Monosodium glutamate.

MSHA   Mine Safety and Health Administration.

MSHRC   Mine Safety and Health Review Commission.

MSK   Musculoskeletal effects.

MSST   Maximum safe storage temperature.

MTBE   Methyl-*tert*-butyl ether.

MTBF   Mean time between failures.

MTCE   Maintenance.

MTD   Maximum tolerated dose; manufacturing technology development.

MUF   Material unaccounted for.

MW   Molecular weight.

MWe   Electrical megawatt.

MWhr   Megawatt-hours.

Mx   Maxwell.

N

v   Lower case Greek letter nu; frequency.

n   Neutron; nano- (10^{-9}).

N   Nitrogen; north; normal solution; uppercase Greek letter nu; newton; (N_A) number of molecules of a gas contained in 22.4 liters, i.e., 6.0×10^{23}.

Na   *Natrium* (Latin); sodium.

NA   Not applicable; not available; nonattainment.

NAAQS   National Ambient Air Quality Standards.

NAC   National Audiovisual Center.

NAE   National Academy of Engineering.

NAFI   National Association of Fire Investigators.

NaK   Sodium potassium alloy.

NAP   National Academy Press.

NAS   National Academy of Sciences.

NASA   National Aeronautics and Space Administration.

n.b.   Latin for *nota bene*, "note well."

Nb   Niobium.

NBBPVI   National Board of Boiler & Pressure Vessel Inspectors.

NBC   National Building Code.

NBFU   National Board of Fire Underwriters.

NBS   National Bureau of Standards; replaced by NIST. *See* NIST.

NCADI   National Clearinghouse for Alcohol and Drug Information.

NCCI   National Council on Compensation Insurance.

NCDRH   National Center for Devices and Radiological Health.

NCI   National Cancer Institute.

NCP   National Contingency Plan.

NCR   Nonconformance report.

NCRP   National Council on Radiation Protection and Measurements.

NCTR   National Center for Toxicological Research.

Nd   Neodymium.

NDA   Nondestructive assay.

NDT   Nondestructive testing.

Ne   Neon.

NEC   National Electrical Code.

NEHA   National Environmental Health Association.

NEMA   National Electrical Manufacturers Association.

NEPA   National Environmental Policy Act.

NESC   National Electrical Safety Code; National Energy Software Center.

NESHAP   National Emission Standard for Hazardous Air Pollutants.

NF   National Formulary.

NFC   National Fire Code.

NFPA   National Fire Protection Association; National Fluid Power Association.

NFSA   National Fire Sprinkler Association.

NHTSA   National Highway Traffic Safety Administration.

Ni   Nickel.

NIEHS   National Institute of Environmental Health Sciences.

NIFS   National Institute for Farm Safety.

NIH   National Institutes of Health.

NIMBY   Not in my backyard.

NIOSH   National Institute for Occupational Safety and Health.

NIST   National Institute of Standards and Technology.

NLM   National Library of Medicine.

NLMA   National Labor Management Association; National Lumber Manufacturers Association.

nm   Nanometer.

Nm   Newton meter; joule (J); moment of force.

N/m   Newton/meter; surface tension.

N/m²   Newton/meter^2; pascal; (Pa).

NMR   Nuclear magnetic resonance.

NMSHA   National Mine Safety and Health Academy.

No   Nobelium.

NO   Normally open; nitric oxide.

NOAA   National Oceanic and Atmospheric Administration.

NOAEL   No observed adverse effect level.

NOC   Not otherwise classified.

NOEL   No observable effect level.

NORM   Naturally occurring radioactive material.

NOS   Not otherwise specified; a term frequently used in shipping but also applied elsewhere.

NO$_x$ Nitrogen oxides.

Np   Neptunium.

NPAA   Noise Pollution and Abatement Act.

NPCA   National Paint & Coatings Association.

NPD   Nitrogen phosphorus detector.

NPDES   National Pollutant Discharge Elimination System.

NPDWS   National Primary Drinking Water Standards.

NPEA   National Petroleum Engineers Association.

NPGA   National Propane Gas Association.

NPL   National Priorities List.

NPRA   National Petroleum Refiners Association.

NPRM   Notice of proposed rulemaking.

NQR   Nuclear quadrupole resonance.

NRC   Nuclear Regulatory Commission; National Response Center; National Research Council.

NRCC   National Registry of Certified Chemists.

NRT   National Response Team.

NSA   National Standards Association; National Slag Association.

NSC   National Safety Council.

NSF   National Science Foundation; National Sanitation Foundation.

NSMA   National Safety Management Association.

NSPB   National Society for the Prevention of Blindness.

NSPE   National Society of Professional Engineers.

NSWMA   National Solid Waste Management Association.

NTIS   National Technical Information Service.

NTP   National Toxicology Program; normal temperature and pressure.

NTS   Not-to-scale; Nevada Test Site.

NTSB   National Transportation Safety Board.

O

Ω Uppercase Greek letter omega.

o   *Ortho.*

O   Oxygen.

O$_2$ The oxygen molecule.

O$_3$ Ozone.

O&M   Operation and Maintenance.

OBA   Operating basis accident.

OC   On center.

OCAW   Oil, Chemical and Atomic Workers (International Union).

OD   Optical density; outside diameter; overdose (usually of a drug).

Oe   Oersted.

OEG   Occupational exposure guideline.

OEL   Occupational exposure limit.

OEP   Office of Emergency Preparedness.

ohm   Unit of electrical resistance.

OHMT   Office of Hazardous Materials Transportation.

OJT   On-the-job training.

OMB   Office of Management and Budget.

OP   Order point.

op. cit.   Latin *opere citato*; "in work cited."

ORD   Office of research and development.

ORM   Operations risk management; other regulated material.

ORNL   Oak Ridge National Laboratory.

ORR   Operational readiness review.

Os   Osmium.

OSA   Operational safety analysis.

OSHA   Occupational Safety and Health Administration.

OSHRC   Occupational Safety and Health Review Commission.

OSS   Office of safeguards & security; off-site shipments.

OSW   Office of Solid Waste.

OSWER   Office of Solid Waste and Emergency Response.

OTA   Office of Technology Assessment.

Ox   Total oxidants.

P

p   Page; pico- (prefix $= 10^{-12}$)

p   *Para.*

P   Phosphorus; peta- (prefix $= 10^{15}$); poise; uppercase Greek letter rho.

Pa   Protactinium, pascal.

PA   Public address; plant air; project administrator.

PABA   *p*-Aminobenzoic acid.

PAH   Polycyclic aromatic hydrocarbon; polynuclear aromatic hydrocarbon.

PAN   Peroxyacetyl nitrate.

PAPR   Powered air-purifying respirator.

PARMA   Public Agency Risk Managers Association.

Pa s   Pascal second; dynamic viscosity.

PAT   Proficiency analytical testing.

PAW   Plasma arc welding.

Pb   *Plumbum* (Latin); lead.

PC   Personal computer; politically correct; production control; programmable controller.

PCA   Portland Cement Association.

PCB   Polychlorinated biphenyl; printed circuit board.

PCC   Poison control center.

PCP   Pentachlorophenol; phencyclidine hydrochloride.

PCV   Pressure control valve.

PCW   Process cooling water.

Pd   Palladium.

PDAS   Process data acquisition system.

PDR   *Physicians' Desk Reference*; preliminary design review; property disposal report.

PE   Professional engineer; project engineer; product engineer; program engineer.

PEL   Permissible exposure limit.

PET   Positron emission tomography.

PHA   Pulse height analyzer; preliminary hazards analysis, process hazard analysis.

φ   Lower case Greek letter phi.

PHS   Public Health Service.

π   Lower case Greek letter pi; an amount equal to about 3.14159.

PI   Principal investigator.

PI&S Product integrity & surveillance.

PIN Product identification number; personal identification number.

PLC Programmable logic controller.

Pm Promethium.

PM Photomultiplier tube; project manager.

PM-10 Particulate matter ten micrometers or less in aerodynamic diameter.

PMCC Pensky-Martens closed cup.

PML Probable maximum loss; physical metrology laboratory.

PMS Preventive maintenance system; performance measurement system; premenstrual syndrome.

PM$_x$ Particulate matter x micrometers or less in aerodynamic diameter.

PNA Polynuclear aromatic hydrocarbon.

Po Polonium.

PO Purchase order; production order; post office.

POC Products of combustion; purgeable organic compounds.

POE Point of exposure.

POGO Privately-owned, government-operated.

POTW Publicly owned treatment works.

ppb, PPB Parts per billion.

PPC Personal protective clothing.

PPE Personal protective equipment.

ppm, PPM Parts per million.

ppt, PPT Parts per trillion; precipitate, prepared.

PQE Procurement quality engineering.

Pr Praseodymium.

PR Purchase request; procurement request.

P$_R$ Rated power.

PRP   Potentially responsible party.

PRV   Pressure relief valve.

PSAR   Preliminary safety analysis report.

PSC   Personnel status change; project status control.

psi   Pounds per square inch.

ϕ Lower case Greek letter psi.

PSV   Pressure safety valve.

Pt   Platinum.

PT   Plant training; liquid penetrant testing; process tool; physical therapist.

PTS   US EPA Pesticides and Toxic Substances Branch.

Pu   Plutonium.

PU&D   Property utilization and disposal.

PUC   Public Utility Commission.

PURPA   Public Utilities Regulatory Policies Act of 1978.

PVC   Polyvinyl chloride.

PW   Process waste.

Q

Q   Volume; coulomb.

QA   Quality assurance.

QAR   Quality assurance record.

QC   Quality control.

Q.E.D.   Latin phrase *quod erat demonstrandum*, "which was to be proved."

QF   Quality factor, also neutron quality factor.

QL   Quality level; quality laboratory.

q.v.   Latin phrase *quod vide*, "which see."

R

ρ Lower case Greek letter rho.

R   The gas constant, $= 0.08206$ L-atm/mole degree; roentgen; resistance; the representation of an organic group in a formula.

Ra   Radium.

RA   Remedial action.

RAC   Recombinant advisory committee.

RACT   Reasonably achievable control technology, reasonable available control technology.

R&D   Research and development.

rad   A unit of absorbed dose of ionizing radiation (100 erg/g $= 1$ gray $= 1$ J/kg); radiation absorbed dose; radian (plane angle).

RAD   Radiation absorbed dose.

rad/s   Radian/second; angular velocity.

rad/s²   Radian/second^2; angular acceleration.

RAM   Random access memory; reliability, availability, maintainability; responsibility assignment matrix.

Rb   Rubidium.

RBC   Red blood cell(s).

RBE   Relative biological effectiveness.

RCA   Radiological control area.

RCRA   Resource Conservation and Recovery Act.

RDA   Recommended dietary allowance.

R&D   Research and development.

Re   Rhenium; Reynolds number.

redox   Oxidation-reduction.

REL   Recommended exposure limit.

REM   Roentgen-equivalent-man.

REP   Roentgen equivalent physical.

rev   Revision; revolution.

rf, RF   Radio frequency.

RFD   Reference dose.

RFETS   Rocky Flats Environmental Technology Site.

RFP   Request for proposals; Rocky Flats Plant (obsolete. *See* RFETS)

Rh   Rhodium.

r.h.   Relative humidity.

Rh Factor   Rhesus factor.

RIA   Radioimmunoassay.

RIMS   Risk and Insurance Management Society.

RM   Raw material; resources management.

rms   Root mean square.

Rn   Radon.

RN   Registered Nurse.

RNA   Ribonucleic acid.

roi   Return on investment; region of interest.

ROM   Read-only memory; range of motion.

RPG   Radiation protection guide.

RQ   Reportable quantity.

RRT   Regional response team.

R$_t$ Reliability.

RS   Registered Sanitarian.

RSD   Relative standard deviation.

RSO   Radiological safety officer.

RT   Radiographic testing.

RTECS   *Registry of Toxic Effects of Chemical Substances*, published by NIOSH.

RTK   Right to know.

RTP   Request for technical proposals.

RTR   Real-time radiography.

Ru   Ruthenium.

S

σ Lower case Greek letter sigma.

s   Second.

S   Sulfur; south; entropy; siemens (A/V).

s^{-1} reciprocal second; hertz (Hz).

Σ Upper case Greek letter sigma.

S&S   Safeguards and Security.

SA   Safety analysis; statistical applications.

SAAM   Selective alpha air monitor.

SAB   Science Advisory Board.

SAE   Society of Automotive Engineers (Inc.).

SAR   Safety analysis report; simultaneous activity request; structure activity relationship.

SARA   Superfund Amendment and Reauthorization Act.

SAS   Society for Applied Spectroscopy.

Sb   *Stibium* (Latin); antimony.

Sc   Scandium.

SCA   Single-channel analyzer.

SCBA   Self-contained breathing apparatus.

SCE   Sister chromatid exchange.

SCEM   Scanning electron microscope.

SCG   Storage compatibility group (chemicals that can be stored together safely).

SCUBA   Self-contained underwater breathing apparatus.

SDWA   Safe Drinking Water Act.

Se   Selenium.

sec   Second; section; secant.

SEI   Safety Equipment Institute.

SEM   Scanning electron microscope (microscopy).

SERI   Solar Energy Research Institute.

SES   Standards Engineers Society.

SETA   Seta flash closed tester.

SG   Specific gravity; strain gauge; safeguards.

Si   Silicon.

SI   Le Système International d'Unités.

SIA   Semiconductor Industry Association.

SIO   Signal input/output.

Sm   Samarium.

SME   Society of Manufacturing Engineers; subject matter expert.

SMSA   Standard Metropolitan Statistical Area (BLS).

Sn   *Stannum* (Latin); tin.

SNLA   Sandia National Laboratories Albuquerque.

SNLL   Sandia National Laboratories Livermore.

SNM   Special nuclear material.

SOCMA   Synthetic Organic Chemical Manufacturers Association.

SOP   Standard operating procedure; specified operating power.

SOT   Society of Toxicology.

SO$_x$ Sulfur oxides.

SPCC   Spill prevention control and counter measure.

SPE   Society of Plastics Engineers.

SPEGL   Short-term public emergency guidance level.

SPR   Simplified Practice Recommendations (of the US Department of Commerce).

SQG   Small-quantity generator.

sr   Steradian (solid angle).

Sr   Strontium.

SRA   Society of Risk Analysis.

SRM   Standard Reference Material.

SRMC   Society of Risk Management Consultants.

SRP   Savannah River Plant.

SRS   Savannah River Site.

SS   Stainless steel; safeguards systems; safe and secure; shift superintendent; safety stock; special source; Social Security.

SSHO   Site safety and health officer.

SSS   Systems Safety Society.

SST   Safe, secure transport; safe, secure trailer.

STEL   Short-term exposure limit.

STEM   Scanning transmission electron microscope (microscopy).

STEV   Short-term exposure value.

STP   Standard temperature and pressure; sewage treatment plant.

SU   Standard unit.

SUS   Saybolt universal seconds.

Sv   Sievert (J/kg).

SWDA   Solid Waste Disposal Act.

SWIMS   Solid waste information management system.

SWMU   Solid waste management unit.

SWRF   Stored waste retrieval facility.

SYS   System, systemic effects.

T

τ Lower case Greek letter tau.

θ Lower case Greek letter theta.

T   Temperature; tritium; ton; tera, (10^{12}); tesla (Wb/m^2); uppercase Greek letter tau.

T-M   Time and materials.

2,4,5-T   2,4,5-Trichlorophenoxyacetic acid.

TA   Teaching assistant; travel authorization (form).

Ta   Tantalum.

TAC   Toxic air contaminant.

TAPPI   Technical Association of Pulp and Paper Industry.

Tb   Terbium; biological half-life.

TB   Tuberculosis.

TBD   To be determined.

TC   Thermocouple.

Tc   Technetium.

TCC   Tag closed cup.

TCDD   Tetrachlorodibenzodioxin; dioxin.

TCDF   Terchlorodibenzofuran.

TCE   Trichloroethylene.

TC$_{Lo}$ Toxic concentration low.

TCLP   Toxicity characteristic leaching procedure.

TCRI   Toxic chemical release inventory.

TD   Toxic dose.

TDL   Toxic dose, lethal; total dust loading.

TD$_{Lo}$ Total dose, lowest for an adverse effect.

TDS   Total dissolved solids.

TE   Test engineer; totally enclosed; technical evaluation.

Te   Tellurium.

TEC   Total estimated cost.

TEFC   Totally enclosed fan cooled.

T_{eff} Effective half-life.

TEM   Transmission electron microscope (microscopy).

TENV   Totally enclosed nonventilated.

TER   Teratogen.

TFX   Toxic effects.

TGA   Thermogravimetric analysis.

Th   Thorium.

THC   Total hydrocarbons.

therm   A heat unit = 100,000 Btu.

Ti   Titanium.

TIRC   Toxicology information response center.

Tl   Thallium.

TLC   Thin-layer chromatography; total lung capacity.

TLCC   Total life cycle costs.

TLD   Thermoluminescent dosimeter.

TLV®   Threshold limit value.

TLV®-Ceiling   Threshold limit value-ceiling.

TLV®-STEL   Threshold limit value-short-term exposure limit.

TLV®-TWA   Threshold limit value-time-weighted average; concentration limit for a normal 8-hour shift.

Tm   Thulium.

TNT   Trinitrotoluene.

TOC   Total organic carbon.

torr   One mm of mercury (1333.2 bars).

TOXLINE   On-line toxicology information from the National Library of Medicine (NLM).

TPQ   Threshold planning quantity.

TQ   Threshold quantity.

TRB   Transportation Research Board.

TRI   Toxics release inventory.

TRU   Transuranium or transuranic.

TRUPACT   Transuranic package transporter.

TS   Technical security.

TSA   Technical safety appraisal.

TSCA   Toxic Substances Control Act.

TSD   Treatment, storage, and disposal.

TSD/DOE   Transportation Safeguards Division/Department of Energy.

TSDF   Treatment, storage, and disposal facility.

TSI   Transportation Safety Institute.

TSO   Time-sharing option.

TSP   Total suspended particulates; trisodium phosphate.

TSS   Total suspended solids.

TTU   Transportable treatment unit.

TW   Tower water; technical writer; technical writing.

TWA   Time-weighted average.

TWX   Teletype transmission.

U

U   Uranium.

UAQI   Uniform air quality index.

UAW   United Auto Workers.

UBC   Uniform Building Code.

UCL   Upper confidence limit; upper control limit.

UEL   Upper explosive limit.

UFA   United Farmworkers of America.

UFC   Uniform Fire Code.

UFL   Upper flammability limit.

UGST   Underground storage tank.

uhf   Ultrahigh frequency.

UL   Upper limit; Underwriters Laboratories Incorporated.

ULD   Upper level discriminator.

UMW   United Mine Workers.

UPS   Uninterruptible power supply.

UPW   Uniform present worth.

URI   Upper respiratory infection.

USC   United States Code.

USCG   United States Coast Guard.

USCS   United States Commercial Standard.

USDA   United States Department of Agriculture.

USGS   United States Geological Survey.

USMA   United States Metric Association.

USP   United States Pharmacopeia.

USRDA   United States Recommended Dietary Allowance.

UST   Underground storage tank.

USTR   United States Transuranic Registry.

USTUR   United States Transuranic and Uranium Registries.

UT   Ultrasonic testing.

UTS   Ultimate tensile strength.

UV   Ultraviolet.

UV-A   The ultraviolet region between 315 and 400nm.

UV-B   The ultraviolet region between 280 and 315 nm.

UV-C   The ultraviolet region between 100 and 280 nm.

V

V   Vanadium; volt.

VA   Veterans Administration; now called Department of Veterans Affairs.

V/A   Ohm.

VAX   Virtual address extension.

vdf   Video frequency.

VDT   Video display terminal.

VE   Visible emission(s).

vhf   Very high frequency.

vibr.   Vibration, vibrator.

vlf   very low frequency.

V/M   Volt/meter; electric field strength.

VOC   Volatile organic compound.

VOS   Veterans of Safety.

VP   Vapor pressure; velocity pressure.

VPP   Voluntary Protection Program.

Vs   Volt second; weber (Wb).

Vs/A   Volt second/ampere; henry (H, Wb/A).

W

W Tungsten, Wolfram (German); watt; west.

WAC Waste acceptance criteria.

WAD Work-authorizing document.

WASP Waste accountability, shipping and packaging.

Wb Weber (Vs).

Wb/A Weber/ampere; henry (H).

Wb/m2 Weber/meter2; tesla (T).

WBC White blood cell.

WBGT Wet bulb globe temperature.

WBS Work breakdown structure.

WC Water column.

WDS Wavelength dispersive spectroscopy.

WEEL® Workplace Environmental Exposure Level.

WERC Wind Energy Research Center.

WG Water gauge.

W/g Watts per gram.

W-HA Walsh-Healy Act.

WHIMS Workplace hazardous materials information system.

WHO World Health Organization.

WIP Work in process.

WIPP Waste Isolation Pilot Plant.

WIPP-WAC Waste Isolation Pilot Plant; Waste Acceptance Criteria.

WLM Working level month(s).

WLN Wiswesser line notation.

W/m²   Watt/meter²; heat flux density; power density.

W/m K   Watt/meter degree Kelvin; thermal conductivity.

W/m²sr   Watt/meter² steradian; radiance.

WO   Work order.

WOG   Water, oil, gas.

WP   Word processing; work package.

WQCC   Water Quality Control Commission.

WRAP   Waste processing and packaging.

WRC   Water Resources Council.

WSF   Water-soluble fraction.

W/sr   Watt/steradian; radiant intensity.

wt   Weight.

www   World Wide Web.

X

X   Uppercase Greek letter chi.

Xe   Xenon.

Y

Y   Yttrium.

Yb   Ytterbium.

YTD   Year-to-date.

Z

Z   Atomic number.

Z_{eff} Effective atomic number.

ZFF   Zinc fume fever.

Z list   OSHA's Toxic and Hazardous Substances Tables Z-1, Z-2, and Z-3 of air contaminants, in 29 CFR 1910.1000.

Zn   Zinc.

Zr   Zirconium.

ZRL   Zero risk level.

Appendix B

Brief Biographies

A

Abel, Sir Fredrick Augustus English chemist (1827–1902), coinventor of cordite and the device, named after him, for determining flash points.

Agricola, Georgius Latin name of Georg Bauer (1494–1555), a German mineralogist, metallurgist, physician, and author of *De Re Metallica* (1556), a record of sixteenth century mining and metalworking in which the use of protective masks for miners and smelters was advocated.

Altman, Sidney Canadian chemist (1939–) who shared the 1989 Nobel Prize in Chemistry for studies of catalytic properties of RNA.

Anfinsen, Christian Boehmer American chemist (1916–) who shared the 1972 Nobel Prize in Chemistry for work on ribonuclease.

Ångstrom, Anders Jonas Swedish astronomer and physicist (1814–1874) who studied light, made spectral analyses in the solar system, and discovered hydrogen in the solar atmosphere; the Ångstrom unit used to measure the length of light waves is named after him.

Arrhenius, Svante August Swedish scientist (1859–1927) awarded the 1903 Nobel Prize in Chemistry for his studies of compound dissociation in solvents; noteworthy is his equation. *See* Arrhenius equation.

Aston, Francis William   English chemist (1877–1945) awarded the 1922 Nobel Prize in Chemistry for isotope studies with his mass spectrograph.

Auger, Pierre-Victor   French physicist (1899–1931) who discovered the photoelectric effect.

Avogadro, Amedeo   Italian chemist (1776–1856) noted for stating the principle that equal volumes of gases at the same temperature and pressure contain the same number of molecules, i.e., 6.023×10^{23} for 22.4 liters of a gas.

B

Balmer, Johann Jakob   Swiss chemist (1825–1898) noted for his work on spectral series and his 1885 formula for the wavelengths of hydrogen.

Bardeen, John   American physicist (1908–1991) who shared two Nobel Prizes in Physics (1956 and 1972).

Barton, Sir Derek Harold Richard   English chemist (1918–) who shared the 1969 Nobel Prize in Chemistry for developing the concept of confirmation and its application to chemistry.

Bauer, Georg   *See* Agricola.

Baumé, Antoine   French chemist (1728–1804) who invented the hydrometer named after him and devised two scales for measuring specific gravity using this device: one for substances heavier than water and one for substances lighter than water.

Beckmann, Ernst   German chemist (1853–1923) who invented a very sensitive type of mercury thermometer used to very accurately measure small changes in temperature.

Becquerel, Antoine Henri   French physicist (1851–1908) who shared the 1903 Nobel Prize in Physics for his studies of radioactivity.

Beer, August   German physicist (1825–1863) who was one of the founders of photometry and after whom Beer's Law is named.

Beilstein, Friedrich Konrad   German chemist born in Russia (1838–1906) who wrote the multi-volume *Handbook of Organic Chemistry* on organic chemical reactions and properties. Developed a test to detect halogens in organic compounds.

Berg, Paul   American chemist (1926–) who shared the 1980 Nobel Prize in Chemistry for studies of the biochemistry of nucleic acids, particularly recombinant DNA.

Bergius, Friedrich   German chemist (1884–1949) who shared the 1931 Nobel Prize in Chemistry for the invention and development of high-pressure chemical methods.

Berthollet, Claude Louis   French chemist (1748–1822) who discovered the use of chlorine for bleaching and, together with Lavoisier, devised a system of chemical nomenclature on which today's system is based.

Berzelius, Jons Jakob   Swedish chemist (1779–1848) who discovered and isolated selenium, cerium, thorium, titanium, zirconium, silicon, and niobium; also introduced the present system of chemical symbols.

Bohr, Niels   Danish physicist (1895–1962) awarded the 1922 Nobel Prize in Physics for his studies of the structure of atoms and the radiations from them.

Boltzmann, Ludwig   Austrian physical chemist (1844–1906) who developed statistical mechanics as applied to the kinetic theory of gases and summarized his findings on their viscosity and diffusion in the Stefan–Boltzmann law.

Bosch, Carl   German chemist (1874–1945) who shared the 1931 Nobel Prize in Chemistry for the invention and development of high-pressure chemical methods.

Bouguer, Pierre   French scientist (1698–1758) who was one of the founders of photometry, the measurement of light intensities.

Boyer, Paul Delos   American chemist (1918–) who shared the 1997 Nobel Prize in Chemistry for elucidation of the enzymatic mechanism underlying the synthesis of adenosine triphosphate.

Boyle, Robert   Irish scientist and theologian (1627–1691) who studied the properties of air and formulated Boyle's law.

Bragg, Sir William Henry   English physicist (1862–1942) who shared with his son the 1915 Nobel Prize in Physics for X-ray crystallography.

Bragg, Sir William Lawrence   Australian-born physicist (1890–1971) who shared with his father the 1915 Nobel Prize in Physics for work on crystallography.

Briggs, Henry   English mathematician (1561–1631) who first published logarithm tables in 1617.

Brown, Herbert Charles   American chemist (1912–) who shared the 1979 Nobel Prize in Chemistry for work with boron containing compounds.

Büchner, Eduard   German chemist (1860–1917), brother of Hans Büchner, awarded the 1907 Nobel Prize in Chemistry for showing that alcoholic fermentation of sugars is caused by enzymes in yeast; studied under Bayer; killed during World War I.

Büchner, Hans   German hygienist and bacteriologist (1850–1902), brother of Eduard Büchner, who showed that there are substances in the blood that protect against infection.

Büchner, Johann Andreas   German pharmacist (1783–1852) who established the scientific basis of pharmacy.

Bunsen, Robert Wilhelm   German chemist (1811–1899) who was the co-discoverer of cesium and rubidium; a pioneer of spectral analysis; invented the Bunsen burner.

Butenandt, Adolf Friederich Johann   German chemist (1903–1995) who shared the 1939 Nobel Prize in Chemistry for work on sex hormones. The authorities of his country forced him to decline the award, but he later received the diploma and the medal.

C

Cannizzaro, Stanislao   Italian chemist (1826–1910) who marched with Garibaldi's Thousand. He was the first to appreciate the importance of Amedeo Avogadro's work in connection with atomic weights and demonstrated how to estimate atomic weights and that molecules of elements may consist of more than one atom. He integrated organic and inorganic chemistry and discovered the reaction named after him.

Carnot, Nicolas Léonard Sadi   French physicist (1796–1832) who was a pioneer in the study of thermodynamics; studied steam engines, and died of cholera.

Cavendish, Henry   English chemist and physicist (1731–1810) who determined the composition of air and water.

Cech, Thomas Robert   American chemist (1947–) who shared the 1989 Nobel Prize in Chemistry for his discovery of catalytic properties of RNA.

Celsius, Andres   Swedish physicist and astronomer (1701–1744) who developed the centigrade or Celsius temperature scale.

Chadwick, Sir James   English physicist (1891–1974) awarded the 1935 Nobel Prize in Physics for discovery of the neutron.

Charles, Jacques Alexandre César   A French physicist, chemist, and inventor (1746–1823) who anticipated Guy-Lussac's studies of expanding gases.

Cheyne, John   Scottish physician (1777–1836) who lived in Dublin, who, along with William Stokes, helped recognize an abnormal breathing pattern in sleeping and unconscious adultsthat is common in infants.

Clapeyron, Benoit Paul Emile   French engineer (1799–1864) who developed a theoretical relationship between the temperature of a liquid and its vapor pressure showing that it is not a straight line.

Clausius, Rudolf Julius Emmanuel   German mathematical physicist (1822–1888) who developed the second law of thermodynamics.

Compton, Arthur Holly   American physicist (1892–1962) who shared the 1927 Nobel Prize in Physics for the discovery of the Compton effect, of importance in the attenuation of X rays and gamma rays because it ejects an excited electron from the atom.

Cottrell, Fredrick Gardner   American chemist (1877–1948) who invented electrostatic precipitation for the removal of suspended particulate matter from air and other gases.

Crick, Francis Henry Compton   British biophysicist (1916–) who, with J. D. Watson, and with the help of X-ray diffraction photographs, constructed a molecular model of DNA and proposed that DNA determines the sequence of amino acids in a polypeptide. He shared the 1962 Nobel Prize in Physiology or Medicine with Watson and Wilkins.

Curie, Marie (née Manya Sklodowska)   French physicist born in Warsaw (1867–1934) who worked with her French husband, Pierre Curie (1859–1906), on magnetism and radioactivity and discovered radium. Pierre and Marie Curie shared the 1903 Nobel Prize in Physics with Becquerel for the discovery of radioactivity. After her husband's death in a horse-drawn carriage accident, she isolated polonium and radium and was awarded the 1911 Nobel Prize in Chemistry.

Curie, Pierre   French chemist (1859–1906) who shared the 1903 Nobel Prize in Physics with his wife, Marie Curie. He and Marie Curie discovered radium and polonium in their investigation of radioactivity.

D

Dakin, Henry Drysdale   An English chemist (1880–1952) who co-authored the *Handbook of Chemical Antiseptics*, developed Dakin's solution for treating wounds during World War I, and was awarded the Davy Medal by the Royal Society in 1941.

Dalton, John   An English chemist and physicist (1766–1844) who discovered the law of partial pressures of gases, discovered the law of multiple proportions, maintained the electric origin of the aurora borealis, and gave the first detailed description of color blindness.

Darcy, Henri-Philibert-Gaspard   French hydraulic engineer (1803–1858) who first derived the equation that governs the laminar flow of fluids in homogeneous, porous media and thereby established the theoretical basis of groundwater hydrology.

Darling, Samuel Taylor   American physician (1872–1925) and chief of laboratories in the Panama Canal Zone, who discovered the pulmonary infection caused by histoplasma, which is a fungus grown in soil contaminated by bird excrement.

Darwin, Charles Robert   English naturalist (1809–1882) who developed the theory of natural selection and the pangenesis hypothesis, i.e., that humans are derived from anthropoids.

Davy, Sir Humphry   English chemist (1778–1829) who discovered the effects of nitrous oxide when inhaled; invented the miners lamp; showed that chlorine was an element, and demonstrated that diamond is carbon.

Debye, Peter Joseph Wilhelm   American physicist (1884–1966) born in the Netherlands, who lived in Germany until 1935 and was awarded the 1936 Nobel Prize in Chemistry for studies of molecular structure through dipole moments and X-ray diffraction; worked on the Manhattan Project in the USA.

Deisenhofer, Johann   German chemist (1943–) who shared the 1988 Nobel Prize in Chemistry for the determination of the three-dimensional structure of a photosynthetic reaction center.

Delaney, Jim   Democratic congressman from Queens, NY, who introduced the amendment (which bears his name) to the 1958 legislation (Section 409) on food additives. It has since been repealed.

Democritus   A fourth and fifth century Greek philosopher, known as the Abderite and laughing philosopher, who adopted and extended the atom theory of Leucippus and used the term "atoms" to describe the smallest indivisible parts of matter.

DeQuervain, Frits   Swiss physician (1868–1940) who discovered that certain types of repetitive hand use cause numbness and tingling in the fingers.

Descartes, René   French scientist, mathematician, and philosopher (1596–1650), resident of Holland, whose Latin name was Renatus Cartesius; developed the Cartesian system of coordinates and mathematical certainty and modern scientific thinking summarized by the words, *cogito, ergo sum* ("I think, therefore I am").

Dirac, Paul Adrien Maurice   English physicist (1902–1984) who shared the 1933 Nobel Prize in Physics for his work in quantum mechanics, notably the book, *The Principles of Quantum Mechanics*, published in 1930.

Döbereiner, Johann Wolfgang   German chemist (1780–1849) who discovered the catalytic properties of platinum and palladium; anticipated the periodic table by observing "triads of elements", e.g., chlorine, bromine, and iodine.

Dollo, Louis (Antoine Marie Joseph)   French paleontologist (1857–1931) who worked in Brussels; stated the law of irreversibility in evolution.

Doppler, Christian Johann   Austrian physicist (1803–1853) known for his work with sound, especially the effect named for him.

Dulong, Pierre Louis   French chemist (1785–1838) noted for Dulong and Petit's law, which relates the specific heat capacity of a solid element to its atomic mass, which for over a century was the way to approximate atomic weights.

E

Ehrlich, Paul   German bacteriologist (1854–1915) awarded the 1908 Nobel Prize in Medicine or Physiology; a pioneer in immunology and chemotherapy; developed methods to stain tubercle bacillus, a remedy for syphilis, and a diphtheria antitoxin.

Eigen, Manfred   German chemist (1927–) awarded the 1967 Nobel Prize in Chemistry for his work on extremely rapid chemical reactions, hydrogen ions, and enzyme control.

Einstein, Albert   German-born (1879–1955) naturalized Swiss and American theoretical physicist awarded the 1922 Nobel Prize in Physics for developing the general and specialized theories of relativity and unified field theory of electromagnetism and gravitation, explaining Brownian movement, and developing the law of photoelectric effect.

Ellenbog, Ulrich   Austrian physician who described lead and mercury poisoning in 1473.

Erlenmeyer, Emil   German chemist (1825–1909) who proposed the formula for naphthalene and designed the flat-bottomed, cone-shaped glass flask named after him.

Ernst, Richard Robert   Swiss chemist (1933–) awarded the 1991 Nobel Prize in Chemistry for developing Fourier transform NMR, which enabled NMR to analyze small quantities of material.

F

Fahrenheit, Gabriel Daniel   German physicist (1686–1736) who lived in Holland and England; used mercury instead of alcohol in the thermometer and introduced the Fahrenheit scale of measurement.

Faraday, Michael   British chemist, physicist, and apprenticed bookbinder (1791–1867) who worked with Sir Humphry Davy; made many contributions to science such as work on the liquefaction of gases, the conservation of force, electrolysis, polarized light, and the relationship between electricity and magnetism.

Fehling, Hermann von   German chemist (1821–1885) who introduced the important oxidizing solution named after him.

Fermi, Enrico   Italian-born American physicist (1901–1954) awarded the 1938 Nobel Prize in Physics; studied statistical mechanics, worked on the Manhattan Project to develop the first US atomic bomb, and was the first to achieve a controlled nuclear reaction.

Fick, Adolph Eugen   German physiologist (1829–1901) who studied and developed rules for diffusion.

Fieser, Louis   American chemist (1899–1977) who synthesized vitamin K_1; did research on aromatic carcinogens and steroids such as cortisone.

Fischer, Edmond Henri   German-born American chemist (1920–) who worked with Krebs on plant enzymes, growth regulators, hormonal mechanisms and metabolism; shared the 1992 Nobel Prize in Physiology or Medicine.

Fischer, Emil Hermann   German chemist (1852–1919) awarded the 1902 Nobel Prize in Chemistry for studies of the chemistry of carbohydrates and proteins.

Fischer, Ernst Otto   German chemist (1918–) who shared the 1973 Nobel Prize in Chemistry for his work with organometallic sandwich compounds.

Fischer, Hans   German chemist (1881–1945) awarded the 1930 Nobel Prize in Chemistry for studies of animal and vegetable pigments, hemoglobin, chlorophylls, porphyrins, and bilirubin.

Fleming, Sir Alexander   A Scottish biochemist and bacteriologist (1881–1955) who developed penicillin from mold, which led to the development of many antibiotics.

Flory, Paul   American chemist (1910–1986) awarded the 1974 Nobel Prize in Chemistry for his work with polymers.

Fourier, Baron Jean Baptiste Joseph   French mathematician (1768–1830) who applied mathematics to heat flow, discovered the equation bearing his name, and demonstrated other applications by which a single variable can be expanded in a series.

Fukui, Kenichi   Japanese professor at Kyoto University (1918–) who shared the 1981 Nobel Prize in Chemistry for his work in quantum mechanics and reactivity.

G

Galen of Pergamum   Greek physician (129–c. 216); a vivisectionist who regarded anatomy as the foundation of medical knowledge; continued Hippocratic ideas; his physiology concepts influenced medicine for 1400 years.

Gauss, Karl Friedrich   German mathematician and astronomer (1777–1855) who developed the first mathematical theory of electricity, participated in geodetic surveys, and developed an absolute system of magnetic units.

Gay-Lussac, Joseph Louis   French chemist and physicist (1778–1850) who made balloon ascents to study earth magnetism and composition of air; described the law of volumes named after him; invented the hydrometer; with von Humboldt, studied the composition of water.

Geiger, Hans   German physicist (1882–1945) who investigated beta-ray radioactivity and worked with Walther Müller to develop a counter that measured it.

Gibbs, Josiah Willard   American physicist (1839–1903) who studied thermodynamics, entropy, and the electrical properties of light; established the phase rule and the field of physical chemistry.

Gilbert, Walter   American biochemist (1932–) who shared the 1980 Nobel Prize in Chemistry for the study of the chemical structure of nucleic acids.

Glauber, Johann Rudolf   German chemist and physician (1604–1668) who lived in Holland; discovered hydrochloric and nitric acid; also studied the decomposition of salt by acids and bases.

Gooch, Frank Austin   American chemist (1852–1929) who developed analytical methods and a particular type of filter and crucible named after him.

Goodyear, Charles   American inventor (1800–1860) who experimented with rubber and developed vulcanization.

Graham, Thomas   Scottish chemist (1805–1869) who pioneered the study of colloids; established the law named after him; discovered dialysis and polybasic acids.

Gram, Hans Christian Joachim   Danish physician (1853–1938) who developed the method and types of stains to identify specific bacteria.

Gray, George William   Scottish chemist (1926–) who made the first stable liquid crystal.

Grignard, Franois Auguste Victor   French chemist (1871–1935) awarded the 1912 Nobel Prize in Chemistry for introducing the use of organo-magnesium compounds (Grignard reagents) that form the basis of many organic synthetic reactions.

Gutzeit, Max   German chemist (1847–1915) who developed the tests for arsenic and antimony.

H

Haber, Fritz   German chemist (1868–1934) awarded the 1918 Nobel Prize in Chemistry for his work in nitrogen fixation.

Hahn, Otto   German physical chemist (1879–1968) awarded the Nobel Prize for Chemistry in 1944 for his pioneering work on atomic fission.

Hall, Charles Morton   American chemist (1863–1914) who discovered how to make aluminum electrolytically; helped found the Aluminum Company of America.

Hamilton, Alice   American physician (1869–1970) regarded as the mother of American occupational medicine; first female faculty member of Harvard; studied lead poisoning, phossy jaw, and many other occupational disorders.

Harden, Sir Arthur   English chemist (1861–1940) who shared the 1929 Nobel Prize in Chemistry for studies on fermentation.

Hassel, Odd   Norwegian chemist (1897–1981) who shared the 1969 Nobel Prize in Chemistry for the development of the concept of confirmation and its application in chemistry.

Hauptman, Herbert Aaron   American biophysicist (1917–) who shared the 1985 Nobel Prize in Chemistry for work with X-ray crystallography used to determine the three-dimensional structure of various biochemicals.

Haworth, Sir Walter Norman   English chemist (1893–1950) who shared the 1937 Nobel Prize in Chemistry for determining the chemical structure of vitamin C.

Heisenberg, Werner Karl   German physicist (1901–1976) awarded the 1932 Nobel Prize in Physics for his uncertainty principle and work in quantum mechanics.

Helmholtz, Baron Hermann Ludwig Ferdinand von   German physician, physicist, mathematician, and philosopher (1821–1894) known for developing the law of conservation of energy.

Henry, Joseph   American physicist (1797–1878) noted for his studies of electromagnetic phenomena; the unit of electrical induction was named for him.

Herschbock, Dudley Robert   American chemist (1932–) awarded the 1986 Nobel Prize in Chemistry for studies on the mechanisms of chemical reactions and detailing the sequence of events and energy release.

Hertz, Gustav Ludwig   German physicist (1887–1975) awarded the 1925 Nobel Prize in Physics for confirming the quantum theory.

Hertz, Heinrich Rudolf   German physicist (1857–1894) who worked on electromagnetic waves; the first to broadcast and receive radio waves. The unit of frequency is named after him.

Herzberg, Gerhard   German-born chemist (1904–1999) who emigrated to Canada and was awarded the 1971 Nobel Prize in Chemistry for the detection of free radicals and studies of the energy levels of atoms.

Hess, Germain Henri   Swiss-born chemist (1802–1850) who did early studies on the conservation of energy in chemical reactions.

Hevesy, Georg Charles de   Hungarian chemist (1885–1966) awarded the 1943 Nobel Prize in Chemistry for his work in developing radioisotope tracer techniques; co-discoverer of hafnium.

Heyrovsky, Jaroslav   Czech chemist (1890–1967) awarded the 1959 Nobel Prize in Chemistry for work with mercury electrodes and development of the electroanalytical chemical technique of polarography.

Hinshelwood, Sir Cyril Norman   English chemist (1897–1968) who shared the 1956 Nobel Prize in Chemistry for work on histamines and on the effect of drugs on bacterial cells; also investigated chemical reaction kinetics.

Hippocrates   Greek physician ($\sim$ 377 BC) known as "the father of medicine," associated with the medical profession's Hippocratic Oath.

Hodgkin, Dorothy Mary Crowfoot   Egyptian-born chemist who lived in England (1910–1994); awarded the 1964 Nobel Prize in Chemistry for the use of X-ray techniques in determining the structure of certain molecules, including penicillin, vitamins B_1 and B_2, and insulin.

Hoffman, Roald   Polish-born chemist (born Roald Safran, 1937–) who lived in the United States and shared the 1981 Nobel Prize in Chemistry for the development of mathematical rules to predict the results of chemical reactions, hence, altering the way chemical experiments were designed.

Hoffman, August Wilhelm von   German chemist (1818–1892) who produced aniline from coal products, discovered formaldehyde and other compounds, and developed a theory for classifying different types of chemicals.

Huber, Robert   German chemist (1937–) who shared the 1988 Nobel Prize in Chemistry for the three-dimensional structure of proteins essential to photosynthesis.

J

Jenner, Edward   English physician (1749–1823) who observed that dairymaids had immunity to smallpox and discovered how to vaccinate against the disease.

Joliot-Curie, Irène   French chemist (1897–1956), daughter of Pierre and Marie Curie, who, together with her husband, Jean Frédéric Joliot-Curie, was awarded the 1935 Nobel Prize in chemistry for their work on the production of artificial radioactive elements.

Joliot-Curie, Jean Frédéric   French chemist (1900–1958), World War II resistance fighter, and communist who worked with Marie Curie and married her daughter Irene, with whom he shared the 1935 Nobel Prize in Chemistry for their work on the production of artificial radioactive elements.

Joule, James Prescott   English physicist (1881–1889) who worked with Dalton and Kelvin; demonstrated that heat is a form of energy, established the theory of the conservation of energy, and studied temperature changes in gases and refrigeration.

K

Kaposi, Moritz   Austrian dermatologist (1837–1902) who named a specific type of malignant tumor commonly associated with AIDS.

Karrer, Paul   Swiss chemist (1889–1971) who shared the 1937 Nobel Prize in Chemistry for his investigations of carotenoids, flavins, and vitamins A and B_2.

Kekule von Stradonitz, Frederich August   German chemist (1829–1911) who began his studies in architecture but was influenced by Liebig to study chemistry; best known for his study and depiction of the structure of benzene and his four-volume work on organic chemistry.

Kelvin, Sir William Thompson   British mathematician and physicist (1824–1907) born in Ireland who studied heat, temperature, magnetism, and electricity. Proposed the absolute Kelvin scale of temperature measurement.

Kendrew, Sir John Cowdery   English chemist (1917–) who shared the 1962 Nobel Prize in Chemistry for studies of the structure of globular proteins.

Kirchhoff, Gustav Robert   German physicist (1824–1887) who was a student of Gauss and made important contributions to the theory of circuits using topology. Best known for his publication of Kirchhoff's law in 1854.

Kistiakowsky, George Bogdan   American chemist (1900–1982) born in Russia who was a member of the Manhattan Project at Los Alamos National Laboratory and was a world-renowned authority on high explosives.

Kjeldahl, Johan Gustav Christoffer Thorsager   Danish chemist (1849–1900) known for his expertise in analytical chemistry, most notably for the method of determining the amount of nitrogen in a substance, which can be converted to the amount of protein present. The special flask used for nitrogen-Kjeldahl determinations is named after him.

Klug, Sir Aaron   Lithuanian-born biochemist (1926–) who emigrated to South Africa then to the United Kingdom; awarded the 1982 Nobel Prize in Chemistry for his studies of X-ray diffraction, modeling, and the structure of viruses in combination with proteins and for the development of crystallographic electron microscopy.

Koch, Heinrich Hermann Robert   German bacteriologist and physician (1843–1910) who discovered tuberculosis bacillus and cholera bacillus; awarded the 1905 Nobel Prize in Medicine or Physiology.

Kohlrausch, Friedrich Wilhelm Georg   German physicist (1840–1910) who studied electrolytic conduction of ions in solution.

Krebs, Sir Hans Adolf   German physiologist (1900–1981) who emigrated to England and was awarded the 1953 Nobel Prize in Medicine or Physiology for his work on metabolism.

Kuhn, Richard   Austrian-born biochemist (1900–1967) who worked in Germany and Switzerland and was awarded the 1938 Nobel Prize in Chemistry for his work on the structure and synthesis of vitamins A and B_6 and carotinoids. Nazi Germany did not allow him to accept the award, which was presented after World War II.

L

Lambert, Johann Heinrich   German physicist and mathematician (1728–1777) who studied light and showed that π was an irrational number.

Langmuir, Irving   American chemist (1881–1957) awarded the 1932 Nobel Prize in Chemistry for his studies of surface chemistry. Invented the mercury pump enabling the attainment of very low pressures needed for vacuum tubes; made theoretical contributions to the study of adsorption and thermonuclear fusion; and coined the term "plasma" for ionized gas Worked on problems of ice formation on aircraft wings that led to the discovery of producing rain by seeding clouds with dry ice and silver iodide.

Lavoisier, Antoine Laurent   French chemist (1743–1794) called the founder of modern chemistry; wrote *Traité élémentaire de chimie*; showed that air is a mixture of gases disproving the phlogiston theory; helped devise a modern method of naming chemical compounds and the metric system; guillotined in Paris during the French Revolution.

Lawrence, Ernest Orlando   American physicist (1901–1958) awarded the 1939 Nobel Prize in Physics for inventing the cyclotron. The element lawrencium is named after him.

Le Chatelier, Henry   French chemist (1850–1936) who discovered the law of reactions and the effects of pressure and temperature on equilibrium called Le Chatelier's principle. Also studied metallurgy, ceramics, and combustion and devised a railway water-brake and an optical pyrometer.

Lewis, Gilbert Newton   American physicist and chemist (1875–1946) who developed the valence theory of chemical reactions, discovered "heavy" water, invented the cyclotron, and established an acid-base theory.

Libby, Willard Frank   American chemist (1908–1980) awarded the 1960 Nobel Prize in Chemistry for his role in developing ^{14}C dating; worked on the atomic bomb and served on the Atomic Energy Commission (AEC).

Liebig, Baron Justus Freiherr von   A German chemist (1803–1873) who worked with Gay-Lussac; studied organic and agricultural chemistry and developed new analytical techniques, such as distillation. Developed the condenser that carries his name.

Lister, Baron Joseph   English surgeon (1827–1912) who studied inflammation of wounds and used phenol to prevent infection. Considered the founder of antiseptic surgery.

Little, Arthur Dehon   American chemical engineer (1863–1935) who was an authority on paper technology and chemistry; founded Arthur D. Little, Inc. in 1909.

Löffler, Friedrich August Johannes   German bacteriologist and surgeon (1852–1915) who was the first to culture diphtheria bacillus, discovered the cause of glanders and swine erysipelas, isolated an organism causing food poisoning, and prepared a vaccine against foot-and-mouth disease.

Long, Crawford Williamson   American surgeon (1815–1878) who claimed to be the first to use diethyl ether as an anesthetic.

M

Marcus, Rudolph Arthur   American chemist (1923–) awarded the 1992 Nobel Prize in Chemistry for his electron transfer theory.

Martin, Archer John Porter   English chemist (1910–1994) who shared the 1952 Nobel Prize in Chemistry for work with partition chromatography.

Maxwell, James Clark   Scottish physicist (1831–1879) who is credited as the father of electromagnetic theory.

McCready, Benjamin W.   American physician who wrote the first book on occupational medicine in 1837.

McMillan, Edwin Mattison   American physicist (1907–) who shared the 1951 Nobel Prize in Chemistry for his work in nuclear physics and the discovery of neptunium and plutonium.

Mendeleyev, Dmitri Ivanovich   Siberian chemist (1834–1907) who in 1869 published the principle of periodicity among the elements.

Merrifield, Robert Bruce   American chemist (1921–) awarded the 1984 Nobel Prize in Chemistry for his work with peptide and protein syntheses.

Michel, Hartmut   German chemist (1948–) who shared the 1988 Nobel Prize in Chemistry for work on the structure of certain proteins essential for photosynthesis.

Midgley, Thomas Jr.   American chemist (1889–1944) noted for his work with synthetic rubber and antiknock gasoline.

Mitchell, Peter Dennis   English biochemist (1920–) awarded the 1978 Nobel Prize in Chemistry for work in cellular energy transfer.

Mohr, Karl Friedrich   German pharmacist (1806–1879) known for his work in volumetric analysis.

Mohs, Friedrich   German mineralogist (1773–1839) who developed the scale for measuring hardness.

Moisson, Henri   French chemist (1852–1907) awarded the 1906 Nobel Prize in Chemistry for first isolating fluorine.

Molina, Mario   Mexican chemist (1943–) who shared the 1995 Nobel Prize in Chemistry for work in atmospheric chemistry.

Moore, Stanford   American biochemist (1913–1982) who shared the 1972 Nobel Prize in Chemistry for enzyme work.

Moseley, Henry   English chemist (1887–1915); killed in World War I; noted for his application of X-ray spectra to more accurately position elements in the periodic table.

Müller, Paul Hermann   Swiss chemist (1899–1965) awarded the 1948 Nobel Prize in Medicine or Physiology for showing the usefulness of DDT as an insecticide.

Mulliken, Robert Sanderson   American scientist (1896–1986) awarded the 1966 Nobel Prize in Chemistry for work with isotope separations.

N

Natta, Giulio   Italian chemist (1903–1979) who shared the 1963 Nobel Prize in Chemistry for work on catalytic polymerization.

Nernst, Walther Hermann   German chemist (1864–1941) awarded the 1920 Nobel Prize in Chemistry for developing the third law of thermodynamics.

Newton, Sir Isaac   English physicist and mathematician (1642–1727) who postulated the law of gravity, studied light, and invented differential calculus.

Nobel, Alfred Bernhard   Swedish chemist (1833–1896) who invented a manageable form of nitroglycerin called dynamite and smokeless gunpowder; founded the Nobel Prizes in Physics, Chemistry, Medicine or Physiology, Literature, and Peace (Economics was added later).

Northrop, John Howard   American chemist (1891–1987) who shared the 1946 Nobel Prize in Chemistry for isolation and crystallization of enzymes.

O

Oersted, Hans Christian   Danish physicist (1771–1851) who discovered the magnetic effect of an electric current.

Ohm, Georg Simon   German physicist (1787–1854) who did pioneering research in electricity, specifically resistance.

Olah, George Andrew   Hungarian-born American chemist (1927–) awarded the 1994 Nobel Prize in Chemistry for work with carbocations (positively charged hydrocarbon molecules), which led to the development of new fuels and ways to raise the octane number of gasoline.

Onsager, Lars   Norwegian-born American chemist (1903–1976) awarded the 1968 Nobel Prize in Chemistry for theories on electrolyte conduction and dielectrics.

Oppenheimer, Julius Robert   Controversial American physicist (1904–1967) who helped develop classical quantum physics, worked on the US Manhattan Project, which made the first atomic bomb; directed the Los Alamos National Laboratory; advisor to the US AEC; awarded the US Enrico Fermi medal.

Ostwald, Wilhelm   German chemist (1853–1932) awarded the 1909 Nobel Prize in Chemistry for catalyst studies. Regarded as the founder of modern physical chemistry.

P

Paracelsus, Phillippus Aureolus   Swiss alchemist and physician (1493–1541) who investigated diseases of miners; taught that diseases were specific and could be treated and cured by specific remedies; used opium sulfur iron and mercury as treatments; emphasized value of observation and experience.

Pascal, Blaise   French mathematician and physicist (1623–1662) who invented a calculating machine, a barometer, a hydraulic press, and the syringe.

Pasteur, Louis   French chemist (1822–1895) noted for heat-treating food to kill or inactivate toxic microorganisms; also popularized inoculations.

Pauli, Wolfgang   Austrian-born American physicist (1900–1958) who studied nuclear physics and postulated the existence of the neutrino; awarded the 1945 Nobel Prize in Physics.

Pauling, Linus Carl   American chemist (1901–1994) awarded the 1954 Nobel Prize in Chemistry for crystallography and bonding theories; an advocate of vitamin C and orthomolecular medicine. Also awarded the Nobel Peace Prize in 1962.

Pedersen, Charles John   Korean-born American (1904–1989) awarded the 1987 Nobel Prize in Chemistry for studies of the mechanisms of molecular recognition.

Perkin, Sir William Henry   English chemist (1838–1907) credited with making the first synthetic dyestuff.

Perutz, Max Ferdinand   Austrian biologist (1914–) who shared the 1962 Nobel Prize in Chemistry for work on crystalline protein structures such as hemoglobin.

Petit, Alexis Thérèse   French physicist (1791–1820) associated with the Dulong and Petit's law that for all elements the product of the specific heat and the atomic weight is the same.

Petri, Julius R.   German bacteriologist (1852–1921) who invented the dish used in microbiology named after him.

Pitot, Henri   French hydraulic and civil engineer (1695–1771) who invented the pitot tube used to measure the relative velocity of a fluid past the orifice of the tube.

Planck, Max Karl Ernst Ludwig   German physicist (1858–1947) who worked on thermodynamics and black body radiation and introduced the quantum theory; awarded the 1918 Nobel Prize for Physics.

Pliny the Elder (Gaius Plinius Secundus)   Roman scholar (23–79) who wrote an encyclopedia on natural history; used animal bladders as filters against inhaling lead dust and fumes, observed the eruption of Mt. Vesuvius and was killed by its fumes.

Poisson, Siméon Denis   French mathematician (1781–1840) who applied mathematics to physics, especially electricity and magnetism; authored works on Fourier series, calculus of variation, and probability.

Polanyi, John Charles   German-born Canadian (1929–) who shared the 1986 Nobel Prize in Chemistry for studies of the dynamics of elementary chemical processes; his work led to the development of lasers.

Pople, John Anthony   English-born American chemist and mathematician (1925–) awarded the 1998 Nobel Prize in Chemistry for developing computational methods in quantum chemistry; a pioneer in nuclear magnetic resonance spectrometry (NMR).

Porter, George   English chemist (1920–) who shared the 1967 Nobel Prize in Chemistry for studies of fast chemical reactions and photosynthesis.

Pott, Sir Percivall   English physician (1714–1788) who established the occupational relationship between scrotal cancer in chimney sweeps and poor personal hygiene; introduced techniques to make surgery less painful; Pott's fracture and Pott's disease are named after him.

Pregl, Fritz   Austrian chemist and physician (1869–1930) awarded the 1923 Nobel Prize in chemistry for his work in developing microanalytical techniques.

Prelog, Vladimir   Swiss chemist born in the former Yugoslavia (1906–1998) who shared the 1975 Nobel Prize in Chemistry for synthesis and stereochemistry of organic compounds.

Priestley, Joseph   English chemist and physicist (1733–1804) who emigrated to America; discovered oxygen and nitrous oxide.

Prigogine, Ilya   Russian-born Belgian chemist (1917–) awarded the 1977 Nobel Prize in Chemistry for nonequilibrium thermodynamics studies; applied his work to living systems; helped establish chaos theory.

Prout, William   British chemist and physiologist (1785–1850) who discovered hydrochloric acid in the stomach,

R

Raman, Sir Chandrasekhara Venkata   Indian physicist (1888–1970) awarded the 1930 Nobel Prize in Physics for his discoveries relating to the scattering of light; the Raman effect is named for him.

Ramazzini, Bernardino   Italian physician (1633–1714) and pioneer of occupational health. In *De Morbis Artificum Diatriba* (Diseases of Workers) he wrote about occupational diseases and environmental hazards such as exposure to lead by potters and painters and made observations on epidemics.

Ramsey, Sir William   British chemist (1852–1916) born in Scotland who was awarded the 1904 Nobel Prize in Chemistry for his discovery of the noble gases.

Rankine, William John Macquorn   British engineer (1820–1872) born in Scotland who studied steam engines, machinery, and shipbuilding; helped develop thermodynamics and the theories of elasticity and waves.

Raoult, François Marie   French chemist (1830–1910) who studied solution chemistry and vapor pressure. Raoult's law is named for him.

Rayleigh, Lord John William Strutt   English physicist (1842–1919) awarded the 1904 Nobel Prize in Physics for investigations of gases and the discovery of argon.

Raynaud, Maurice   French physician (1834–1881) who described the disease named after him in his 1862 MD thesis entitled, "Local Asphyxia and Symmetrical Gangrene of the Extremities."

Reynolds, Osborne   English engineer, physicist, and educator (1842–1912) known for his work in hydraulics and heat transfer and the parameters (Reynolds stress and Reynolds number) named after him.

Richards, Theodore William   The first American (1868–1928) to win a Nobel Prize in Chemistry in 1914 for work in the determination of the atomic weights of about 60 elements indicating the existence of isotopes.

Robinson, Sir Robert   English chemist (1886–1975) awarded the 1947 Nobel Prize in chemistry for work on plants with biochemical significance.

Roentgen, Wilhelm Konrad   A German physicist (1845–1923) awarded the first Nobel Prize in Physics in 1901 for the discovery of X rays in 1895.

Rowland, Frank Sherwood   American chemist (1927–) who shared the 1995 Nobel Prize in Chemistry for work in atmospheric chemistry.

Rutherford, Sir Ernest (First Baron Rutherford of Nelson)   New Zealand-born physicist (1871–1937) who lived in England; awarded the 1908 Nobel Prize in Chemistry for establishing the theory of a nuclear atom.

Ruzicka, Leopold Stephen   Croatian-born chemist (1887–1976) who moved to Switzerland; shared the 1939 Nobel Prize in Chemistry for organic syntheses of ringed molecules, terpenes (found in the essential oils of many plants), and sex hormones.

S

Sanger, Frederick   English biochemist (1918–) awarded the 1958 Nobel Prize in Chemistry for work on protein structures, particularly insulin.

Schawlow, Arthur Lawrence   American physicist (1921–1999) who shared the 1981 Nobel Prize in Physics for contributions to laser spectroscopy.

Scheele, Carl Wilhelm   German-born Swedish chemist (1742–1786) who discovered oxygen, chlorine, glycerine, hydrogen sulfide, and several types of acid.

Schrödinger, Erwin   Austrian physicist (1887–1961) who shared the 1933 Nobel Prize in Physics for the discovery of new productive forms of atomic theory, e.g., wave mechanics.

Seaborg, Glen Theodore   American chemist (1912–1999) who shared the 1951 Nobel Prize in Chemistry for discoveries in the transuranium elements.

Semenov (Semyonov), Nikolay Nikolayevich   Russian chemist (1896–1986) who shared the 1956 Nobel Prize in Chemistry for research into the mechanism of chemical reactions.

Shockley, William Bradford   English-born American physicist (1910–1989) who shared the 1956 Nobel Prize in Physics for inventing transistors.

Smalley, Richard Erret   American chemist (1943–) who shared the 1996 Nobel Prize in Chemistry for the discovery of fullerenes, closed spheroidal aromatic molecules with an even number of carbon atoms..

Smith, Michael   Canadian chemist (1932–) who shared the 1993 Nobel Prize in Chemistry for developing site-directed mutagenesis.

Soddy, Frederick   English physicist (1877–1965) awarded the 1921 Nobel Prize in Chemistry for studies of the chemistry of radioactive substances.

Stanley, Wendall Meredith   American biochemist (1904–1971) who shared the 1946 Nobel Prize in Chemistry for virus and enzyme protein research.

Staudinger, Hermann   German chemist (1881–1965) awarded the 1953 Nobel Prize in Chemistry for his discoveries in the field of macromolecular chemistry that led to the development of plastics.

Stein, William Howard   American biochemist (1911–1980) who shared the 1972 Nobel Prize in Chemistry for contributions to the understanding of the connection between chemical structure and catalytic activity of the active center of ribonuclease.

Stokes, Sir George Gabriel   Irish physicist and mathematician (1819–1903) who used spectroscopy to determine the chemical compositions of the sun and stars; studied diffraction and formulated Stokes' law for the force opposing a small sphere in its passage through a viscous fluid.

Sumner, James Batcheller   American biochemist (1887–1955) who shared the 1946 Nobel Prize in Chemistry for his discovery that enzymes can be crystallized and confirmed the existence of protein.

Svedberg, Theodor   Swedish chemist (1884–1971) awarded the 1926 Nobel Prize in Chemistry for studies of colloids.

Synge, Richard Laurence Millington   Irish physicist (1914–) who shared the 1952 Nobel Prize in Chemistry for his studies of the physical chemistry of proteins; devised gas chromatography.

T

Taube, Henry   Canadian born chemist (1915–) awarded the 1983 Nobel Prize in Chemistry for studies of electron transfer reactions.

Tesla, Nikola   Croatian-born American electrical engineer (1856–1943) who worked on electromagnetic motors that became the basis for alternating-current machinery; high-frequency electricity; developed the Tesla coil, a resonant air-core transformer, and hydroelectricity, wireless communication, solar power, and the forerunner of radar.

Thomson, Sir Joseph John   British physicist (1856–1940) noted for his theory of atomic structure, which earned him the first Nobel Prize in Physics in 1901.

Tiselius, Arne Wilhelm Kaurin   Swedish biochemist (1902–1971) awarded the 1948 Nobel Prize in Chemistry for his research with electrophoresis and adsorption separations.

Todd, Sir Alexander Robertus   Scottish chemist (1907–1997) awarded the 1957 Nobel Prize in Chemistry for his studies on nucleotides, co-enzymes, vitamins B and E.

Torricelli, Evangelista   Italian mathematician and physicist (1608–1647) who invented the mercury barometer.

Townes, Charles Hard   American physicist (1915–) who shared the 1964 Nobel Prize in Physics for work in quantum electronics that led to masers and lasers.

Tyndall, John   Irish physicist (1820–1893) who did research on heat radiation, acoustics, and light scattering.

Typhoid Mary   Mary Mallon, a New York City cook determined to be a carrier of typhoid in 1907.

U

Urey, Harold Clayton   American chemist (1894–1981) awarded the 1934 Nobel Prize in Chemistry for his discovery of the heavy isotopes of hydrogen and oxygen.

van der Waals, Johannes Diderik   Dutch physicist (1837–1923) awarded the 1910 Nobel Prize in Physics for applying the ideal gas laws to real gases and deriving the van der Waals equation of state.

V

van't Hoff, Jacobus Henricus   Dutch chemist (1852–1911) awarded the first Nobel Prize in Chemistry in 1901 for his discovery of the laws of chemical dynamics and osmotic pressure.

Venturi, Giovanni Battista   Italian physicist (1746–1822) who did research on fluids; discovered the Venturi effect, i.e., the decrease in pressure of a fluid in a pipe when the diameter is reduced by a gradual taper, which led to the carburetor and flow-collecting devices.

Volta, Alessandro Giuseppe Antonio Anastasio   Italian physicist (1745–1827) who invented a device to generate static electricity, discovered methane gas, and developed the first electric battery.

von Baeyer, Johann Friedrich Wilhelm Adolf   German chemist (1835–1917) awarded the 1905 Nobel Prize in Chemistry for work with dyes, photosynthesis, and aromatic compounds.

W

Walker, John Ernest   English chemist (1941–) who shared the 1997 Nobel Prize in Chemistry for his elucidation of the enzymatic mechanism underlying the synthesis of adenosine triphosphate (ATP).

Wallach, Otto   German chemist (1847–1931) awarded the 1910 Nobel Prize in Chemistry for his work with alicyclic compounds. Also did pioneering work on essential oils, sex hormones, and vitamins.

Watson, James Dewey   American geneticist (1928–) who shared the 1962 Nobel Prize in Medicine or Physiology with Crick and Wilkins for determining the structure of DNA and RNA.

Watt, James   Scottish engineer (1736–1819) who invented the modern steam engine and the centrifugal governor.

Werner, Alfred   Swiss chemist (1866–1919) awarded the 1913 Nobel Prize in Chemistry for developing the coordination theory of atoms in molecules, especially inorganic molecules.

Wieland, Heinrich Otto   German chemist (1877–1957) awarded the 1927 Nobel Prize in Chemistry for his work on bile acids, cholesterol, toxins, and oxidation.

Wien, Wilhelm Carl Werner Otto Fritz Franz   German physicist (1864–1928) awarded the 1911 Nobel Prize in Physics for his studies of heat radiation; he stated his displacement law of blackbody radiation spectra at different temperatures in 1893.

Wilkins, Maurice Hugh Frederick   New Zealand-born biophysicist (1916–) living in Great Britain who shared the 1962 Nobel Prize in Medicine or Physiology with Crick and Watson for determining the structure of DNA and RNA.

Wilkinson, Geoffry   British chemist (1921–) who shared the 1973 Nobel Prize in chemistry for his studies of organometallics, particularly the "sandwich" compounds, e.g., ferrocene.

Willstater, Richard Martin   German chemist (1872–1942) awarded the 1915 Nobel Prize in Chemistry for his work with plant pigments, chlorophyll, cocaine, and atropine.

Windaus, Adolf Otto Reinhold   German chemist (1876–1959) awarded the 1928 Nobel Prize in Chemistry for his work with sterols and their connections with vitamins.

Wittig, Georg   German chemist (1897–1987) who shared the 1979 Nobel Prize in Chemistry for his development of phosphorus-containing compounds into important reagents in organic synthesis and pharmaceuticals.

Wohler, Friedrich   German chemist (1800–1882) who made urea from inorganic ammonium cyanate, thus proving that organic and inorganic chemistry were linked, not separate.

Woodward, Robert Burns   American chemist (1917–1979) awarded the 1965 Nobel Prize in Chemistry for his outstanding work in organic syntheses, e.g., quinine, cholesterol, and chlorophyll.

Z

Zewail, Ahmed   Egyptian-born American chemist (1946–) awarded the 1999 Nobel Prize in Chemistry for his studies of chemical reaction transition states utilizing femtosecond spectroscopy.

Ziegler, Karl   A German chemist (1898–1973) who shared the 1963 Nobel Prize in Chemistry for his development of plastics.

Zsigmondy, Richard Adolf   An Austrian chemist (1865–1929) awarded the 1925 Nobel Prize in Chemistry for work on colloids and invention of the ultramicroscope.

Appendix C

Health and Safety Related Organizations

A

Acoustical Society of America (ASA)   2 Huntington Quadrangle, Melville NY 11747-4502.

Agency for Toxic Substances and Disease Registry (ATSDR)   1600 Clifton Rd., NE, Atlanta GA 30333.

Air and Waste Management Association (AWMA)   1 Gateway Center, Pittsburgh PA 15222.

Air Pollution Control Association (APCA)   PO Box 2861, Pittsburgh PA 15230.

Air Transport Association of America (ATAA)   1301 Pennsylvania Ave., NW, Washington DC 20004-1707.

Alliance for Acid Rain Control and Energy Policy (AARCEP)   444 Capitol St., NW, Washington DC 20001.

Alliance of American Insurers (AAI)   1501 Woodfield Rd., Schaumburg IL 60173-4980.

American Academy for Laboratory Accreditation (AALA)   656 Quince Orchard Rd., Gaithersburg MD 20878-1409.

American Academy of Environmental Engineers (AAEE)   130 Holiday Ct., Annapolis MD 21401.

American Academy of Industrial Hygienists (AAIH)   302 South Waverly Rd., Lansing MI 48917.

American Academy of Occupational Medicine (AAOM)   2340 Arlington Heights Rd., Suite 400, Arlington Heights IL 80005.

American Association for Accreditation of Laboratory Animal Care (AAALAC)   11300 Rockville Pike, Rockville MD 21401.

American Association for the Advancement of Science (AAAS)   1333 H St., NW, Washington DC 20005.

American Association of Occupational Health Nurses (AAOHN)   50 Lennox Pointe, Atlanta GA 30324.

American Association of Poison Control Centers (AAPCC)   20 North Pine, Baltimore MD 21201.

American Association of State Highway and Transportation Officials (AASHTO)   444 Capitol St., NW, Washington DC 20001.

American Association of Textile Chemists and Colorists (AATCC)   PO Box 12215, 1 Davis Dr., Research Triangle Park NC 27709.

American Astronautical Society (AAS)   6852 Rolling Mill Pl., Springfield VA 22152.

American Biological Safety Association (ABSA)   1202 Allanson Rd., Mundelein IL 60060.

American Board of Industrial Hygiene (ABIH)   6015 W. St. Joseph, Suite 102, Lansing MI 48917-3980.

American Cancer Society (ACS)   1599 Clifton Rd., NE, Atlanta GA 30329.

American Chemical Society (ACS)   1155 Sixteenth St., NW, Washington DC 20036.

American Chemistry Council (ACC)   1300 Wilson Blvd., Arlington VA 22209.

American College of Laboratory Animal Medicine (ACLAM)   96 Chester St., Chester NH 03036.

American College of Occupational Medicine (ACOM)   55 West Seegers Rd., Arlington Heights IL 60005.

American College of Toxicology (ACT)   9650 Rockville Pike, Bethesda MD 20814.

American Conference of Governmental Industrial Hygienists (ACGIH) 1330 Kemper Meadow Dr., Cincinnati OH 45240.

American Council on Science and Health (ACSH)   1995 Broadway, New York, NY 10023-5860.

American Dental Association (ADA)   211 East Chicago Ave., Chicago IL 60611.

American Federation of Labor-Congress of Industrial Organizations (AFL-CIO)   815 16th St., NW, Washington DC 20006.

American Foundrymen's Society (AFS)   505 State St., Des Plaines IL 60016-8399.

American Gas Association (AGA)   1515 Wilson Blvd., Arlington VA 22209-2470.

American Hospital Association (AHA)   325 Seventh St., NW, Washington DC 20004.

American Industrial Hygiene Association (AIHA)   2700 Prosperity Ave., Suite 250, Fairfax VA 22031.

American Institute of Architects (AIA)   1735 New York Ave., NW, Washington DC 20006.

American Institute of Chartered Casualty Property Underwriters (AICPCU)   720 Providence Rd., Malvem PA 19355.

American Institute of Chemical Engineers (AIChE)   345 E. 47th St., New York NY 10017.

American Institute of Chemists (AIC)   7315 Wisconsin Ave., Bethesda MD 20814.

American Institute of Electrical and Electronics Engineers (AIEEE)   345 E. 47th St., New York NY 10017.

American Institute of Mining, Metallurgical, and Petroleum Engineers (AIME)   345 E. 47th St., New York NY 10017.

American Institute of Physics (AIP)   335 E. 47th St., New York NY 10017.

American Insurance Services Group (AISG)   85 John St., New York NY 10038.

American Iron and Steel Society (AISS)   1133 15th St., NW, Washington DC 20005.

American Leather Chemists' Association (ALCA)   University of Cincinnati, Cincinnati OH 45221.

American Medical Association (AMA)   535 North Dearborn St., Chicago IL 60610.

American Meteorological Society (AMS)   45 Beacon St., Boston MA 02108.

American National Standards Institute (ANSI)   1819 L St., NW, Washington DC 20036.

American Nuclear Society (ANS)   555 N. Kensington Ave., LaGrange Park IL 60525.

American Occupational Medical Association (AOMA)   2340 S. Arlington Heights Rd., Arlington Heights IL 60005.

American Oil Chemists Society (AOCS)   PO Box 3489, Champaign IL 61826-3489.

American Petroleum Institute (API)   1220 L St., NW, Washington DC 20005.

American Pharmaceutical Association (APhA)   2215 Constitution Ave., NW, Washington DC 20037-2985.

American Pollution Control Association (APCA)   PO Box 2861, Pittsburgh PA 15230.

American Public Health Association (APHA)   1015 15th St., NW, Washington DC 20005.

American Red Cross (ARC)   430 17th St., NW, Washington DC 20006.

American Society for Heating, Refrigeration and Air Conditioning Engineers (ASHRAE)   1791 Tullie Circle, NE, Atlanta GA 30329.

American Society for Mechanical Engineers (ASME)   Three Park Ave., New York NY 10016-5990.

American Society for Metals International (ASM)   9639 Kinsman Rd., Materials Park OH 44073-0002.

American Society for Nondestructive Testing (ASNT)   1711 Arlingate Ln., PO Box 28518, Columbus OH 43228-0518.

American Society for Testing Materials (ASTM)   100 Barr Harbor Dr., West Conshohocken PA 19428-2959.

American Society for Training & Development (ASTD)   Box 1443, 1640 King St., Alexandria VA 22313.

American Society of Civil Engineers (ASCE)   1015 15 St., NW, Suite 600, Washington DC 20005.

American Society of Microbiology (ASM)   1752 N St., NW, Washington DC 20036-2804.

American Society of Safety Engineers (ASSE)   1800 E. Oakton, Des Plaines IL 60018.

American Society of Sanitary Engineers (ASSE)   PO Box 40362, Bay Village OH 44140.

American Standards Association (ASA)   11 W. 42nd St., New York NY 10036.

American Supply Association (ASA)   222 Merchandise Mart, Suite 1360, Chicago IL 60654.

American Vacuum Society (AVS)   335 E. 45th St., New York NY 10017.

American Veterinary Medical Association (AVMA)   1931 North Meacham Rd., Schaumburg IL 60173.

American Water Works Association (AWWA)   6666 West Quincy Ave., Denver CO 80235.

American Welding Society (AWS)   PO Box 351040, Miami FL 33135.

Argonne National Laboratory (ANL)   DOE, 9700 S. Cass Ave., Argonne IL 60440.

Asbestos Information Association (AIA)   1745 Jefferson Davis Highway, Arlington VA 22202.

Association of Environmental Consulting Firms (AECF)   1 E. Wacker Dr., Chicago IL 60601.

Association of Federal Safety and Health Professionals (AFSHP)   7549 Wilhelm Dr., Lanham MD 20706.

Association of Groundwater Scientists and Engineers (AGSE)   PO Box 182039, Columbus OH 43218.

Association of Official Analytical Chemists (AOAC)   111 N. 19th St., Arlington VA 22209.

Aviation Safety Institute (ASI)   PO Box 304, Worthington OH 43085.

B

Battelle Memorial Institute   Headquarters, 505 King Ave., Columbus OH 43201-2693.

Battelle Pacific Northwest Laboratories (BNWL)   Also known as Pacific Northwest National Laboratory; PO Box 999, 902 Battelle Blvd., Richland WA 99352.

Board of Certified Hazard Control Management (BCHCM)   8009 Carita Ct., Bethesda MD 20817.

Board of Certified Product Safety Management (BCPSM)   8009 Carita Ct., Bethesda MD 20817.

Board of Certified Safety Professionals (BCSP)   208 Burwash Ave., Savoy IL 61874-9571.

Brick Institute of America (BIA)   11490 Commerce Park Dr., Reston VA 20191.

Brookhaven National Laboratory (BNL)   DOE, PO Box 5000, Upton NY 11973-5000.

Building Officials and Code Administrators International (BOCA)   4051 W. Flossmoor Rd., Country Club Hills IL 60478.

Bureau of Labor Statistics (BLS)   Department of Labor, 441 G St., NW, Washington DC 20212.

Bureau of Mines (BOM)   Department of the Interior, 2401 E St., NW, Washington DC 20241.

C

Campus Safety Association (CSA)   c/o National Safety Council, 1121 Spring Lake Dr., Itasca IL 60143-3201.

Centers for Disease Control and Prevention (CDC)   US DHHS, PHS, 1600 Clifton Rd., NE, Atlanta GA 30333.

Chemical Abstracts Service (CAS)   2540 Olentangy River Rd., PO Box 3012, Columbus OH 43210-0012.

Chemical Industry Institute of Toxicology (CIIT)   PO Box 12137, 6 Davis Dr., Research Triangle Park NC 27709.

Chemical Manufacturers Association (CMA)   1300 Wilson Blvd., Arlington VA 22209. *See* American Chemistry Council (ACC).

Chemical Waste Transportation Council (CWTC)   c/o National Solid Waste Management Association, 1730 Rhode Island Ave., NW, Washington DC 20006.

Chemical Workers Union (CWU)   1655 W. Market St., Akron OH 4313.

Compressed Gas Association (CGA)   1725 Jefferson Davis Highway, Suite 1004, Arlington VA 22202-4102.

Consumer Products Safety Commission (CPSC)   5401 Westbard Ave., Bethesda MD 20816.

Corps of Engineers (US Army COE)   20 Massachusetts Ave., NW, Washington DC 20314.

Council on Environmental Quality (CFQ)   722 Jackson Pl., NW, Washington DC 20006.

D

Department of Agriculture (USDA)   14th & Independence, SW, Washington DC 20250.

Department of Commerce (DOC)   Constitution Ave. & E St., NW, Washington DC 20230.

Department of Defense (DOD)   The Pentagon, Washington DC 20301

Department of Energy (DOE)   1000 Independence Ave., SW, Washington DC 20585.

Department of Health and Human Services (DHHS)   200 Independence Ave., SW, Washington DC 20201.

Department of the Interior (DOI)   1849 C St., NW, Washington DC 20240.

Department of Justice (DOJ)   950 Pennsylvania Ave., NW, Washington DC 20530.

Department of Labor (DOL)   200 Constitution Ave., NW, Washington DC 20210.

Department of Transportation (DOT)   400 Seventh St., SW, Washington DC 20590.

E

Edison Electric Institute (EEI)   701 Pennsylvania Ave., NW, Washington DC 20004.

Electric Power Research Institute (EPRI)   PO Box 10412, Palo Alto CA 94303.

Environmental Protection Agency (EPA)   401 M St., SW, Washington DC 20460.

F

Factory Mutual Engineering Research Organization (FMER)   1151 Boston-Providence Turnpike, PO Box 9102, Newton MA 02062.

Federal Aviation Administration (FAA)   600 Independence Ave., SW, Washington DC 20591.

Federal Communications Commission (FCC)   445 12th St., SW, Washington DC 20554.

Federal Emergency Management Agency (FEMA)   500 C St., SW, Washington DC 20472.

Federal Energy Regulatory Commission (FERC)   825 North Capitol St., NE, Washington DC 20426.

Federal Highway Administration (FHWA)   400 7th St., Washington DC 20590.

Federal Maritime Commission (FMC)   1100 L St., NW, Washington DC 20573.

Federal Mine Safety and Health Review Commission (FMSHRC)   1730 K St., NW, Washington DC 20006.

Federal Railroad Administration (FRA)   400 7th St., SW, Washington DC 20590.

Federal Trade Commission (FTC)   CRC-240, Washington DC 20580.

Food and Drug Administration (US FDA)   5600 Fishers Ln., Rockville MD 20857.

Food Safety and Inspection Service (FSIS)   14th St. & Independence Ave., SW, Washington DC 20250.

G

Government Printing Office (GPO)   North Capitol & H St., NW, Washington DC 20402.

H

Hazardous Waste Treatment Council (HWTC)   1919 Pennsylvania Ave., NW, Washington DC 20006.

Health Physics Society (HPS)   1313 Dolly Madison Blvd., McLean VA 22101-3926.

Human Factors and Ergonomics Society (HFES)   PO Box 1369, Santa Monica CA 90406.

I

Idaho National Engineering Lab (INEL)   PO Box 1625, Idaho Falls ID 83415.

Illuminating Engineering Society of North America (IESNA)   120 Wall St., New York NY 10005.

Industrial Health Foundation (IHF)   43 Penn Circle W, Pittsburgh PA 15206.

Industrial Risk Insurers (IRI)   85 Woodland St., PO Box 5010, Hartford CT 06102.

Industrial Safety Equipment Association (ISEA)   1910 N. Moore St., Arlington VA 22209.

Institute of Electrical and Electronics Engineers (IEEE)   345 E. 47th St., New York NY 10017.

Institute of Environmental Sciences (IES)   940 East Northwest Highway, Mt. Pleasant IL 60056.

Institute of Gas Technology (IGT)   1700 South Mount Prospect Rd., Des Plaines IL 60018.

Institute of Hazardous Materials Management (IHMM)   11900 Parklawn Dr., Rockville MD 20852.

Institute of Industrial Engineers (IIE)   25 Technology Park, Atlanta GA 30092.

Institute of Makers of Explosives (IME)   1120 19th St., NW, Washington DC 20036.

Institute of Medicine (IOM)   2101 Constitution Ave., NW, Washington DC 20418.

Institute of Nuclear Material Management (INMM)   60 Revere Dr., North-brook IL 60062.

Institute of Transportation Engineers (ITE)   525 School St. SW, Washington DC 20024.

Instrument Society of America (ISA)   *See* International Society for Measurement and Control.

Interagency Testing Committee (ITC)   An independent advisory committee to the US EPA created in 1976 under TSCA; includes representatives of Council on Environmental Quality (CEQ), DOC, EPA, NCI, NIEHS, NIOSH, NSF, OSHA; liaison members include ATSDR, CPSC, DOD, DOI, FDA, NLM, NTP, and USDA.

International Agency for Research on Cancer (IARC)   150 Cours Albert Thomas, F-69372 Lyon CEDEX 08, France.

International Air Transportation Association. (IATA)   IATA Centre, Route de l'Aeroport 33, PO Box 416, 15 - Airport, CH-1215 Geneva, Switzerland.

International Association of Industrial Accident Boards and Commissions (IAIABC)   1575 Aviation Ctr. Parkway, Daytona FL 32826.

International Association of Mechanics and Aerospace Workers (IAM)   9000 Machinists Pl., Upper Marlboro MD 20722.

International Atomic Energy Agency (IAEA)   Karbster Ring 11, PO Box 590, A-1011, Vienna, Austria.

International Commission of Radiological Units and Measurements, Inc. (ICRU)   791 Woodmont Ave., Bethesda MD 20314.

International Commission on Non-Ionizing Radiological Protection (ICNRP)   Bundesamt für Strahlenschutz, Institut für Strahlenhygiene, Ingols. der Landstrasse 1, 85764 Oberschleiheim, Germany.

International Commission on Radiological Protection (ICRP)   SE 171, 16 Stockholm, Sweden.

International Conference of Building Officials (ICBO)   5360 South Workman Mill Rd., Whittier CA 90601.

International Occupational Hygiene Association (IOHA)   Georgian House, Great Northern Rd., Derby DE1 1LT, United Kingdom.

International Organization for Standardization (ISO)   1, rue de Varembe, Case postale 56, CH-1211 Geneva, Switzerland.

International Society for Measurement and Control   Also known as the Instrument Society of America (ISA), 67 Alexander Dr., Research Triangle Park NC 27709.

International Transportation Safety Board (ITSB)   800 Independence Ave., SW, Washington DC 20594.

International Union of Pure and Applied Chemistry (IUPAC)   IUPAC Secretariat, PO Box 13757, Research Triangle Park NC 27709-3757.

Interstate Commerce Commission (ICC)   Congress abolished the ICC in 1995 and created a Surface Transportation Board within DOT to perform the small number of regulatory tasks that had remained in the ICC.

L

Laser Institute of America (LSA)   12424 Research Parkway, Orlando FL 32826.

Lawrence Livermore National Laboratory (LLNL)   7000 East Ave., Livermore CA 94550-9234.

Lead Industries Association (LIA)   292 Madison Ave., New York NY 10017.

Library of Congress (LOC)   101 Independence Ave., SW, Washington DC 20540.

Los Alamos National Laboratory (LANL)   PO Box 1663, Los Alamos NM 87545-1362.

M

Maritime Administration (MA)   US DOT, 400 7th St., SW, Washington DC 20590.

Metal Casting Society (MCS)   455 State St., Des Plaines IL 60016.

Mine Safety and Health Administration (MSHA)   US DOL, 4015 Wilson Blvd., Arlington VA 22203.

Mine Safety and Health Review Commission (MSHRC)   1730 K St., NW, Washington DC 20006.

N

National Academy of Engineering (NAE)   2101 Constitution Ave., Washington DC 20418.

National Academy Press (NAP)   2101 Constitution Ave. NW, Washington DC 20418.

National Academy of Sciences (NAS)   2101 Constitution Ave. NW, Washington DC 20418.

National Aeronautics and Space Administration (NASA)   600 Independence Ave. SW, Washington DC 20546.

National Association of Fire Investigators (NAFI)   PO Box 957257, Hoffman Estates, IL 60195.

National Audiovisual Center (NAC)   8700 Edgeworth Dr., Capitol Heights MD 20743.

National Board of Boiler & Pressure Vessel Inspectors (NBBPVI)   1055 Crupper Ave., Columbus OH 43229.

National Bureau of Standards (NBS)   *See* National Institute of Standards and Technology.

National Cancer Institute (NCI)   9000 Rockville Pike, Bethesda MD 20892.

National Center for Devices and Radiological Health (NCDRH)   5600 Fishers Ln., Rockville MD 20857.

National Center for Toxicological Research (NTCR)   Highway 365, Jefferson AR 72079.

National Clearinghouse for Alcohol and Drug Information   PO Box 2345, Rockville MD 20847.

National Council on Compensation Insurance (NCCI)   200 East 42nd St., New York NY 10017.

National Council on Radiation Protection and Measurements (NCRP)   7910 Woodmont Ave., Bethesda MD 20814.

National Electrical Manufacturers Association (NEMA)   210 L St., Washington DC 20037.

National Energy Software Center (NESC)   US EPA, Enterprise Technology Services Division, MD 34, Research Triangle Park NC 27711.

National Environmental Health Association (NEHA)   720 South Colorado Blvd., Denver CO 80222.

National Fire Protection Association (NFPA)   1 Batterymarch Park, PO Box 9101, Quincy MA 02269-9101.

National Fire Sprinkler Association (NFSA)   PO Box 1000, Robin Hill Corporation Pk., Patterson NY 12563.

National Highway Traffic Safety Administration (NHTSA)   DOT, 400 7th St., SW, Washington DC 20590.

National Institute of Environmental Health Sciences (NIEHS)   PO Box 12233, Research Triangle Park NC 27709.

National Institute for Farm Safety (NIFS)   2601 Rose Ct., Columbia MO 65202.

National Institute of Occupational Safety and Health (NIOSH)   1600 Clifton Ave., NE, Atlanta GA3033.

National Institute of Standards and Technology (NIST)   Formerly the National Bureau of Standards (NBS) Department of Commerce, Quince Orchard Rd., Gaithersburg MD 20899.

National Institutes of Health (NIH)   9000 Rockville Pike, Bethesda MD 80222.

National Labor Management Association (NLMA)   PO Box 819, Jamestown NY 14702-0819.

National Library of Medicine (NLM)   8600 Rockville Pike, Bethesda MD 20209.

National Mine, Safety and Health Academy   PO Box 1166, Beckley WV 25801.

National Oceanic and Atmospheric Administration (NOAA)   600 Independence Ave., SW, Washington DC 20546.

National Paint & Coatings Association (NPCA)   1500 Rhode Island Ave., NW, Washington DC 20005.

National Pesticide Information Retrieval System   Purdue University, Entomology Hall, West Lafayette IN 47907.

National Petroleum Engineers Association (NPEA)   1899 L St., NW, Washington DC 20036.

National Petroleum Refiners Association (NPRA)   1899 L St., NW, Washington DC 20036.

National Propane Gas Association (NPGA)   1600 Eisenhower Ln., Lisle IL 60532.

National Registry of Certified Chemists (NRCC)   815 Fifteenth St., NW, £508, Washington DC 20005.

National Renewable Energy Laboratory (NREL)   1617 Cole Blvd., Golden CO 80401-3393.

National Research Council (NRC)   2101 Constitution Ave., Washington DC 20418.

National Response Center (NRC)   2100 Second Ave., SW, Washington DC 20593.

National Safety Council (NSC)   1121 Spring Lake Dr., Itasca IL 60143-3201.

National Safety Management Association (NSMA)   3871 Piedmont Ave., Oakland CA 94611.

National Sanitation Foundation (NSF)   PO Box 130140, Ann Arbor MI 48113.

National Science Foundation (NSF)   1800 G St., NW, Washington DC 20550.

National Slag Association (NSA)   110 W. Lancaster Ave., Wayne PA 19087-4043.

National Society for the Prevention of Blindness (NSPB)   500 East Remington Rd., Schaumberg IL 60173.

National Society of Professional Engineers (NSPE)   1420 King St., Alexandria VA 22314.

National Solid Waste Management Association (NSWMA)   1730 Rhode Island Ave., NW, Washington DC 20036.

National Standards Association (NSA)   5161 River Rd., Bethesda MD 20816.

National Technical Information Service (NTIS)   5285 Port Royal Rd., Springfield VA 22161.

National Toxicology Program (NTP)   PO Box 12233, Research Triangle Park NC 27709.

National Transportation Safety Board (NTSB)   DOT, 490 L'Enfant Plaza, SW, Washington DC 20594.

Nuclear Regulatory Commission (NRC)   One White Flint North, 11555 Rockville Pike, Rockville MD 20852-2738.

O

Oak Ridge National Laboratory (ORNL)   DOE, PO Box 2800, Oak Ridge TN 37830.

Occupational Safety and Health Administration (OSHA)   DOL, 200 Constitution Ave., NW, Washington DC 20210.

Occupational Safety and Health Review Commission (OSHRC)   1825 K St., NW, Washington DC 20006.

Office of Emergency Preparedness (OEP)   US DHHS, US PHS, Office of Emergency Preparedness, National Disaster Medical System, 12300 Twinbrook Parkway, Suite 360, Rockville MD 20850.

Office of Hazardous Materials Transportation (OHMT)   DOT, 400 7th St., SW, Washington DC 20590.

Office of Research and Development (ORD)   401 M St., SW, Washington DC 20460.

Office of Solid Waste and Emergency Response (OSWER)   US EPA, Ariel Rios (5101), 1200 Pennsylvania Ave., NW, Washington DC 20460.

Office of Technology Assessment (OTA)   600 Pennsylvania Ave., SW, Washington DC 20510.

Oil, Chemical and Atomic Workers International Union (OCAW)   255 Union Blvd., Lakewood CO 80228.

P

Pacific Northwest National Laboratory   *See* Batelle Pacific Northwest National Laboratories.

Portland Cement Association (PCA)   5420 Old Orchard Rd., Skokie IL 60077.

Public Agency Risk Managers Association (PARMA)   5750 Almaden Expressway, San Jose CA 95118.

Public Health Service (PHS)   200 Independence Ave., SW, Washington DC 20201.

R

Risk and Insurance Management Society (RIMS)   205 E. 42nd St., New York NY 10017.

Rocky Flats Environmental Technology Site (RFETS)   PO Box 464, Golden CO 80402-0464.

S

Safety Equipment Institute (SEI)   1307 Dolly Madison Blvd., Suite 3A, McLean VA 22101.

Sandia National Laboratories New Mexico (SNLNM)   DOE, PO Box 5800, 7011 East Ave., Albuquerque NM 87185-5800.

Sandia National Laboratories, California (SNLCA)   DOE, PO Box 969, Livermore CA 94551.

Savannah River Site (SRS)   DOE, PO Box A, Aiken SC 29802.

Semiconductor Industry Association (SIA)   4300 Stevens Creek Blvd., San Jose CA 95129.

Society for Applied Spectroscopy (SAS)   201 B Broadway St., Frederick MD 21701-6501.

Society of Automotive Engineers (SAE)   400 Commonwealth Dr., Warendale PA 15096.

Society of Manufacturing Engineers (SME)   PO Box 930, Dearborn MI 48121.

Society of Plastics Engineers (SPE)   14 Fairfield Dr., Brookfield Center CT 06805.

Society of Risk Analysis (SRA)   1313 Dolly Madison Blvd., McLean VA 22101.

Society of Risk Management Consultants (SRMC)   3255 Fritchie Dr., Baton Rouge LA 70809.

Society of Toxicology (SOT)   1767 Business Center Dr., Reston VA 22190.

Standards Engineers Society (SES)   13340 SW 96th Ave., Miami FL 33176.

Synthetic Organic Chemical Manufacturers Association (SOCMA)   1330 Connecticut Ave., NW, Suite 300, Washington DC 20036.

Systems Safety Society (SSS)   Technology Trading Park, 5 Escort Dr., Sterling VA 21210.

T

Technical Association of Pulp and Paper Industries (TAPPI)   PO Box 105113, Atlanta GA 31348.

Teratology Society   9650 Rockville Pike, Bethesda MD 20814.

Toxicology Information Response Center (TIRC)   ORNL, 1060 Commerce Park, MS 6480, Oak Ridge TN 37830.

Transportation Research Board (TRB)   2101 Constitution Ave., NW, Washington DC 20418.

Transportation Safety Institute (TSI)   DOT, 6500 S. MacArthur Blvd., Oklahoma City OK 73125.

U

Underwriters Laboratories, Inc. (UL)   333 Pfingsten Rd., Northbrook IL 60062.

United Auto Workers (UAW)   8000 E. Jefferson Ave., Detroit MI 48214.

United Farmworkers of America (UFW)   PO Box 62, Keene CA 93570.

United Mine Workers of America (UMW)   900 15th St., NW, Washington DC 20005.

United States Coast Guard (USCG)   DOT, 2100 Second St., SW, Washington DC 20593.

United States Department of Agriculture (USDA)   *See* Department of Agriculture.

United States Metric Association (USMA)   10245 Andasol Ave., Northridge CA 91325.

United States Pharmacopeia (USP)   12601 Twinbrook Parkway, Rockville MD 20852; A private organization that establishes drug and pharmaceutical standards.

United States Transuranic and Uranium Registries (USTUR)   Washington State University, at Tri-Cities, 2710 University Dr., Richland WA 99352.

V

Veterans Affairs, Department of (VA)   810 Vermont Ave., NW, Washington DC 20420.

Veterans of Safety (VOS)   203 North Wabash Ave., Chicago IL 60601.

W

World Health Organization (WHO)   Avenue Appia 20, 1211 Geneva 27 Switzerland.